Springer Proceedings in Physics

Volume 154

For further volumes:
http://www.springer.com/series/361

Efstathios K. Polychroniadis
Ahmet Yavuz Oral · Mehmet Ozer
Editors

International Multidisciplinary Microscopy Congress

Proceedings of InterM, Antalya, Turkey, October 10–13, 2013

Springer

Editors
Efstathios K. Polychroniadis
Department of Physics
Aristotle University of Thessaloniki
Thessaloniki
Greece

Mehmet Ozer
Department of Physics
Istanbul Kultur University
Istanbul
Turkey

Ahmet Yavuz Oral
Department of Materials Science
Gebze Institute of Technology
Gebze
Kocaeli
Turkey

ISSN 0930-8989 ISSN 1867-4941 (electronic)
ISBN 978-3-319-04638-9 ISBN 978-3-319-04639-6 (eBook)
DOI 10.1007/978-3-319-04639-6
Springer Cham Heidelberg New York Dordrecht London

Library of Congress Control Number: 2014935986

Printed on acid-free paper

Springer is part of Springer Science+Business Media (www.springer.com)

Preface

Since the seventeenth century, when the modern optical microscope was invented, scientists have variously attempted to increase their ability to get "better pictures." Many techniques have been created and considerable efforts have been made toward increasing the resolution. Many instruments have been designed and improved during the past years, now allowing the sub-Ångstrom resolution.

Independently of the techniques, microscopy is widely used in many areas in both materials science and the life sciences. This finally has made microscopy a multidisciplinary field attracting the interest of many researchers in science and technology.

The first International Multidisciplinary Microscopy Congress (INTERM 2013) provided all these scientists an opportunity to meet, present their work, discuss, and mutually interact in order to enhance and promote their research work.

This volume, published by Springer, includes all the papers presented at this Congress, held in Antalya, Turkey, in October 10–13, 2013.

On behalf of the organizing committee we would like to thank all the plenary and invited speakers for their valuable contribution and especially Professor Gustaaf Van Tendeloo (EMAT, University of Antwerp, Belgium) for his excellent opening lecture.

We would also like to thank TURA Tourism for their support in the organization of the Congress, as well as our sponsors: Doga Nanobiotech, JEOL Ltd., and NanoMEGAS. Finally, we would like to thank the publishers for the quality of this edition.

Thessaloniki, Greece — Efstathios K. Polychroniadis
Gebze, Turkey — Ahmet Yavuz Oral
Istanbul, Turkey — Mehmet Ozer

Organization

International Scientific Committee

- Gustaaf Van Tendeloo (University of Antwerp, Belgium)
- Athanassios S. Galanis (Application Specialist, NanoMEGAS R&D Department)
- Konstantinos Giannakopoulos (Res. Assoc., Invited Res. at NCSR Demokritos, Greece)
- Wolfgang Jaeger (University of Kiel, Germany)
- Lorenz Kienle (Christian-Albrechts-University Kiel, Germany)
- Stavros Nicolopoulos (Director of the company NanoMEGAS, Belgium)
- Tetsuo Oikawa (JEOL Ltd., Japan)
- Ahmet Yavuz Oral (Gebze Institute of Technology, Turkey)
- Saiedeh Saghafi (Bioelectronics Department, Vienna University of Technology)
- Efstathios K. Polychroniadis (Aristotle University of Thessaloniki, Greece)

Organizing Committee

- Z. Banu Bahsi (Gebze Institute of Technology, Turkey)
- Nermin Demirkol (Kocaeli University, Turkey)
- Ersin Kayahan (Kocaeli University, Turkey)
- Ahmet Yavuz Oral (Gebze Institute of Technology, Turkey)
- Mehmet Ozer (Istanbul Kultur University, Turkey)
- Efstathios K. Polychroniadis (Aristotle University of Thessaloniki, Greece)

Sponsors

- Doga Nanobiotech

- The Company—JEOL Ltd.

- The company NanoMEGAS.

Conference Organizing Company

TURA TOURISM LTD.

Cumhuriyet Cad. No: 109/A Elmadag—Sisli/Istanbul
E-mail: interem2013@turaturizm.com.tr
Phone: +90-212-241 27 00 Fax: +90-212-241 29 89
Web: http://www.turaturizm.com.tr

Contents

Contributors

Eray Abakay Engineering Faculty, Department of Metallurgy and Materials Engineering, Sakarya University, 54187 Serdivan, Sakarya, Turkey, e-mail: eabakay@sakarya.edu.tr

Carmen N. Afonso Laser Processing Group, Instituto de Optica, CSIC, Serrano 121, 28006 Madrid, Spain, e-mail: cnafonso@io.cfmac.csic.es

Gencer Ağdaş Engineering Faculty, Department of Metallurgy and Materials Engineering Esentepe Campus, Sakarya University, 54187 Sakarya, Turkey

R. Ahmed College of Engineering, Alfaisal University, P.O. Box 50927, Riyadh 11533, Saudi Arabia; School of Engineering and Physical Sciences, Heriot-Watt University, Edinburgh EH14 4AS, UK, e-mail: rahmed@alfaisal.edu

Sinan Aksöz Department of Mechanics and Materials Technology, Atatürk High Vocational School, University of Gazi, Ankara, Turkey, e-mail: sinanaksoz@gazi.edu.tr

Boris Alchagirov Kh.M. Berbekov Kabardino-Balkarian State University, Nalchik, Russia 360004, e-mail: alchg@kbsu.ru

Edreese H. Alsharaeh Department of Chemistry, Alfaisal University, P.O. Box 50927, Riyadh 11533, Saudi Arabia, e-mail: ealsharaeh@alfaisal.edu

Ariadne Andreadou Department of Physics, Aristotle University of Thessaloniki, 54124 Thessaloniki, Greece, e-mail: aria@auth.gr

Aliye Arabacı Faculty of Engineering, Department of Metallurgical Engineering, Istanbul University, 34320 Avcilar, Istanbul, Turkey, e-mail: aliye@istanbul.edu.tr

Ruslan Arkhestov Kh.M. Berbekov Kabardino-Balkarian State University, Nalchik, Russia 360004, e-mail: ruslan.arhestov@mail.ru

M. A. Ashraf College of Engineering, Alfaisal University, P.O. Box 50927, Riyadh 11533, Saudi Arabia

Salim Aslanlar Department of Metallurgical and Materials Engineering, University of Sakarya, Sakarya, Turkey, e-mail: aslanlar@sakarya.edu.tr

V. Y. Bairamukov National Research Centre "Kurchatov Institute", B.P. Konstantinov Petersburg Nuclear Physics Institute, Orlova Roshcha, Gatchina, Leningrad District, Russia 188300, e-mail: vbayramukov@gmail.com

N. Barbier Département de Physique, Université Orsay Paris Sud, Paris, France

Iwona Bednarczyk Institute of Materials Science, Silesian University of Technology, Krasińskiego 8, 40-019 Katowice, Poland

J. Bermudez INFN-Laboratori Nazionali di Legnaro, Legnaro, PD, Italy

Jürgen Beyerer Fraunhofer Institute of Optronics, System Technologies and Image Exploitation IOSB, Fraunhoferstr. 1, 76131 Karlsruhe, Germany

Christopher Bohr Department of Otorhinolaryngology, Head and Neck Surgery, Erlangen University Hospital, Erlangen, Germany

Alina S. Borisova Doctor of Engineering Science, Omsk State Technical University, Omsk, Russia, e-mail: lianilla@mail.ru

Bülent Bostan Department of Materials and Metallurgy Engineering, University of Gazi, Ankara, Turkey, e-mail: bostan@gazi.edu.tr

S. Bruni ENEA-UTSISM and Laboratorio Laerte, Via Martiri di Montesole 4, 40128 Bologna, Italy

Fabiola Bubel Laboratory of Biotechnology, Institute of Opencast Mining, Poltegor-Institute, Parkowa 25, 51-625 Wroclaw, Poland, e-mail: fabiola.bubel@igo.wroc.pl

Berrak Bulut Faculty of Chemical and Metallurgical Engineering, Metallurgical and Materials Engineering Department, Istanbul Technical University, Istanbul, Turkey, e-mail: berrakbulut@hotmail.com

S. Carturan INFN-Laboratori Nazionali di Legnaro, Legnaro, Padua, Italy; Department of Physics and Astronomy, University of Padua, Padua, Italy

Danilo A. Cavalcante Cell Biology Department, University of Brasília, Brasilia, Brazil, e-mail: danilo.ac9@gmail.com

Narendraraj Chandran Department of Physics, Aristotle University of Thessaloniki, 54124 Thessaloniki, Greece, e-mail: naren@physics.auth.gr

Eun-Seob Cho National Fisheries Research and Development Institute, 216 Gijanghaean-ro, Gijang-eup, Gijang-gun, Busan, Korea

M. Cinausero INFN-Laboratori Nazionali di Legnaro, Legnaro, PD, Italy

G. Collazuol Department of Physics and Astronomy, University of Padua, Padua, Italy

E. L. Algarra Cruz ENEA-UTSISM and Laboratorio Laerte, Via Martiri di Montesole 4, 40128 Bologna, Italy; Scuola di Lettere e Beni Culturali, Università di Bologna, office of Ravenna, Via Mariani 5, 48100 Ravenna, Italy

M. Degerlier Science and Art Faculty Physics Department, Nevsehir Haci Bektas University, Nevsehir, Turkey, e-mail: meltem.degerlier@lnl.infn.it

A. Şükran Demirkıran Engineering Faculty, Department of Metallurgy and Materials Engineering, Esentepe Campus, Sakarya University, 54187 Sakarya, Turkey, e-mail: dkiran@sakarya.edu.tr

Nermin Demirkol Metallurgical and Materials Engineering Department, Faculty of Chemical and Metallurgical Engineering, Istanbul Technical University, Istanbul, Turkey, e-mail: nermin.demirkol@kocaeli.edu.tr

Marlene T. De-Souza Cell Biology Department, University of Brasília, Brasilia, Brazil, e-mail: marlts@unb.br

Zbigniew Dobrzański Department of Animal Hygiene and Animal Welfare, Wroclaw University of Environmental and Life Sciences, Chełmońskiego 38C, 50-631 Wroclaw, Poland, e-mail: zbigniew.dobrzanski@up.wroc.pl

V. A. Drozdov Omsk Scientific Center, Siberian Branch, Russian Academy of Sciences, 54, Neftezavodskaya st., Omsk, Russia 644040

Saeideh Ebrahimiasl Department of Chemistry, Islamic Azad University, Ahar, Iran, e-mail: ebrahimi.saeideh@yahoo.com

Mehmet Ekici Vocational School of Yalova, University of Yalova, Yalova, Turkey, e-mail: mekici@yalova.edu.tr

M. El-Ameen College of Engineering, Alfaisal University, P.O. Box 50927, Riyadh 11533, Saudi Arabia

A. M. El-Sherik Research and Development Centre, Saudi Aramco, Dhahran 31311, Saudi Arabia

Nuri Ergin Engineering Faculty, Department of Metallurgy and Materials Engineering, Sakarya University, 54187 Sakarya, Turkey, e-mail: nergin@sakarya.edu.tr

Ziya Engin Erkmen Faculty of Chemical and Metallurgical Engineering, Metallurgical and Materials Engineering Department, Marmara University, Istanbul, Turkey, e-mail: eerkmen@marmara.edu.tr

N. H. Faisal College of Engineering, Alfaisal University, P.O. Box 50927, Riyadh 11533, Saudi Arabia; School of Engineering, Robert Gordon University, Garthdee Road, Aberdeen AB10 7GJ, UK

Wojciech Franus Division of Geotechnics, Lublin University of Technology, Nadbystrzycka 40, 20-618 Lublin, Poland, e-mail: w.franus@pollub.pl

Peter Frühberger Fraunhofer Institute of Optronics, System Technologies and Image Exploitation IOSB, Fraunhoferstr. 1, 76131 Karlsruhe, Germany, e-mail: peter.fruehberger@iosb.fraunhofer.de

Rachid Gheriani Laboratory of Radiation and Plasmas and Physics of Surfaces (LRPPS), University of Ouargla, Ouargla, Algeria, e-mail: ragheriani@yahoo.fr

M. F. A. Goosen Office of Research and Graduate Studies, Alfaisal University, P.O. Box 50927, Riyadh 11533, Saudi Arabia

F. Gramegna INFN-Laboratori Nazionali di Legnaro, Legnaro, Padua, Italy

Jakub Grzesiak Electron Microscopy Laboratory, Wroclaw University of Environmental and Life Sciences, Kożuchowska 5b, 50-631 Wroclaw, Poland; Department of Animal Hygiene and Animal Welfare, Wroclaw University of Environmental and Life Sciences, Chełmońskiego 38C, 50-631 Wroclaw, Poland, e-mail: jakub.grzesiak@up.wroc.pl

M. Ipek Engineering Faculty, Department of Metallurgy and Materials Engineering, Esentepe Campus, Sakarya University, 54187 Sakarya, Turkey, e-mail: mipek@sakarya.edu.tr

Elena A. Ivankina Institute of Molecular and Cellular Biology, Sib.Br. RAS, 8, Acad. Lavrentiev ave, Novosibirsk, Russia 630090

Magdalena Barbara Jabłońska Institute of Materials Science, Silesian University of Technology, Krasińskiego 8, 40-019 Katowice, Poland, e-mail: magdalena.jablonska@polsl.pl

Saeed Jafarirad Inorganic Chemistry Department, University of Tabriz, Tabriz, Iran

Nadezhda J. Japaridze I. Beritashvili Center of Experimental Biomedicine, Tbilisi, Georgia, e-mail: japaridze.nadia@gmail.com

Nuray Karakuş Engineering Faculty, Department of Metallurgy and Materials Engineering, Sakarya University, 54187 Sakarya, Turkey, e-mail: nurayc@sakarya.edu.tr

Barasbi Karamurzov Kh.M. Berbekov Kabardino-Balkarian State University, Nalchik, Russia 360004

Eyup Sabri Kayali Faculty of Chemical and Metallurgical Engineering, Metallurgical and Materials Engineering Department, Istanbul Technical University, Istanbul, Turkey, e-mail: kayali@itu.edu.tr

Zareta Kegadueva Kh.M. Berbekov Kabardino-Balkarian State University, Nalchik, Russia 360004

Bülent Kilinc Arifiye Vocational High School, Sakarya University, 54580 Arifiye, Sakarya, Turkey, e-mail: bklilinc@sakarya.edu.tr

Eun-Soo Kim Department of Electronic Engineering, HoloDigilog Human Media Research Center (Holodigilog), 3D Display Research Center (3DRC), Kwangwoon University, 20 Kwangwoon-ro, Nowon-gu, Seoul, Korea, e-mail: eskim@kw.ac.kr

Byung-Mok Kim Department of Electronic Engineering, HoloDigilog Human Media Research Center (Holodigilog), 3D Display Research Center (3DRC), Kwangwoon University, 20 Kwangwoon-ro, Nowon-gu, Seoul, Korea

Edmund Klaus Fraunhofer Institute of Optronics, System Technologies and Image Exploitation IOSB, Fraunhoferstr. 1, 76131 Karlsruhe, Germany

Jung-Sik Koo HoloDigilog Human Media Research Center (Holodigilog), 3D Display Research Center (3DRC), Department of Electronic Engineering, Kwangwoon University, 20 Kwangwoon-ro, Nowon-gu, Seoul, Korea

Domna G. Kotsifaki Department of Physics, School of Applied Mathematical and Physical Sciences, National Technical University of Athens, Zografou Campus, Heroon Polytechniou 9, 15780 Athens, Greece, e-mail: dkotsif@central.ntua.gr

N. N. Koval′ Siberian Branch, Institute of High Current Electronics, Russian Academy of Sciences, Tomsk, Russia 634055

Yu. G. Kryazhev Omsk Scientific Center, Siberian Branch, Russian Academy of Sciences, 54, Neftezavodskaya st., Omsk, Russia 644040

Mariam G. Ksovreli I. Beritashvili Center of Experimental Biomedicine, Tbilisi, Georgia, e-mail: mari.qsovreli@yahoo.com

D. V. Lebedev National Research Centre "Kurchatov Institute", B.P. Konstantinov Petersburg Nuclear Physics Institute, Orlova Roshcha, Gatchina, Leningrad district, Russia 188300

V. A. Likholobov Omsk Scientific Center, Siberian Branch, Russian Academy of Sciences, 54, Neftezavodskaya st., Omsk, Russia 644040

Iakov A. Litvin Department of Kinetics and Catalysis, Semenov Institute of Chemical Physics, Kosigina Street 4, Moscow, Russian Federation 119991

G. Maggioni INFN-Laboratori Nazionali di Legnaro, Legnaro, Padua, Italy; Department of Physics and Astronomy, University of Padova, Padova, Italy

Andreas Maier Pattern Recognition Lab, Department of Computer Science, Friedrich-Alexander University Erlangen-Nuremberg, Martensstr. 3, 91058 Erlangen, Germany; Erlangen Graduate School in Advanced Optical Technologies (SAOT), Friedrich-Alexander University Erlangen-Nuremberg, Erlangen, Germany

G. Maino ENEA-UTSISM and Laboratorio Laerte, Via Martiri di Montesole 4, 40128 Bologna, Italy; Scuola di Lettere e Beni Culturali, Università di Bologna, office of Ravenna, Via Mariani 5, 48100 Ravenna, Italy

Mersini Makropoulou Department of Physics, School of Applied Mathematical and Physical Sciences, National Technical University of Athens, Zografou Campus, Heroon Polytechniou 9, 15780 Athens, Greece, e-mail: mmakro@central.ntua.gr

Alkyoni Mantzari Department of Physics, Aristotle University of Thessaloniki, 54124 Thessaloniki, Greece, e-mail: am@auth.gr

T. Marchi INFN-Laboratori Nazionali di Legnaro, Legnaro, PD, Italy

G. Marghella ENEA-UTSISM and Laboratorio Laerte, Via Martiri di Montesole 4, 40128 Bologna, Italy, e-mail: giuseppe.marghella@enea.it

Maya Marinova Laboratoire de Physique des Solides, Université Paris Sud, Bât 510, 91405 Orsay, France, e-mail: maya.marinova-atanassova@u-psud.fr

Krzysztof Marycz Electron Microscopy Laboratory, Wroclaw University of Environmental and Life Sciences, Kożuchowska 5b, 50-631 Wroclaw, Poland; Department of Animal Hygiene and Animal Welfare, Wroclaw University of Environmental and Life Sciences, Chełmońskiego 38C, 50-631 Wroclaw, Poland, e-mail: krzysztof.marycz@up.wroc.pl

Raouf Mechiakh University El-Hadj Lakhdar, 5000 Batna, Algeria, e-mail: raouf_mechiakh@yahoo.fr

Firas Mualla Pattern Recognition Lab, Department of Computer Science, Friedrich-Alexander University Erlangen-Nuremberg, Martensstr. 3, 91058 Erlangen, Germany, e-mail: firas.mualla@cs.fau.de

I. V. Muromtsev Institute of Hydrocarbon Processing, Siberian Branch, Russian Academy of Sciences, 54, Neftezavodskaya st., Omsk, Russia 644040

Helmut Neumann Department of Medicine I, Friedrich-Alexander University Erlangen-Nuremberg, Erlangen, Germany

Jakub Nicpoń Faculty of Veterinary Medicine, Department of Surgery, Wroclaw University of Environmental and Life Sciences, pl. Grunwaldzki 51, 50-366 Wrocław, Poland, e-mail: jakub.nicpon@up.wroc.pl

Leonid V. Omelyanchuk Institute of Molecular and Cellular Biology, Sib.Br. RAS, 8, Acad. Lavrentiev ave, Novosibirsk, Russia 630090

Juliana C. Orem Cell Biology Department, University of Brasília, Brasilia, Brazil, e-mail: julianacapella@live.com

Yıldız Yaralı Özbek Engineering Faculty, Department of Metallurgy and Materials Engineering, Esentepe Campus, Sakarya University, 54187 Sakarya, Turkey, e-mail: yyarali@sakarya.edu.tr

Ozkan Ozdemir Technology Faculty, Department of Metallurgy and Materials Engineering, Sakarya University, 54187 Sakarya, Turkey, e-mail: oozdemir@sakarya.edu.tr

Ugur Ozsarac Department of Metallurgical and Materials Engineering, University of Sakarya, Sakarya, Turkey, e-mail: ozsarac@sakarya.edu.tr

Irina G. Palchikova Technological Design Institute of Scientific Instrument Engineering, Sib.Br. RAS, 41, Russkaya str, Novosibirsk, Russia 630058, e-mail: palchikova@tdisie.nsc.ru

M. Dalla Palma INFN-Laboratori Nazionali di Legnaro, Trieste, Italy; Department of Industrial Engineering, University of Trento, Trento, Italy

Rafal Panek Division of Geotechnics, Lublin University of Technology, Nadbystrzycka 40, 20-618 Lublin, Poland, e-mail: r.panek@pollub.pl

Alekxey V. Panichkin Doctor of Physics and Mathematics, Omsk Branch of the Institute of Mathematics of the SB RAS. S.L. Soboleva, Omsk, Russia, e-mail: panich@ofim.oscsbras.ru

E. Papalazarou Départemement de Physique, Université Orsay Paris Sud, Paris, France

R. J. Peláez Laser Processing Group, Instituto de Optica, CSIC, Serrano 121, 28006 Madrid, Spain, e-mail: ramon.j.pelaez@gmail.com

Efstathios K. Polychroniadis Department of Physics, Aristotle University of Thessaloniki, 54124 Thessaloniki, Greece, e-mail: polychr@auth.gr

Dimitris Polyzos Department of Physics, School of Applied Mathematical and Physical Sciences, National Technical University of Athens, Zografou Campus, Heroon Polytechniou 9, 15780 Athens, Greece, e-mail: dpolyzos@mail.ntua.gr

O. V. Protasova Institute of Hydrocarbon Processing, Siberian Branch, Russian Academy of Sciences, 54, Neftezavodskaya st., Omsk, Russia 644040

A. Quaranta INFN-Laboratori Nazionali di Legnaro, Legnaro, Padua, Italy; Department of Industrial Engineering, University of Trento, Trento, Italy

Kinga Rodak Institute of Materials Science, Silesian University of Technology, Krasińskiego 8, 40-019 Katowice, Poland

Simon Schöll Pattern Recognition Lab, Department of Computer Science, Friedrich-Alexander University Erlangen-Nuremberg, Martensstr. 3, 91058 Erlangen, GermanyASTRUM IT GmbH, Am Wolfsmantel 2, 91058 Erlangen, Germany; Erlangen Graduate School in Advanced Optical Technologies (SAOT), Friedrich-Alexander University Erlangen-Nuremberg, Erlangen, Germany, e-mail: simon.schoell@cs.fau.de

Valery F. Semeshin Institute of Molecular and Cellular Biology, Sib.Br. RAS, 8, Acad. Lavrentiev ave, Novosibirsk, Russia 630090

Ugur Sen Engineering Faculty, Department of Metallurgy and Materials Engineering, Sakarya University, 54187 Sakarya, Turkey, e-mail: ugursen@sakarya.edu.tr

Saduman Sen Engineering Faculty, Department of Metallurgy and Materials Engineering, Sakarya University, 54187 Sakarya, Turkey, e-mail: sdmnsen@sakarya.edu.tr

Kwang-Beom Seo Department of Electronic Engineering, HoloDigilog Human Media Research Center (Holodigilog), 3D Display Research Center (3DRC), Kwangwoon University, 20 Kwangwoon-ro, Nowon-gu, Seoul, Korea

Yung-Sang Seo National Fisheries Research and Development Institute, 216 Gijanghaean-ro, Gijang-eup, Gijang-gun, Busan, Korea

Alexandros A. Serafetinides Department of Physics, School of Applied Mathematical and Physical Sciences, National Technical University of Athens, Zografou Campus, Heroon Polytechniou 9, 15780 Athens, Greece, e-mail: aseraf@central.ntua.gr

G. M. Seropyan F. M. Dostoevsky State University, Omsk, Russia 644077

Aleksey A. Skoblin Department of Kinetics and Catalysis, Semenov Institute of Chemical Physics, Kosigina street 4, Moscow, Russian Federation 119991

Agnieszka Śmieszek Electron Microscopy Laboratory, Wroclaw University of Environmental and Life Sciences, Kożuchowska 5b, 50-631 Wroclaw, Poland; Department of Animal Hygiene and Animal Welfare, Wroclaw University of Environmental and Life Sciences, Chełmońskiego 38C, 50-631 Wroclaw, Poland, e-mail: agnieszka.smieszek@up.wroc.pl

Eugeny S. Smirnov Technological Design Institute of Scientific Instrument Engineering, Sib.Br. RAS, 41, Russkaya str, Novosibirsk, Russia 630058, e-mail: ome@mcb.nsc.ru

B. Sommerfeldt Institute of Bioprocess Engineering, Friedrich-Alexander-Universität Erlangen-Nürnberg, Erlangen, Germany

Grazyna Stanczak Geophysics and Environmental Protection, University of Science and Technology, Krakow, Poland, e-mail: gstanczak@geol.agh.edu.pl

S. Steidl Pattern Recognition Lab, Department of Computer Science, Friedrich-Alexander-Universität Erlangen-Nürnberg, Martensstr. 3, 91058 Erlangen, Germany

Sergey V. Stovbun Department of Kinetics and Catalysis, Semenov Institute of Chemical Physics, Kosigina street 4, Moscow, Russian Federation 119991, e-mail: s.stovbun@chph.ras.ru

A. D. Teresov Institute of High Current Electronics, Siberian Branch, Russian Academy of Sciences, Tomsk, Russia 634055

V. I. Tikhonov National Research Centre "Kurchatov Institute", B.P. Konstantinov Petersburg Nuclear Physics Institute, Orlova Roshcha, Gatchina, Leningrad district, Russia 188300

Evgeny V. Trapeznikov Omsk State Technical University, Omsk, Russia, e-mail: evtrapeznikov@yandex.ru

M. V. Trenikhin Institute of Hydrocarbon Processing, Siberian Branch, Russian Academy of Sciences, 54, Neftezavodskaya st., Omsk, Russia 644040, e-mail: tremv@yandex.ru

Giorgos Tsigaridas Department of Physics, School of Applied Mathematical and Physical Sciences, National Technical University of Athens, Zografou Campus, Heroon Polytechniou 9, 15780 Athens, Greece, e-mail: gtsigaridas@teilam.gr

Larisa G. Varepo Doctor of Engineering Science, Omsk state technical university, Omsk, Russia, e-mail: larisavarepo@yandex.ru

Faruk Varol Vocational School of Karasu, University of Sakarya, Sakarya, Turkey, e-mail: fvarol@sakarya.edu.tr

L. Volpe Tecnopolo dell'Università di Ferrara, Dipartimento di Fisica e Scienze della terra, Università di Ferrara, Via Saragat 1, 44122 Ferrara, Italy

Magdalena Wdowin Mineral and Energy Economy Research Institute of the Polish Academy of Sciences, Wybickiego 7, 31-261 Kraków, Poland, e-mail: wdowin@meeri.pl

Gokhan Yoruk Technology Faculty , Department of Metallurgy and Materials Engineering, Sakarya University, 54187 Sakarya, Turkey

Seok-Hyun Youn National Fisheries Research and Development Institute, 216 Gijanghaean-ro, Gijang-eup, Gijang-gun, Busan, Korea

Amir Zarei Department of Chemistry, Islamic Azad University, Ahar, Iran

Igor F. Zhimulev Institute of Molecular and Cellular Biology, Sib.Br. RAS, 8, Acad. Lavrentiev ave, Novosibirsk, Russia 630090

Mzia G. Zhvania Ilia State University, Tbilisi, Georgia; I. Beritashvili Center of Experimental Biomedicine, Tbilisi, Georgia, e-mail: mzia_zhvania@iliauni.edu.ge

Part I
Applications of Microscopy in the Physical Sciences

Chapter 1
Micro and Nano Structural Characterization of SiC

Narendraraj Chandran, Ariadne Andreadou, Alkyoni Mantzari, Maya Marinova and Efstathios K. Polychroniadis

Abstract For the last decades, substantial research has been focused on the development of Silicon Carbide electronic devices thanks to its outperformance over Silicon. A review of the author's recent results on the microstructure of SiC layers studied by Transmission Electron Microscopy is presented in the current work. Structural defects appearing in epilayers grown by different methods will be presented. The studied samples were grown by different means-from vapour phase techniques such as Physical Vapour Transport, Chemical Vapour Deposition and Sublimation Epitaxy.

1.1 Introduction

Due to a wide bandgap energy, high-saturation electron velocity, high breakdown field and high thermal conductivity, Silicon carbide (SiC)-based electronics are suitable for high-power and high frequency applications, as well as for severe environments, including high temperature and high radiation [1]. The large Si–C binding energy (about 5.0 eV) makes it intrinsically resistant to radiation fields, high temperature

N. Chandran (✉) · A. Andreadou · A. Mantzari · E. K. Polychroniadis
Department of Physics, Aristotle University of Thessaloniki, GR54124 Thessaloniki, Greece
e-mail: naren@physics.auth.gr

A. Andreadou
e-mail: aria@auth.gr

A. Mantzari
e-mail: am@auth.gr

E. K. Polychroniadis
e-mail: polychr@auth.gr

M. Marinova
Laboratoire de Physique des Solides, Universitė Paris Sud, Bât 510, 91405 Orsay, France
e-mail: maya.marinova-atanassova@u-psud.fr

E. K. Polychroniadis et al. (eds.), *International Multidisciplinary Microscopy Congress*, Springer Proceedings in Physics 154, DOI: 10.1007/978-3-319-04639-6_1,

and corrosion. Moreover, because of its high thermal conductivity, it constitutes an excellent heat sink. The combination of such excellent thermal, electrical and mechanical properties makes it also an interesting material for sensor applications in non-standard environmental conditions [2].

SiC crystallizes in many different forms, called polytypes, which differ only in the successive stacking of n elemental Si–C bilayers. Since they have different bandgap energies and near band edge electronic structures, they should allow the design of new hetero-polytypic devices with unique optical and electrical properties. Unfortunately, up to now, only two SiC polytypes have a well-defined application field. They are hexagonal 4H-SiC for active electronic device layers and 6H-SiC as substrate for Group III–nitride optoelectronics. In this paper, we have studied different SiC polytypes, such as 15R-SiC, 4H-SiC and 3C-SiC, all grown by vapor based growth techniques namely physical vapor transport (PVT), chemical vapor deposition (CVD) and sublimation epitaxy (SE).

15R-SiC has been an attractive material for decades followed by 4H-SiC, and 6H-SiC as a potential candidate for MOS devices [3]. However, no bulk 15R-SiC crystals of good quality have ever been reported. Polytype inclusions and extended defects are the major concerns for crystal growth. In the current study, polytypic transitions were observed during bulk growth. We mainly focused on the interface and formation of associated defects.

Incorporation of group III and IV elements in SiC made astonishing progress in modern power generation technology which requires better semiconductor devices that are capable of operating at higher temperature and at high frequency radiations [4]. Ge and B show relatively high solubility of $3.8 \times 10^{20}\ cm^{-3}$ and $1.9 \times 10^{20} cm^{-3}$ by replacing Si atoms at 2500 K in SiC [5]. However these results differ in experimental studies depending on the growth technique and temperature. In our case Ge was added as a dopant during CVD 4H-SiC growth.

3C-SiC is regarded advantageous over the other hexagonal polytypes because it shows the highest electron mobility [6] and lower density of the near-interface-traps in the SiO_2/3C-SiC system due to the smaller band gap. A crucial problem hindering the future development of a 3C-SiC technology is the unavailability of 3C-SiC substrates. Frequently, in order to overcome problems related to mismatch, as in the case of Si, the growth is performed on substrates from a hexagonal polytype (4H- or 6H-SiC). In our case, 3C-SiC nucleates spontaneously, during SE growth, on (0001) 6H-SiC surfaces or grows homoepitaxially on 3C-SiC buffer layers priory grown by Vapor-Liquid-Solid (VLS) mechanism.

In all above cases cross-sectional Transmission Electron Microscopy (XTEM) as well as High Resolution Transmission Electron Microscopy (HRTEM) study was performed to gain information on the nano-structural quality of the materials and their defects.

1.2 15R-SiC Grown by Physical Vapor Transport

15R-SiC seeds were used to grow single crystals by PVT method. C face 3° off-axis and Si face on-axis seeds were employed. Common growth conditions were applied, i.e. 2200 °C for the growth temperature and a pressure comprised within a range of 15–20 mbar. The thickness of the grown crystals was around 3 mm and the polytype was determined by Raman Spectroscopy. A mixture of 15R/4H-SiC and 15R/6H-SiC crystals were obtained on C-face and Si-face seeds respectively. The main defects observed were stacking faults (SFs) quite abundant in 15R-SiC and normal to [0001] direction. Apart from SFs, a high density of dislocations (D) was observed such as Screw Dislocations (SD), Threading Screw Dislocations (TSD), Threading Edge Dislocations (TED) and Basal Plane Dislocations (BPD). Their density was found to be higher in 15R-SiC area compared to 4H-SiC and 6H-SiC inclusions. Few TEDs and BPDs are identified in TEM images due its well-known conversion mechanism and nature of propagation. Threading dislocations are often introduced at the initial stage of growth and propagate along c-axis and normally terminate or convert into either SFs or BPDs [7].

Polytype transitions depend on various factors such as seed polytype and polarity, supersaturation [3, 8] etc. SiC polytype transitions have been extensively investigated for decades and few authors proposed models for polytypic transition on H-SiC to C-SiC based on asymmetry in the mobility of partial dislocations [9] and 15R-SiC to 6H-SiC based on inhomogeneous density of screw dislocations [10]. On the other hand models are also developed for twins in 15R-SiC such as "Friedelian" reticular merohedric twinning [11]. These studies give better insight for the current study since in case of the C face seeded growth, along with the stacking fault, a (23)/(32) twin sequence was formed in 15R-SiC just below the interface sequence (32)/(22) which is structurally and energetically favored (Fig. 1.1a). This was confirmed by having Fast Fourier Transform (FFT's) images at these regions. Whereas in Si face seeded growth the polytype transition is more complex due to high SF density (Fig. 1.2b). This implies that twin associated polytype transition is coherent than the other form. Further detailed study is required to extract conclusions on the impact of twins in polytypic transition.

1.3 Formation of Ge Droplets on 4H-SiC by Chemical Vapor Transport

Epitaxial 4H-SiC growth was performed at 1500 °C on Si face 8° off-axis 4H-SiC seeds. 0.02 sccm GeH_4 was added to the $SiH_4 + C_3H_8$ precursors' mixture during the deposition run. Ge droplets formed on surface by accumulation of non-incorporated Ge atoms during cooling process.

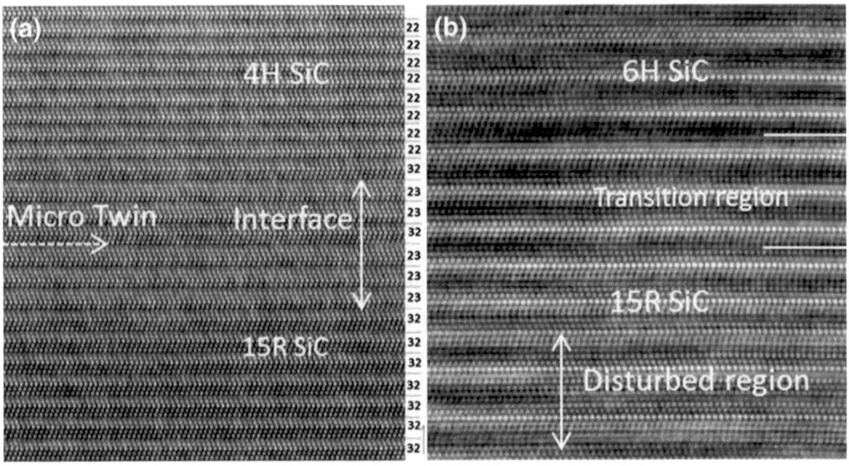

Fig. 1.1 **a** filtered HRTEM micrograph showing a coherent transition region between 15R-SiC and 4H-SiC interface with alternating (32)/(23) transforming to (32)/(22) atomic sequence. The Zhdanov notation of the stacking sequence, listed at *right*, was measured as indicated by the *arrow*. **b** HRTEM micrograph of 15R-SiC/6H-SiC interface appearing more complex with an extent transition region due the presence of numerous stacking faults

HRTEM was employed to investigate the Ge droplets (Fig. 1.2a) due to recent research on the growth of low-dimensional Ge islands on SiC with an increasing interest to develop optoelectronic SiC devices [12, 13]. The droplets were polycrystalline and their size varied between 20 and 300 nm. Nano-crystallites within the droplets ranged from 5 to 15 nm or in some cases even smaller.

Nano-crystallites were oriented in different directions forming droplets on 4H-SiC. The samples are imaged in two different zone axes i.e. SiC [11–20] (Fig. 1.2a, b) and SiC [1–100] (Fig. 1.2c, d) in order to establish the epitaxial relation. HRTEM micrograph of Fig. 1.2b shows a Ge nano-crystal oriented along Ge{001}//4H-SiC{0001}. At the SiC/Ge interface few atomic layers exhibited higher contrast due to atom exchange. The epitaxial orientation of atomic layers was not very clear in SiC [11–20] zone direction. In SiC[1–100] zone direction, it was evident that most of Ge crystals were oriented along Ge{111}// SiC {0001} direction with few misfit dislocations due to large lattice mismatch of about 30 % (Fig. 1.2d). However the step structure of off-axis substrate favored the epitaxial growth considerably reducing the lattice mismatch. TEM characterization at the interface region showed that Ge islands can be grown at low temperature on off axis SiC substrates. Moreover, no Ge inclusions have been detected within the layer.

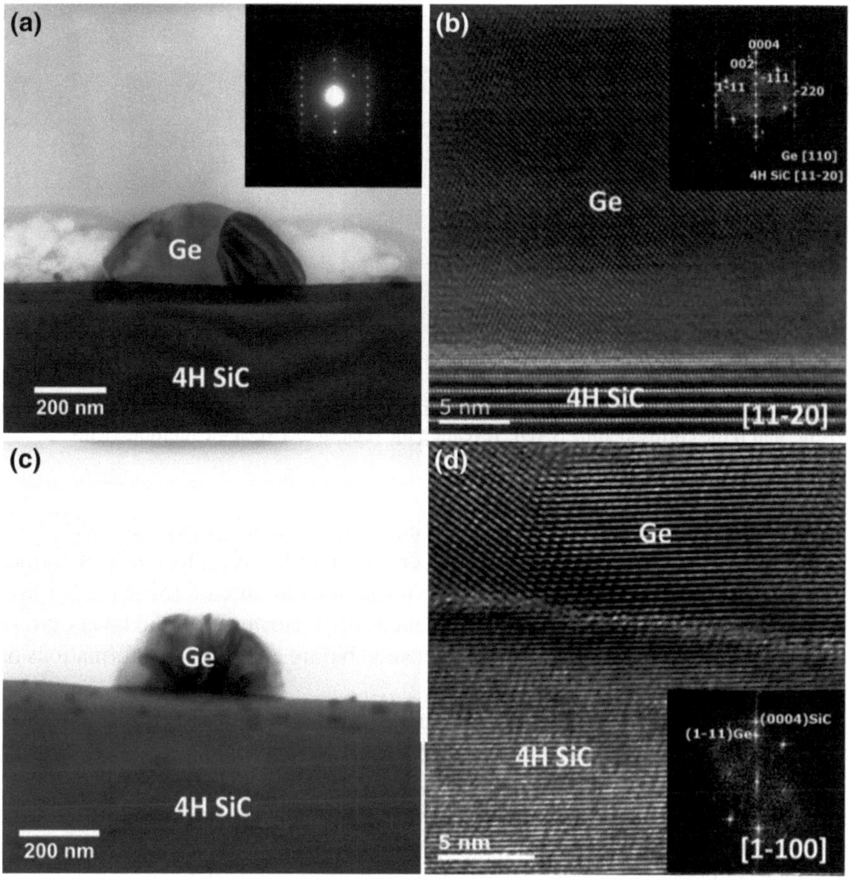

Fig. 1.2 **a** Low Magnification TEM micrograph along [11–20] zone direction and **b** HRTEM micrograph and corresponding FFT (insert) revealing the epitaxial growth of Ge {002}// 4H-SiC{0004}. Low magnification and HRTEM micrograph **c**, **d** along [1–100] zone direction showing that the SiC off-axis step structure favored the epitaxial growth of Ge {1–11} // SiC {0001}

1.4 3C-SiC Grown by Sublimation Epitaxy

In this case one set of samples was grown on the Si-face of 6H-SiC on-axis substrates in vacuum conditions for 30 min at a temperature of 2000 °C and the temperature gradient ranged from 5 to 8 °C/mm [14, 15] and another set was grown at a source temperature of 1775 °C on differently treated prior growth substrates (e.g. as-received by supplier on-axis, Si-face 6H-SiC substrates, additionally polished 6H-SiC substrates or subjected to a deposition of ~1.5 μm (111) 3C-SiC buffer layer, grown by the VLS mechanism) [16]. A general observation concerning the growth on 6H-SiC substrates by SE is that the 3C-SiC epilayer/substrate interface was not always abrupt. During the initial ramp up the growth can continue by competition

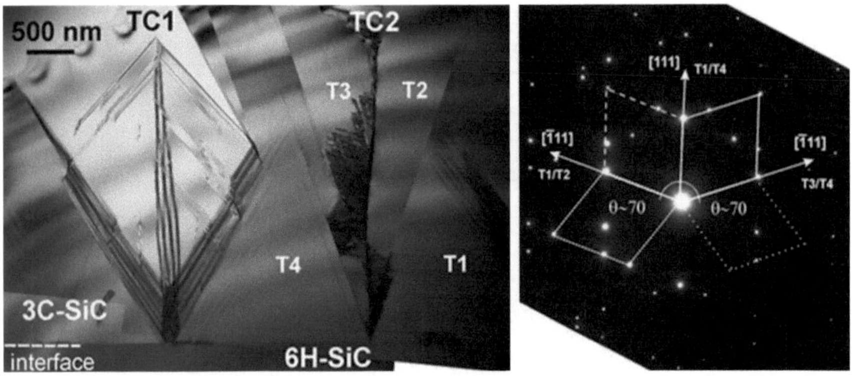

Fig. 1.3 **a** the formation of fourfold twin structures with nucleation cores on the interface with the 6H-SiC and **b** the corresponding SADP, revealing the relation between the multiple twins

between 3C-SiC heteroepitaxial and 6H-SiC homoepitaxial growth. In some cases a transition zone between the 3C-SiC layer and the 6H-SiC substrate was formed where polytypic transformations occurred through admixture of 15R-SiC, 6H-SiC, 3C-SiC and other irregular stacking sequences [16]. However, in the layers grown when a (111) 3C-SiC buffer layer was deposited before SE, such transformations on the 3C-/6H-SiC interface did not occur.

Typical defects appearing in the layers grown on 6H-SiC substrates, apart from SFs and dislocations, were twins and mainly characteristic triangular defects which are fourfold twin complexes (TC) as shown in Fig. 1.3a. These twin structures have their nucleation cores at the substrate/overgrown interface and once formed they propagate through the whole layer terminating on the surface. When the growth temperature is ~2000 °C [15] and close to the interface with the 6H-SiC they appeared in their simplest form. The analysis of the selected area diffraction pattern (SADP) given in Fig. 1.3b shows that the defect TC2 consists of four grains—two of them twinned along the (111) planes without having a common plane (T1 and T4) and another two (T2 and T3) twinned along the $(\bar{1}11)$ planes of T1 and T4, forming two coherent $\Sigma 3$ interfaces (T1/T2 and T3/T4). Consequently the T2/T3 couple is bounded by one $\Sigma 27$ interface. It is reasonable to assume that the twin complex denoted as TC1 in Fig. 1.3a consists of twins along the $(1\bar{1}1)$ and $(11\bar{1})$ planes. Regarding growth performed at lower temperature and/or on 3C-SiC VLS seeds the complexity of the twinning increases. It is accompanied by formation of other high angle boundaries associated with mis-orientation along the $[0\bar{1}1]$ [16].

The alteration of temperature gradients affects the crystalline quality of overgrown layers in two ways. First, by lowering the temperature gradient the amount of 6H-SiC in the layer increases. Detailed analysis of the structure of the observed different 3C-/6H-SiC transformation interfaces can be found in [17]. Second, the lowering of the temperature gradient favours strongly the reduction of SFs in the grown 3C-SiC domains [15].

1.5 Conclusions

Due to its structural qualities, SiC appears particularly interesting from a crystallographic as well as structural defects point of view. At the same time, the above characteristics affect its growth by making it considerably difficult at certain cases. To make an improvement on the wafer technology, in this work, we investigated SiC crystals grown by three different vapor phase methods. Thus, Physical Vapor Transport (PVT), Chemical Vapor Deposition (CVD) and Sublimation Epitaxy (SE) were used for growing 15R-SiC, Ge doped 4H-SiC and 3C-SiC crystals respectively.

PVT grown 15R-SiC suffers with high order defect density. The planar defects affect the main SiC polytype, which are purely depended on stacking fault formation energies. This fact suggest that may be higher energy is required to form micro-twins which are observed near 15R-SiC/4H-SiC interface. TEM investigation on Ge doped 4H-SiC revealed that CVD technique can be considered suitable growth method for Ge Islands. Growth direction SiC{0001}//Ge {111} found to be the preferred one while the off-axis SiC substrate seems to have favored the epitaxial growth. In case of 3C-SiC layers grown by SE, apart from the typical stacking faults and dislocations some characteristic fourfold twin complexes appeared. Additionally, the thermal gradient decrease led to significant lower stacking fault density but higher amount of 6H-SiC overgrown.

Acknowledgments This work was supported by the European Commission through the NetFISiC and MANSiC projects (Grant no. PITN-GA-2010-264613 and Grant no. MRTN-CT-2006-035735). The authors would like to thank Dr. M. Syväjärvi and professor R.Yakimova from the Department of Physics, Chemistry and Biology in Linköping University, Dr. G. Ferro from Laboratoire des Multimateriaux et Interfaces in Claude Bernard University in Lyon and D. Chaussende from Laboratoire des Materiaux et Genie, INPG Grenoble, for providing the samples.

References

1. J.A. Coopper Jr., Opportunities and technical strategies for silicon carbide device development. Mater. Sci. Forum **15**, 389–393 (2002)
2. C.I. Harris, S. Savage, A. Konstantinov, M. Bakowski, P. Ericsson, Progress towards SiC products. Appl. Surf. Sci. **184**, 393 (2001)
3. N. Schulze, D. Barrett, G. Pensl, Controlled growth of 15R-SiC single crystals by the modified Lely method. Phys. Status Solidi A **178**(2), 645 (2000)
4. T. Kups, M. Voelskow, W. Skorupa, M. Soueidan, G. Ferro, J. Pezoldt, Lattice location determination of Ge in SiC by ALCHEMI, in *Microscopy of Semiconducting Materials 2007*, vol. 120 (2008), pp. 353–358
5. S.A. Reshanov, I.I. Parfenova, V.P. Rastegaev, Group III-V impurities in β-SiC: lattice distortions and solubility. Diamond Relat. Mater. 1278–1282 (2001)
6. M. Bhatnagar, B.J. Baliga, Comparison of 6H-SiC, 3C-SiC, and Si for power devices. IEEE Trans. Devices **40**, 645 (1993)
7. S. Chung, V. Wheeler, R. Myers-Ward, C.R. Eddy, D.K. Gaskill, P. Wu, Y.N. Picard, M. Skowronski, Direct observation of basal-plane to threading-edge dislocation conversion in 4H-SiC epitaxy. J. Appl. Phys. **109**(9), 094906 (2011)

8. L. Barrett Donovan, N. Schulze, G. Pensl, S. Rohmfeld, M. Hundhausen, Near-thermal equilibrium growth of SiC by physical vapor transport. Mater. Sci. Eng. **44**, B61–62 (1999)
9. P. Pirouz, J.W. Yang, Polytypic transformations in SiC: the role of TEM. Ultramicroscopy **51**(1), 189 (1993)
10. Yu. Zhang, Chen Hui, Choi Gloria, Raghothamachar Balaji, Dudley Michael, H. Edgar James, K. Grasza, E. Tymicki, L. Zhang, D. Su, Y. Zhu, Nucleation mechanism of 6H-SiC polytype inclusions inside 15R-SiC crystals. J. Electron. Mater. **39**(6), 799 (2010)
11. G. Agrosì, G.C. Capitani, E. Scandale, G. Tempesta, Near-atomic images of interfaces between twin-related lamellae in a synthetic 6H-SiC sample. Phys. Chem. Miner. **38**(2), 101 (2010)
12. B. Schröter, K. Komlev, U. Kaiser, G. Hess, G. Kipshidze, W. Richter, Germanium on SiC (0001): surface structure and nanocrystals. Mater. Sci. Forum **247**, 353–356 (2001)
13. K. Aït-Mansour, D. Dentel, L. Kubler, M. Diani, J.L. Bischoff, D. Bolmont, Epitaxy relationships between Ge-islands and SiC(0001). Appl. Surf. Sci. **241**, 403 (2005)
14. M. Beshkova, M. Syväjärvi, R. Vasiliauskas, J. Birch, R. Yakimova, Structural properties of 3C-SiC grown by sublimation epitaxy. Mater. Sci. Forum **181**, 615–617 (2009)
15. M. Marinova, A. Mantzari, M. Beshkova, M. Syväjärvi, R. Yakimova, E.K. Polychroniadis, the influence of the temperature gradient on the defect structure of 3C-SiC grown heteroepitaxially on 6H-SiC by sublimation epitaxy. Mater. Sci. Forum 645–648, 367–370 (2010)
16. R. Vasiliauskas, M. Marinova, M. Syväjärvi, A. Mantzari, A. Andreadou, J. Lorenzzi, G. Ferro, E.K. Polychroniadis, R. Yakimova, Sublimation growth and structural characterization of 3C-SiC on hexagonal and cubic SiC seeds. Mater. Sci. Forum 645–648, 175–178 (2010)
17. M. Marinova, A. Mantzari, M. Beshkova, M. Syväjärvi, R. Yakimova, E.K. Polychroniadis, TEM investigation of the 3C/6H-SiC transformation interface in layers grown by sublimation epitaxy. Solid State Phenomena **163**, 97–100 (2010)

Chapter 2
MWI Synthesis and Characterization of RGO-AgNPs/Polymethyl Methacrylate Nanocomposites

Edreese H. Alsharaeh

Abstract Polymethyl methacrylate and reduced graphene oxide/silver (PMMA/RGO/(RGO))AgNPs) nanocomposites were prepared via in situ bulk polymerization method using two different preparation techniques. In the first approach, mixture of silver nitrate, Hydrazine hydrate and polymethyl methacrylate containing graphene oxide (PMMA/GO) were reduced by microwave irradiation (MWI) to obtain R-(GO-PMMA)/AgNPs nanocomposites. In the second approach, a mixture of (RGO/AgNPs) nanocomposite, which was produced by microwave, and MMA monomers were polymerized using in situ bulk polymerization method to obtain PMMA-RGO/AgNPs nanocomposites. The two nanocomposites were compared and characterized by FTIR, XRD, XPS, HRTEM, and TGA. The results indicate that the nanocomposites obtained using the first approach, which involved MWI, had a better morphology and dispersion with enhanced thermal stability compared with the nanocomposites pre-pared without MWI.

2.1 Introduction

Graphene (GR) is an excellent two-dimensional filler material for polymer composites for application in many technological fields [1–5]. Good dispersion is crucial for achieving the desired enhancement in the final physical and chemical properties of the composites [6]. However, one of the challenges is achieving good dispersion of the nanoscale filler in the composites, especially for GR, which has a strong tendency to agglomerate through the intrinsic van der Waals forces and $\pi-\pi$ stacking due to its hydrophobic nature [7]. Various techniques have been developed for the synthesis of these polymer-graphene composites, including solution mixing; melt blending and in-situ polymerization [8–10]. Recently, a novel approach for the production of metal

E. H. Alsharaeh (✉)
Department of Chemistry, Alfaisal University, P.O. Box 50927, Riyadh 11533, Saudi Arabia
e-mail: ealsharaeh@alfaisal.edu

E. K. Polychroniadis et al. (eds.), *International Multidisciplinary Microscopy Congress*,
Springer Proceedings in Physics 154, DOI: 10.1007/978-3-319-04639-6_2,

nanoparticles (MNPs) supported on reduced graphene oxide nanosheets (RGO) via MWI method is reported, the resulting product is composed of metal nanoparticle supported on graphene sheets with polar functional groups even after the reduction [11]. Polymethylmethacrylate (PMMA) is an important polymeric material that has been used in medicine (bone cement), dentistry, radiation shield, the paper, paint and automobile industries and in many other applications [11–16]. Several studies have reported the successful incorporation of GR nanosheets into the PMMA matrix with different preparation techniques using various methods of GR preparation, such as from reduced GO, functionalized RGO sheets or chemical vapor deposition [17–20], however, there is no published reports on such nanocomposites that is functionalized by the in situ bulk polymerization with (RGO/AgNPs) as a filler. In this letter, we report two different synthesis techniques of (PMMA/RGO/AgNPs) nanocomposites by employing in situ bulk polymerization via MWI.

2.2 Experimental Methods

2.2.1 Preparation of RGO/AgNPs

A mixture of 25 ml of 0.02 M of AgNO3 and 0.40 g GO [21], were stirred and sonicated for 1 h. Then the solution was reduced via conventional microwave oven after adding 400 μl of the hydrazine hydrate (HH) for 1.2 min. For comparison, AgNPs were pre-pared using the same method but without the addition of GO.

2.2.2 In Situ Preparation of PMMA-RGO/AgNPs Nanocomposites

A mixture of 2.0 g MMA, 0.04 g of (RGO/AgNPs), and 0.1 g of benzoyl peroxide (BP) initiator was sonicated for 1 h, and then the mixture were maintained at 60 °C for 20 h to promote *in situ* free radical bulk polymerization. The neat PMMA was prepared for comparison using the same procedure, but no RGO/AgNPs were added.

2.2.3 Preparation of R-(PMMA-GO)/AgNPs Nanocomposite via MWI

A mixture of 2.0 g of MMA, 0.04 g of GO, and 0.1 g of (BP) was sonicated for 1 h, then it was polymerized by the in situ bulk polymerization method as in the previous section, to produce GO-PMMA composites. Then a mixture of 0.40 g (GO-PMMA), 400 μl HH, and 0.08 g of $AgNO_3$ were then sonicated for 1 h and then reduced using MWI.

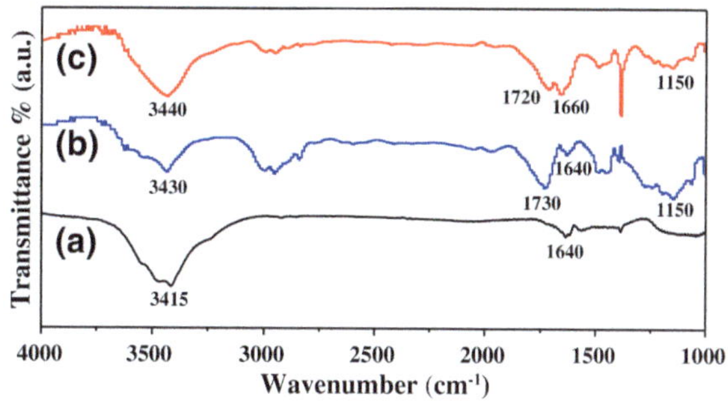

Fig. 2.1 FTIR spectra of **a** RGO/AgNPs, **b** PMMA-RGO/AgNPs, and **c** R-(PMMA-GO)/AgNPs nanocomposites

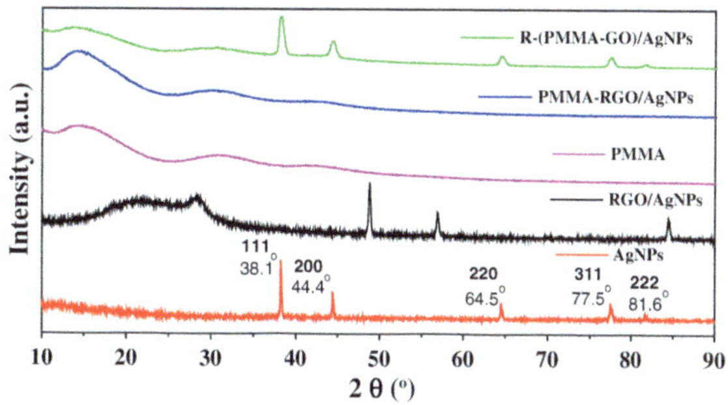

Fig. 2.2 XRD patterns of PMMA/RGO/AgNPs nanocomposites

2.3 Results and Discussion

Figure 2.1 shows the FTIR spectra of the all PMMA/RGO/AgNPs nanocomposites. For the PMMA-RGO/AgNPs (Fig. 2.1b), the spectrum shows the presence of the characteristic bands of PMMA/RGO. The bands at 3430, 1730, 1640, 1150 cm^{-1} that correspond with the O–H, C=O, C=C, C–O–C groups, respectively. This confirms the presence of PMMA/RGO composite. Interestingly, for R-(PMMA-GO)/AgNPs, (Fig. 2.1c) the spectra show that some of these characteristics peaks are much shifted to 3440 and 1660 cm^{-1} in comparison with (Fig. 2.1b) when the microwave was not used, this may suggest that the acrylate group of PMMA and RGO sheets opened by the MWI, which will induce more electron chain transfer sites, and therefore will promote more interfacial interactions between RGO and AgNPs within PMMA matrix [20]. Moreover, (Fig. 2.1c) shows that in the C=O stretching (at 1720 cm^{-1}), the aliphatic C–H, and $-CH_2$ (at 2950, 2840 cm^{-1}), and the C–O–C stretching (at 1150 cm^{-1}) regions, the bands were reduced in intensity compared to (Fig. 2.1b).

XRD analyses of PMMA/RGO/AgNPs composites were conducted. The XRD pattern of the AgNPs (Fig. 2.2) reflects a peaks at 2θ values of about 38.1°, 44.4°, 64.5°, 77.5°, and 81.6° representing the (111), (200), (220), (311), and (222) indicating the formation of metallic silver (Ag) with face-centered cubic (fcc) symmetry [22]. For PMMA-RGO/AgNPs, it is difficult to find the AgNPs in the XRD pattern,

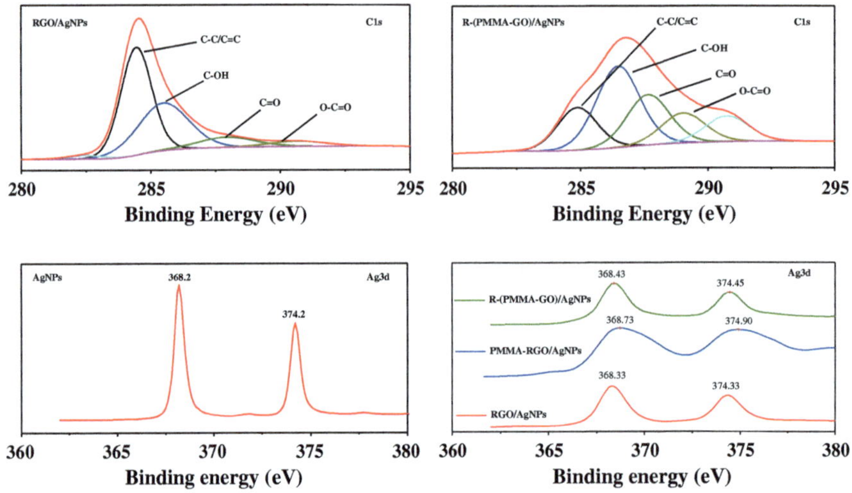

Fig. 2.3 C1s and Ag3d XPS spectra of PMMA/RGO/AgNPs nanocomposites

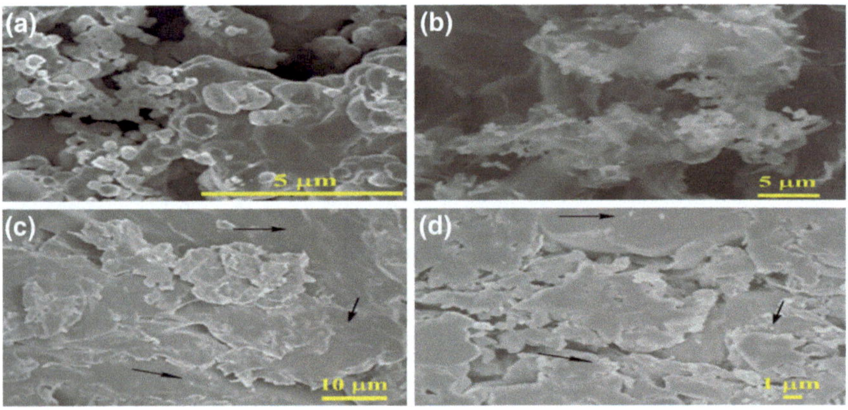

Fig. 2.4 The SEM images of **a** AgNPs, **b** RGO/AgNPs, **c** PMMA-RGO/AgNPs, **d** R-(PMMA-GO)/AgNPs nanocomposites

which might suggest that the AgNPs are intercalated within the PMMA/RGO composites.

The formation of AgNPs-RGO/PMMA nanocomposites were further confirmed by XPS and is summarized in Fig. 2.3. Note that the oxygenated functionalities were fully reduced when the MWI is used indicating the formation of RGO-AgNPs within the PMMA composites.

The morphology of the nanocomposites was studied by SEM as displayed in Fig. 2.4. The SEM image of the RGO/AgNPs-PMMA nanocomposites prepared via in situ method (Fig. 2.4c) shows that the RGO/AgNPs wrinkled and crumpled profile was observed and well dispersed within the PMMA matrix. When the MWI were used

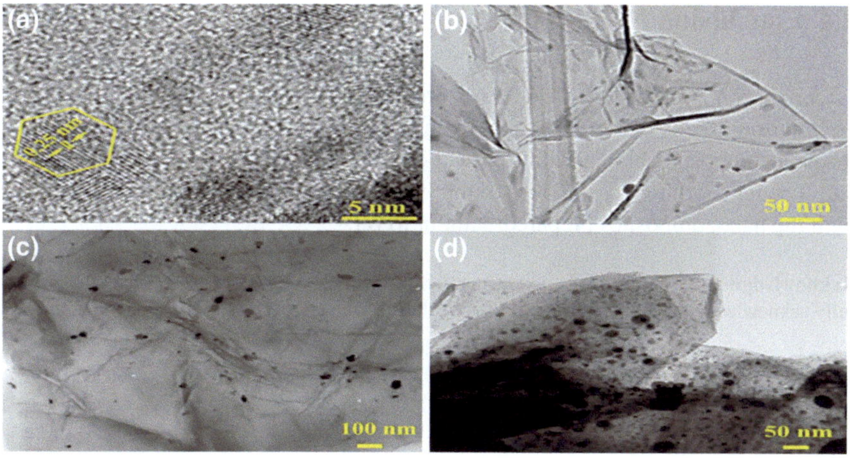

Fig. 2.5 The HRTEM images of **a** AgNPs, **b** RGO/AgNPs, **c** PMMA-RGO/AgNPs, **d** R-(PMMA-GO)/AgNPs nanocomposites

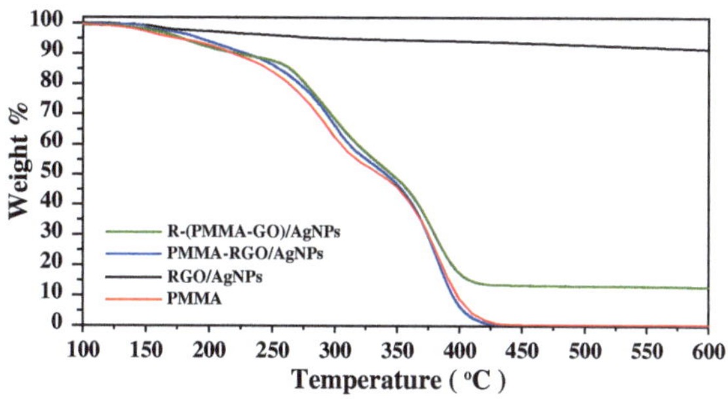

Fig. 2.6 TGA thermograms of PMMA/RGO/AgNPs nanocomposites

R-(PMMA-GO)/AgNPs (Fig. 2.4d), the AgNPs (white spots) were well dispersed within PMMA matrix, and embedded within the matrix.

The morphology of the nanocomposites was also studied by using HRTEM as displayed in Fig. 2.5. Note that RGO-AgNPs sheets are weakly dispersed within PMMA matrix (Fig. 2.5c). However, when the MWI were used R-(PMMA-GO)/AgNPs (Fig. 2.5d), the image clearly shows the RGO/AgNPs is well dispersed within PMMA matrix, and embedded within the matrix.

Thermal gravimetric analyses (TGA) were used to investigate the thermal stability and the interfacial interaction between RGO/AgNPs and the PMMA matrix. Interestingly, in the case of the R-(PMMA-GO)/AgNPs, the curve clearly shows strong enhancement in the thermal stability in comparison with the neat PMMA and PMMA-RGO/AgNPs. We attributed this result to the homogenous dispersion and random alignment of the RGO-AgNPs filler within the PMMA matrix. Therefore, incorporation of RGO/AgNPs by MWI can improve the thermal stability of the composites (Fig. 2.6).

2.4 Conclusion

In conclusion, we report a new *in situ* method to incorporate RGO/MNPs within polymers matrix by the in situ bulk polymerization via MWI. The results indicate that the nanocomposite obtained using our experimental approach, which involved MWI, had a better morphology and dispersion with enhanced thermal stability compared with the nanocomposites prepared without MWI.

Acknowledgments This study is part of research Project No. 405010111127. The authors gratefully acknowledge the continued support from Alfaisal University and its Office of Research.

References

1. A.A. Balandin, S. Ghosh, W.Z. Bao, I. Calizo, D. Teweldebrhan, F. Miao, C.N. Lau, Superior thermal conductivity of single-layer graphene. Nano Lett. **8**(3), 902–907 (2008). doi:10.1021/nl0731872
2. S. Park, R.S. Ruoff, Chemical methods for the production of graphenes. Nat. Nanotechnol. **4**(4), 217–224 (2009). doi:10.1038/nnano.2009.58
3. S. Stankovich, D.A. Dikin, G.H.B. Dommett, K.M. Kohlhaas, E.J. Zimney, E.A. Stach, R.D. Piner, S.T. Nguyen, R.S. Ruoff, Graphene-based composite materials. Nature **442**(7100), 282–286 (2006). doi:10.1038/nature04969
4. T. Ramanathan, A.A. Abdala, S. Stankovich, D.A. Dikin, M. Herrera-Alonso, R.D. Piner, D.H. Adamson, H.C. Schniepp, X. Chen, R.S. Ruoff, S.T. Nguyen, I.A. Aksay, R.K. Prud'homme, L.C. Brinson, Functionalized graphene sheets for polymer nanocomposites. Nat. Nanotechnol. **3**(6), 327–331 (2008). doi:10.1038/nnano.2008.96
5. D.A. Dikin, S. Stankovich, E.J. Zimney, R.D. Piner, G.H.B. Dommett, G. Evmenenko, S.T. Nguyen, R.S. Ruoff, Preparation and characterization of graphene oxide paper. Nature **448**(7152), 457–460 (2007). doi:10.1038/nature06016
6. D.R. Paul, L.M. Robeson, Polymer nanotechnology: nanocomposites. Polymer **49**(15), 3187–3204 (2008). doi:10.1016/j.polymer.2008.04.017
7. J.R. Potts, D.R. Dreyer, C.W. Bielawski, R.S. Ruoff, Graphene-based polymer nanocomposites. Polymer **52**(1), 5–25 (2011). doi:10.1016/j.polymer.2010.11.042
8. C. Lee, X.D. Wei, J.W. Kysar, J. Hone, Measurement of the elastic properties and intrinsic strength of monolayer graphene. Science **321**(5887), 385–388 (2008). doi:10.1126/science.1157996
9. H.M.A. Hassan, V. Abdelsayed, A. Khder, K.M. AbouZeid, J. Terner, M.S. El-Shall, S.I. Al-Resayes, A.A. El-Azhary, Microwave synthesis of graphene sheets supporting metal nanocrystals in aqueous and organic media. J. Mater. Chem. **19**(23), 3832–3837 (2009). doi:10.1039/b906253j
10. J.Y. Jang, M.S. Kim, H.M. Jeong, C.M. Shin, Graphite oxide/poly(methyl methacrylate) nanocomposites prepared by a novel method utilizing macroazoinitiator. Compos. Sci. Technol. **69**(2), 186–191 (2009). doi:10.1016/j.compscitech.2008.09.039
11. S. Pasupuleti, G. Madras, Ultrasonic degradation of poly (styrene-co-alkyl methacrylate) copolymers. Ultrason. Sonochem. **17**(5), 819–826 (2010). doi:10.1016/j.ultsonch.2010.02.003
12. M. Coşkun, P. Seven (2011) Synthesis, characterization and investigation of dielectric properties of two-armed graft copolymers prepared with methyl methacrylate and styrene onto PVC using atom transfer radical polymerization. React. Funct. Polym. **71** (4), 395–401. doi: http://dx.doi.org/10.1016/j.reactfunctpolym.2010.12.012

13. M.S. El-Shall, V. Abdelsayed, A.E.R.S. Khder, H.M.A. Hassan, H.M. El-Kaderi, T.E. Reich, Metallic and bimetallic nanocatalysts incorporated into highly porous coordination polymer MIL-101. J. Mater. Chem. **19**(41), 7625–7631 (2009). doi:10.1039/b912012b
14. J.R. Potts, S.H. Lee, T.M. Alam, J. An, M.D. Stoller, R.D. Piner, R.S. Ruoff, Thermomechanical properties of chemically modified graphene/poly(methyl methacrylate) composites made by in situ polymerization. Carbon **49**(8), 2615–2623 (2011). doi:10.1016/j.carbon.2011.02.023
15. R.D. Priestley, C.J. Ellison, L.J. Broadbelt, J.M. Torkelson, Structural relaxation of polymer glasses at surfaces, interfaces, and in between. Science **309**(5733), 456–459 (2005). doi:10.1126/science.1112217
16. M. Fang, K. Wang, H. Lu, Y. Yang, S. Nutt, Single-layer graphene nanosheets with controlled grafting of polymer chains. J. Mater. Chem. **20**(10), 1982–1992 (2010). doi:10.1039/b919078c
17. X.Y. Yuan, L.L. Zou, C.C. Liao, J.W. Dai, Improved properties of chemically modified graphene/poly(methyl methacrylate) nanocomposites via a facile in-situ bulk polymerization. Exp. Polym. Lett. **6**(10), 847–858 (2012). doi:10.3144/expresspolymlett.2012.90
18. J.C. Wang, H.T. Hu, X.B. Wang, C.H. Xu, M. Zhang, X.P. Shang, Preparation and mechanical and electrical properties of graphene nanosheets-poly(methyl methacrylate) nanocomposites via in situ suspension polymerization. J. Appl. Polym. Sci. **122**(3), 1866–1871 (2011). doi:10.1002/app.34284
19. S. Gilje, S. Han, M. Wang, K.L. Wang, R.B. Kaner, A chemical route to graphene for device applications. Nano Lett. **7**(11), 3394–3398 (2007). doi:10.1021/nl0717715
20. M. Aldosari, A. Othman, E. Alsharaeh, Synthesis and characterization of the in situ bulk polymerization of PMMA containing graphene sheets using microwave irradiation. Molecules **18**(3), 3152–3167 (2013)
21. W.S. Hummers, R.E. Offeman, Preparation of graphitic oxide. J. Am. Chem. Soc. **80**(6), 1339–1339 (1958). doi:10.1021/ja01539a017
22. L.-B. Luo, S.-H. Yu, H.-S. Qian, T. Zhou, Large-scale fabrication of flexible silver/cross-linked poly(vinyl alcohol) coaxial nanocables by a facile solution approach. J. Am. Chem. Soc. **127**(9), 2822–2823 (2005). doi:10.1021/ja0428154

Chapter 3
The Effect of Pulse Plasma Treatment on the Corrosion Resistance of AISI 4140 Steel

Gencer Ağdaş, A. Şükran Demirkıran and Yıldız Yaralı Özbek

Abstract Pulse plasma treatment is a surface modification treatment process, which is widely used to improve wear, fatigue and corrosion resistance of the industrial parts. The paper presents the results obtained from an experimental work investigating the changes in corrosion and microstructural surface properties of AISI 4140 steel. For the pulse plasma treatment, a battery capacity of 800 μF, and three different sample plasma gun nozzle distances were used. The samples were subjected to micro-hardness measurements and it was found that the hardness values of modified surfaces are four times higher than untreated samples. The modified surface layers were examined using an optical microscope and X-ray analyses were carried out all samples. The new phases were found on surface. The samples were exposed to corrosion test in 0.5 M NaCl aqueous solution. The corrosion behavior of samples was determined. An attempt was made to establish an empirical relationship between corrosion resistance and parameters of pulse plasma. The presence of modified layer has improved to corrosion resistance, and it was observed that corrosion cracks were initiated dominantly under the modified layer by pit formation mechanism.

3.1 Introduction

The improvement of mechanical properties and wear resistance of hard alloy is a crucial problem. Different types of treatment techniques were used to find a solution for this issue: coatings deposition, treatment by pulsed high-power ion beam [1, 2], electron beam [3, 4], pulsed laser irradiation [5] and others.

The interaction of intense-pulsed energetic beams such as ion, electron and laser beams with materials and its application in industry have attracted much attention

G. Ağdaş · A. Şükran Demirkıran · Yıldız Yaralı Özbek (✉)
Engineering Faculty, Department of Metallurgy and Materials Eng. Esentepe Campus, Sakarya University, 54187 Sakarya, Turkey
e-mail: yyarali@sakarya.edu.tr

E. K. Polychroniadis et al. (eds.), *International Multidisciplinary Microscopy Congress*, Springer Proceedings in Physics 154, DOI: 10.1007/978-3-319-04639-6_3,

over the past few decades. Indeed, each pulse of concentrated energy flux causes super-fast thermal cycles and modify the surface in a way to impart the material with improved physicochemical and strength properties that are often unattainable with other more conventional surface treatment methods. Pulse plasma process is a relatively new technique developed for the surface modification of metallic materials.

Its main feature is the generation of an electron beam with low energy short pulse and high peak current density. The interaction of the pulsed electron beam with the material can introduce very fast heating and melting in the surface layer of the material that can be followed by a very fast cooling process due to the heat conduction toward the cold substrate. These non-equilibrium processes can easily change the microstructure, chemical composition, phase structure and stress state at the surface.

The pulse plasma process is used to improve surface properties of workpieces of tool steels [3–5]. When irradiated by pulse treatment the near surface layer of targets undergoes a rapid melt and solidification with heating and cooling rates typically 10^7–10^{10} K/s. These rates can promote mixing, rapid diffusion, formation of amorphous or microcrystalline surface layers, hence can improve mechanical performance of material surfaces [6].

In this study, different plasma parameters were applied to the AISI 4140 steel surfaces to investigate the effect of pulse number. After treatment, the cross-sectional microstructures of specimens were observed by means of an optical microscope (Zeiss). The phases in the modified layers were identified by a Rigaku X-ray diffractometer using Cu Kα radiation. The corrosion test was done. Consequently, corrosion surfaces were inspected by using a Jeol scanning electron microscope (SEM JSM-5600).

3.2 Experimental Procedure

In this work, the AISI 4140 steel was used for surface modification. The composition of this steel was given in Table 3.1. The specimens were cut out from square bars and subsequently machined and grinded in order to obtain the desired shape and surface roughness. Before pulse plasma treatments, all samples were polished progressively to 1200-grit with grinding paper to obtain identical surface conditions. The pulse plasma treatment was done only work surface of samples. The parameters of pulse plasma were given in Table 3.2.

The microstructures of the base material and the nitrided layers were examined using an Zeiss optical microscope (OM) and a JEOL JSM-5600 scanning electron microscope (SEM); X-ray diffraction (XRD) investigations of the compound layer were carried out using a Rigaku X-ray diffractometer with Cu Kα radiation. The microhardness of the modified surface was measured using a Future-Tech automatic tester with a small load of 50 g for 15 s.

The samples were exposed to corrosion test in 0.5 M NaCl aqueous solution. The corrosion behavior of samples was determined. An attempt was made to establish an

Table 3.1 The chemical composition of AISI 4140 steel

% Elements	C	Si	Mn	P	S	Cr	Mo
AISI 4140	0.39	0.30	0.80	0.035	0.015	0.90	0.13

Table 3.2 The parameters of pulse plasma treatment

No	Distances (mm)	Electrode	Number of pulse	Battery capacity (mf)
1	40	Mo	5	800
2	40	Mo	10	800
3	40	Mo	15	800
4	50	Mo	5	800
5	50	Mo	10	800
6	50	Mo	15	800
7	60	Mo	5	800
8	60	Mo	10	800
9	60	Mo	15	800
10	70	Mo	10	800
11	70	Mo	15	800

empirical relationship between corrosion resistance and parameters of pulse plasma. The corrosion regions were investigated by scanning electron microscope (SEM) and energy dispersive spectrometer (EDS).

3.3 Results and Discussion

Modified layer's morphology was investigated by using optical microscope. Figure 3.1 shows the optical micrograph of the modified steel samples. Microstructural analysis has shown that pulse plasma application resulted in formation of a uniform structure, different contrast layers. The only difference between sample 2 and sample 5 specimens (Fig. 3.1a and b) is short specimens-nozzle distance. The thickness of modified layer was changed with number of pulse. Due to decreasing nozzle distance in 1 (Fig. 3.1b), grain coarsening is visible compared with sample 1 and sample 3. Decreasing nozzle distance is thought to cause annealing due to excessive heating at the surface and this situation may be leading to recrystallization and grain coarsening (Fig. 3.1c and d) [7–9].

The pulse plasma system has affected grain size in modified layer [4]. The grain structure of outer surface is very fine and dense in modified layer (Fig. 3.1b). Previous studies showed that the thermal stability, mechanical and tribological properties and corrosion behaviour of materials are greatly influenced by the grain size [2, 3, 6].

Figure 3.2 shows SEM results cross-section of modified layer. It can be observed that a diffusion zone with a compound layer present on the surface of the specimen

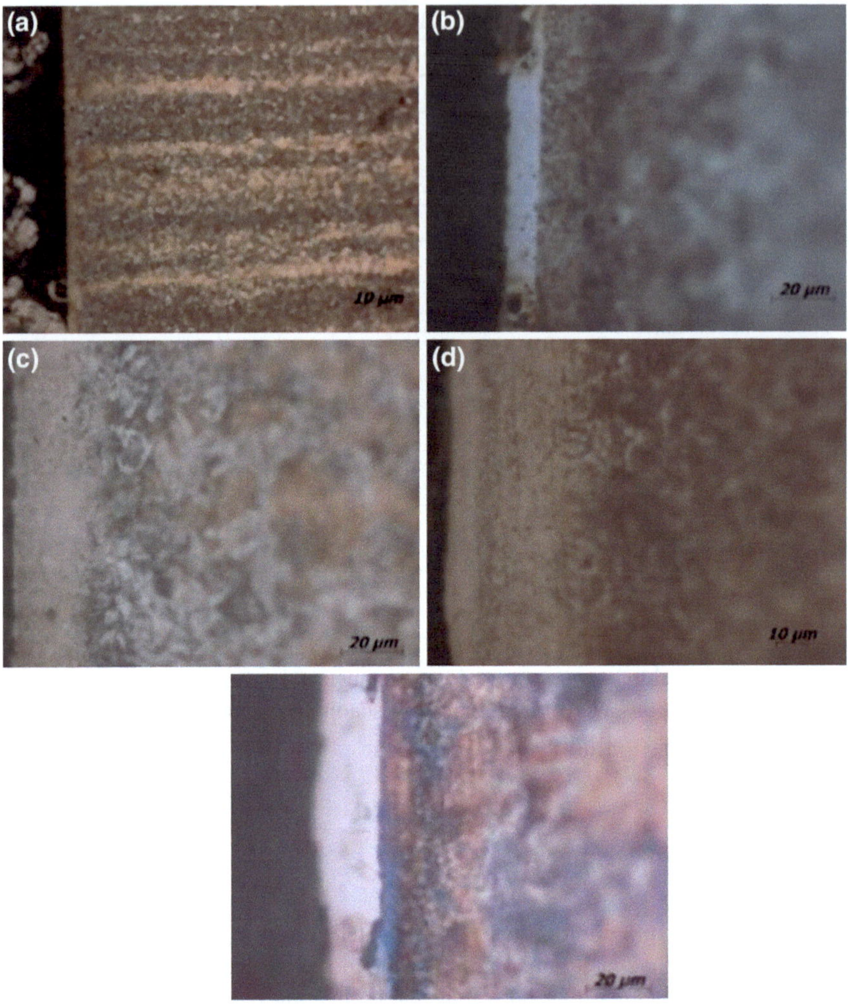

Fig. 3.1 The optic micrograph of **a** sample 1, **b** sample 3, **c** sample 6, **d** sample 9, **e** sample 11

(Fig. 3.2a). In the diffusion zone, some precipitations on the grain boundaries are also visible. The dendritic structure was formed.

Figure 3.3 shows the XRD patterns obtained from different numbers of pulse were applied to samples. The un-treatment specimen showed only α-Fe indicating that there was no retained austenite on the microstructure of the received state. On the other hand, different phases could be identified in the modified layer: the α-Fe phase and phases with molybdenum. On increase distance of nozzle, the intensity of the α-Fe peaks progressively decreased and the intensity of the new phase's peaks progressively increased [3].

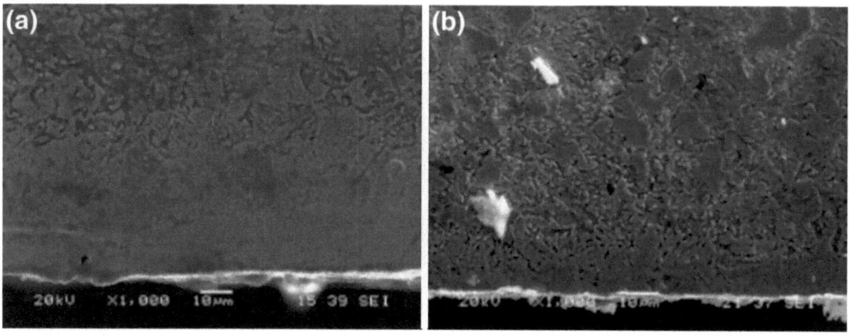

Fig. 3.2 The SEM micrographs of cross-section for **a** sample 6, **b** sample 11

Fig. 3.3 XRD results of distance effect to amount of phases on surface

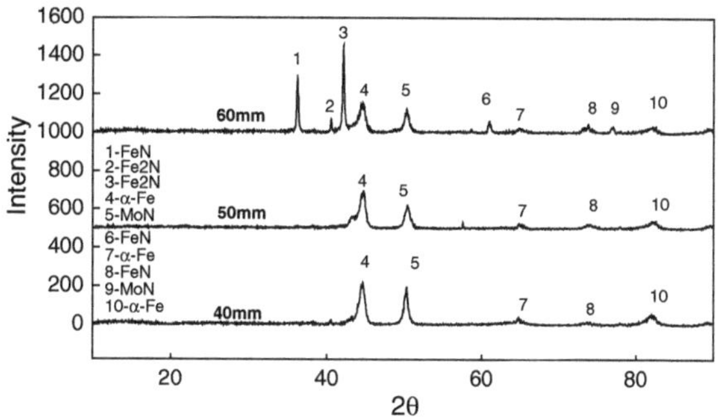

Also, the distance of nozzle-sample is very important to surface properties. This factor affects to amount of forming phases. When the distance was increase to 60 mm, the amount of phases was increase and different phases formed on surface. The broading of peak increases by decreasing the size of grain in 60 mm distance.

The microhardness values of modified samples increased from 180 to 950 HV. This increasing is very important to mechanical improving [6, 9].

The SEM of corroded surface for untreatment samples was given Fig. 3.4a. There were very big cracks on untreatment surface. The intergranular corrosion was occurred on surface. Figure 3.4b–d were given micrograph of surface after corrosion test. The amounts of crack reduce on treated surface [7, 9].

Also, the cracks are bigger and all located at the center of surface. For the 5-pulsed sample, only one pit could be found on the whole tested surface, as shown in Fig. 3.4c. This means that the possibility of cracks was significantly reduced after 15 pulses of pulse plasma treatment.

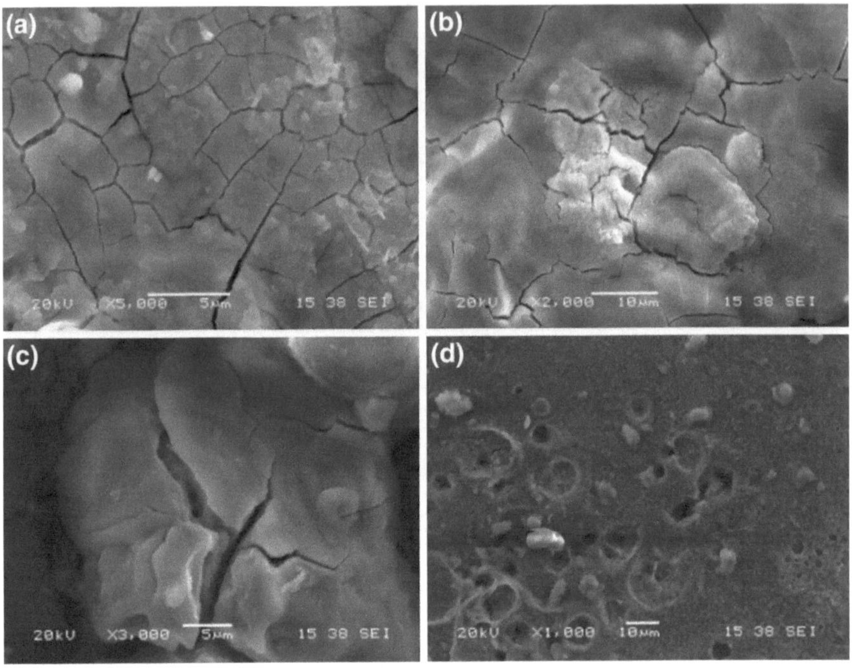

Fig. 3.4 The SEM micrograph **a** untreatment sample, **b** sample 3, **c** sample 6, **d** sample 9

3.4 Conclusion

In this work, the effect of pulse plasma treatment on AISI 4140 steel surface was investigated. Pulse plasma process resulted high surface properties in short processing times with a rather simple operation. Indeed, each pulse of concentrated energy flux causes super-fast thermal cycles and modify the surface in a way to impart the material with improved physicochemical and strength properties that are often unattainable with other more conventional surface treatment methods. The microhardness values increased 4–5 times according to untreatment samples.

The pulse plasma surface modified process has produced a different microstructure and notable grain refinement. It has been observed that the pulse plasma parameters (e.g. number of pulse) play an important role in generating the different surface properties.

The corrosion resistance of steel after pulse plasma treatment was improved.

References

1. N.Y. Tyurın, O.V. Kolısnıchenko, N.G. Tsygankov, Pulse plasma hardening of tolls. The Paton Welding J. **1**, 38–43 (2004)
2. C. Kwietniewski, W. Fontana, C. Moraes, A.S. Rocha, T. Hirsch, A. Reguly, Nitrided layer embrittlement due to edge effect on duplex treated AISI M2 high-speed steel. Surf. Coat. Technol. **179**, 27–32 (2004)
3. K.M. Zhang, J.X. Zou, B.Bolle, T. Grosdidier, Crater-formation-induced metastable structure in an AISI D2 steel treated with a pulsed electron beam, Vacuum, 60–68 (2013)
4. Y. Hao, B. Gao, G.F. Tu, H. Cao, S.Z. Hao, C. Dong, Surface modification of Al-12.6Si alloy by high current pulsed electron beam, Appl. Surf. Sci., 2052–2056 (2012)
5. X. Wang, M.K. Lei, J.S. Zhang, Surface modification of 316L stainless steel with high-intensity pulsed ion beams. Surf. Coat. Technol. **201**, 5884–5890 (2007)
6. M. Soltanieh, H. Aghajani, F. Mahboubi, KhA Nekouee, Surface characterization of multiple coated H11 hot work tool steel by plasma nitriding and hard chromium electroplating processes. Vacuum **86**, 1470–1476 (2012)
7. V.V. Uglov, V.M. Anishchik, N.N. Cherenda, YuV Sveshnikov, V.M. Astashynski, E.A. Kostyukevich, A.M. Kuzmitski, V.V. Askerko, The formation of a tungsten containing surface layer in a carbon steel by compression plasma flow. Surf. Coat. Technol. **202**, 2439–2442 (2008)
8. Y.Y. Ozbek, M. Durman, H. Akbulut, Wear behavior of AISI 8620 steel modified by a pulse-plasma technique. Tribol. Trans. **52**, 213–222 (2009)
9. Dong-Cherng Wen, Plasma nitriding of plastic mold steel to increase wear- and corrosion properties. Surf. Coat. Technol. **204**(4), 511–519 (2009)

Chapter 4
Production of Ni_3Al and Ti_3Al Based Coating by Using Pressure Assisted Combustion Synthesis

Nuri Ergin, Gokhan Yoruk and Ozkan Ozdemir

Abstract In the present work, Ti_3Al and Ni_3Al coatings were carry out on AISI 1010 steel by pressure assisted combustion synthesis method in open atmosphere under an uniaxial pressure of 150 MPa at 1050 °C for 60 min. Starting powders were used titanium powder with 45 μm size, aluminium powder with 15 μm and nickel powder with 4–7 μm size having 99.5, 99 and 99.8 % purity, respectively. Microstructural examinations showed that the Ti_3Al coating has multi-phase microstructure with porosity. The XRD patterns of the test sample indicated that the TiAl, Ti_3Al, Ti_3O and Ti phases. The hardness value of Ti_3Al coating layer was about $520 \pm 6.9\,Hv_{0.05}$. Microstructural and XRD examinations showed that the Ni_3Al coating has only single-phase microstructure and Ni_3Al phase is present. The hardness value of Ni_3Al coating layer was about $377.7 \pm 20.6\,Hv_{0.05}$.

4.1 Introduction

In recent years, several intermetallic compound coatings have been developed and successfully employed to produce ferrous and non-ferrous surfaces. This is due to the superior properties that these coatings possess (e.g., excellent dependability at elevated temperatures, low density, resistance to oxidation, corrosion, and wear), which make them suitable for use in a plethora of applications such as internal combustion and jet engines, gas turbine blades, engine cylinders, metal cutting, welding, aerospace applications, and in repairing the dimensions of worn-out high temperature parts [1–6].

N. Ergin (✉) · G. Yoruk · O. Ozdemir
Department of Metallurgy and Materials Engineering, Technology Faculty,
Sakarya University, 54187 Sakarya, Turkey
e-mail: nergin@sakarya.edu.tr

O. Ozdemir
e-mail: oozdemir@sakarya.edu.tr

E. K. Polychroniadis et al. (eds.), *International Multidisciplinary Microscopy Congress*,
Springer Proceedings in Physics 154, DOI: 10.1007/978-3-319-04639-6_4,

Currently, aluminides coating can be manufactured by using several techniques such as PVD, reactive sintering, reactive casting etc. Among these, combustion syntheses are characterized by high-temperatures, fast heating rates and short reaction times. This process is an attractive technique for synthesizing a wide variety of advanced materials, including powders and near net-shape product of ceramics, intermetallics, composites and functionally graded materials. Combustion synthesis concerned with the ignition of a composed powder mixture in air or inert atmosphere and produce a chemical reaction with a sufficient heat released (exothermic reaction) that it because self sustaining [7–10]. The main aim of the present study is to characterize Ni_3Al and Ti_3Al coating produced on AISI 1010 steel substrate from Ni, Ti, and Al precursors by pressure assisted combustion synthesis technique.

4.2 Experimental Procedure

In order to manufacture Ni_3Al and Ti_3Al-based intermetallic coatings as starting materials Carbonyl Ni powders (99.8 % purity, 4–7 μm), Ti (99.5 % purity, less than 44 μm) and gas-atomized aluminum powder (99 % purity and 15 μm grain size) were used. The starting powders were mixed in stoichiometric ratio corresponding to the Ni_3Al and Ti_3Al intermetallic phases, in a molar proportion of 3:1. Commercial carbon steel AISI 1010 was used as the substrate, which was machined as a disk 10 mm diameter and 5 mm in thickness. Prior to coating process, all substrates were ground using 600 grid emery papers to obtain a good surface finish. The mixed coating powders were placed into the substrate specimen in the mold. The production process was carried out in electrical resistance furnace in open atmosphere under an uniaxial pressure of 150 MPa at 1050 °C for 60 min. After sintering, the specimens were unloaded and cooled to room temperature in an open atmosphere. The morphologies of the samples and the presence of phases were examined by using scanning electron microscopy (SEM), optical microscopy (OM) and X-ray diffraction (XRD) analysis. Microhardnesses of the test materials were measured by using a Vickers indentation technique with a load of 0.5 N.

4.3 Result

The microstructure of Ni_3Al- and Ti_3Al-based intermetallic coatings produced by pressure assisted combustion synthesis method was examined by optical (OM) and scanning electron microscopy (SEM) (Fig. 4.1a, b). As shown in this figure, phase transformation of the NiAl intermetallic coating layer has been completed at a process temperature of 1050 °C for process time of 60 min. It's understood from microstructure studies that Ni_3Al intermetallics coating layer has single phase and dense structure. Ti_3Al intermetallic coating layer possessed a coarser grain structure than Ni_3Al

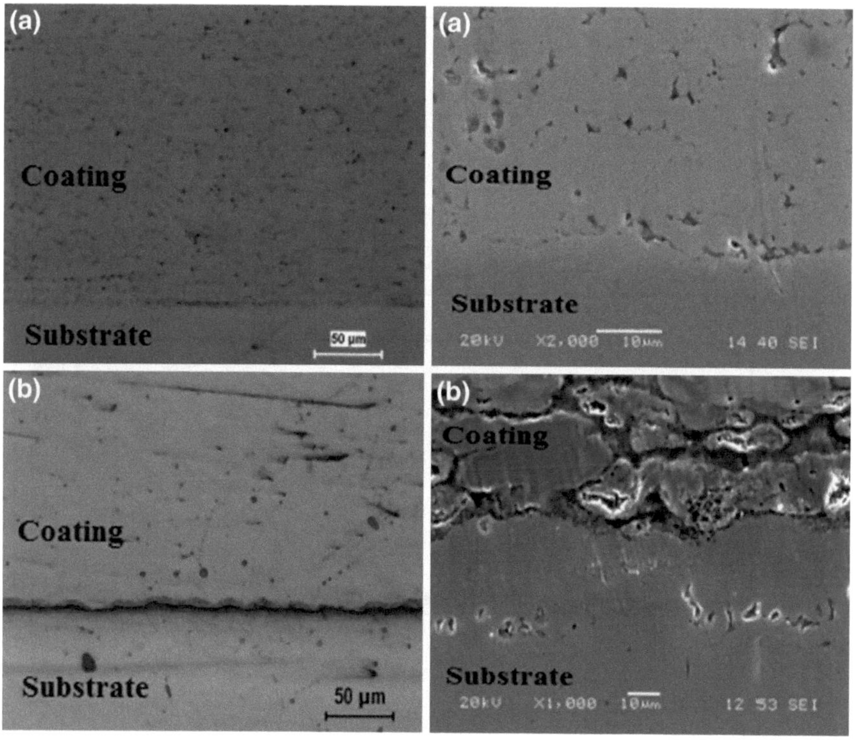

Fig. 4.1 OM and SEM images of coated samples: **a** Ni_3Al and **b** Ti_3Al

intermetallic coating layer, and some pores were present among the grains. Ti_3Al intermetallics coating layer has multi-phase structure.

X-ray diffraction (XRD) technique was used in order to identify the phases in the coated samples. The XRD patterns of Ni_3Al intermetallics coating layer reveal that Ni_3Al single phase (Fig. 4.2a). It can be seen that the Ti_3Al-based coating consists of Ti_3Al as a major phase, along with TiAl, Ti_3O and Ti trace phases (Fig. 4.2b). In a previous work [11], we presented the results of examination of dense Ni_3Al produced from Al and Ni powders by pressure-assisted combustion synthesis. The studies done by Sierra and Vazquez [12] showed that there are pore barriers (large volume pores) on the surface of NiAl coating on carbon steel substrate carried out by self-propagating high-temperature synthesis. In the present study, large volume pores due to the pressure used was not detected in the produced Ni_3Al coating layer.

The distribution of alloying elements on the cross section of Ni_3Al-, Ti_3Al-intermetallic coated samples were determined by energy dispersive spectroscopy (EDS) analysis as shown in Figs. 4.3 and 4.4, respectively.

While, iron atoms diffuse to the coating layer from the matrix, also to Ni atoms diffuse to the steel substrate matrix from the powder mixture at Ni_3Al-coated samples, (Fig. 4.3). It is possible that the diffusion of iron into the coating and diffusion of

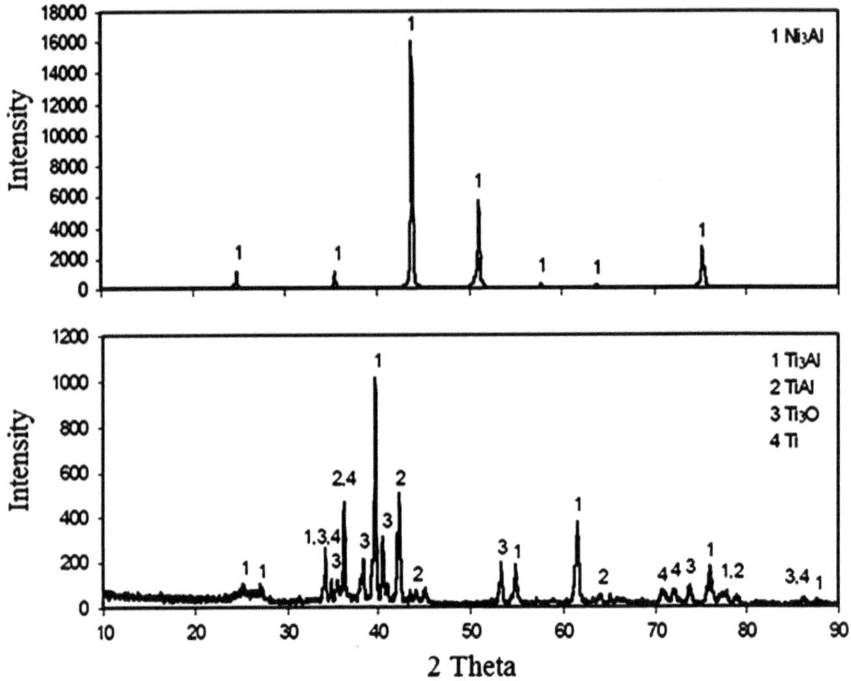

Fig. 4.2 X-ray diffraction patterns of coated samples: **a** Ni_3Al and **b** Ti_3Al

Fig. 4.3 EDS analysis of Ni_3Al coated sample

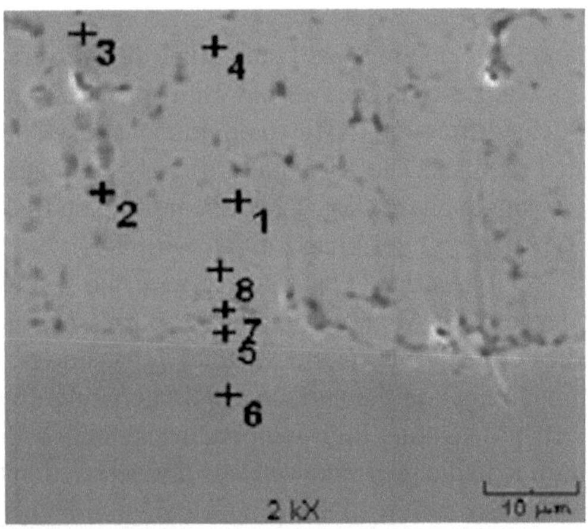

Element	%wt.								
	1	2	3	4	5	6	7	8	
O	-	3.1	-	-	-	-	-	2.1	
Al	10	6.1	9.3	9.7	1.9	-	4.5	6.4	
Fe	2.1	7	-	-	39.9	86.9	22.9	10.6	
Ni	87.9	83.8	90.7	90.3	58.2	13.1	72.6	80.9	
								Total	100

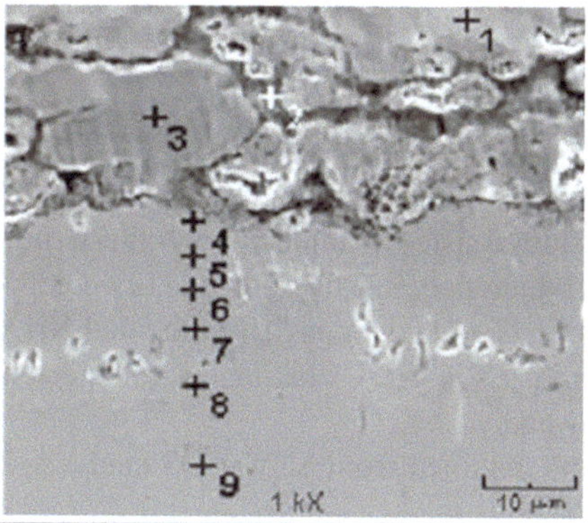

Element	%wt.									
	1	2	3	4	5	6	7	8	9	
Al	18.1	40.2	13.1	4.5	5.1	4.7	4.9	4.8	4.9	
Ti	81.9	21.5	86.9	2.4	1.1	0.5	0,7	-	-	
O	-	38.3	-	-	-	-	-	-	-	
Fe	-	-	-	93.1	93.8	94.8	94.4	95.2	95.1	
									Total	100

Fig. 4.4 EDS analysis of Ti_3Al coated sample

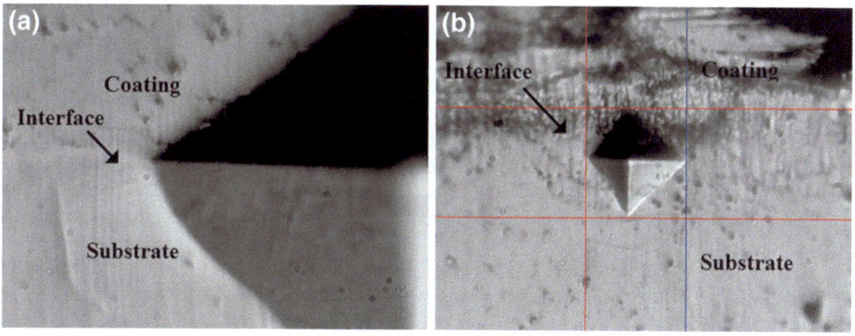

Fig. 4.5 Vickers indentations marks applied to interface of coating layers of **a** Ni_3Al and **b** Ti_3Al

Ni into the substrate formed a metallurgical bond between the coating layer and the substrate. At Ti_3Al coated sample, near the interface, Ti appeared in the substrate, but Fe was not detected in the coating layer (Fig. 4.4). Both of Ti_3Al and Ni_3Al coatings were well adhere to the steel samples and there is no any crack at interface. In addition, Vickers indenter was applied to the interface of steel matrix and coating layer under the load of 10 N and any cracks were not formed at interface as well (Fig. 4.5a, b). The hardness of the Ni_3Al- and Ti_3Al-based coatings and AISI 1010 substrate was approximately 377.7 ± 20.6, 520.0 ± 6.9 and 157 $HV_{0.05}$, respectively.

4.4 Conclusions

The following results can be obtained from the present study:

(a) Ni_3Al- and Ti_3Al-based intermetallic coatings on the AISI 1010 carbon steel produced by pressure-assisted combustion synthesis have a low porosity and seem to have good adherence. Because, no cracks were observed at the interface.
(b) The phases formed in Ni_3Al coatings layer included Ni_3Al single phase. Ti_3Al, TiAl, Ti_3O and Ti phases at Ti_3Al coatings were detected. These findings were confirmed by XRD analysis.
(c) According to results obtained from SEM-EDS analysis, iron atoms are present in Ni_3Al coatings.
(d) The Vickers micro-hardness values for the Ni_3Al, Ti_3Al-based intermetallic coatings and AISI 1010 substrate was approximately 377.7 ± 20.6, 520.0 ± 6.9 and 157 $HV_{0.05}$, respectively.

Acknowledgments This work was supported by the Sakarya University Scientific Research Foundation (Project Number: 2008-50-01-013).

References

1. M. Salehi, F. Karimzadeh, A. Tahvilian, Formation of Ti-Ni intermetallic coatings on carbon tool steel by a duplex process. Surf. Coat. Technol. **148**, 55–60 (2001)
2. Y. Wang, Z. Wang, Y. Yang, W. Chen, The effects of ceria on the mechanical properties and thermal shock resistance of thermal sprayed NiAl intermetallic coatings. Intermetallics **16**, 682–688 (2008)
3. M. Alaeddine, R. Ranganathan, T. Ando, C.C. Doumanidis, Modeling the intermetallic coating growth during reactive thermal processing of layered precursors. Surf. Coat. Technol. **200**, 5181–5192 (2006)
4. C. Sa'nchez Bautista, A. Ferriere, G.P. Rodrı'guez, M. Lo'pez-Almodovar, A. Barba, C. Sierra, A.J. Va'zquez., NiAl intermetallic coatings elaborated by a solar assisted SHS process. Intermetallics **14**, 1270–1275 (2006)
5. Y. Yu, J. Zhou, J. Chen, H. Zhou, C. Guo, B. Guo, Preparation, microstructure and tribological properties of Ni_3Al intermetallic compound coating by laser cladding. Intermetallics **18**, 871–876 (2010)
6. T.C. Totemeier, R.N. Wright, W.D. Swank, FeAl and Mo-Si-B intermetallic coatings prepared by thermal spraying. Intermetallics **12**, 1335–1344 (2004)
7. N. Ergin, G. Yoruk, O. Ozdemir, Characterization of Ni_3Al and Ni_3Al coatings produced by electric current activated sintering method. Acta Physıca Polonıca A No. 2. 245–247 (2013)
8. A.G. Merzhanov, The chemistry of self propagating high-temprature synthesis. J. Mater. Chem. 1779–1786 (2004)
9. A. Varma, A. Mukasyan, Combustion synthesis of intermetallic compound. *Self-Propagating High-temperature Synthesis of Materials*, vol. 5 (1988), pp. 1–34
10. C.L. Yeh, W.Y. Sung, Combustion synthesis of Ni_3Al by SHS boron additions. J. Alloys Compd. **390**, 74–81 (2005)
11. O. Ozdemir, S. Zeytin, C. Bindal, Characterization of NiAl with cobalt produced by combustion synthesis. Vacuum **84**, 430–437 (2010)
12. C. Sierra, A.J. Vazquez, Dry sliding wear behaviour of nickel aluminides coatings produced byself-propagating high-temperature synthesis. Surf. Coat. Technol. **200**, 4383–4388 (2006)

Chapter 5
Scanning Electron Microscopy Analysis of Titanium Coatings on Steel Substrates

Rachid Gheriani and Raouf Mechiakh

Abstract Thin titanium films were deposited on steel substrates by magnetron sputtering method with RF power of 200 W. During the deposition the substrates were heated to 200 °C. The samples were subjected to high vacuum annealing during 30 mins at various temperatures between 400 and 1000 °C. The heat treatments results the interaction between the two parts of the system which was controlled by scanning electron microscopy analysis. The results show the development of the interdiffusion of the chemical compounds of the system with the annealing temperature, manly the diffusion of the carbon toward the titanium films leading to the formation and growth of the hard and refractory metallic carbide TiC confirmed by XRD spectrum registered on treated samples. The intense diffusion of the substrate chemical elements to the external layers and the formation of the oxides at high temperatures lead to the failure of the mechanical properties of the studied samples.

5.1 Introduction

Transition metals carbides and nitrides thin films are of considerable interest for many applications; they make an important part in interstitial compounds. They are formed by the incorporation of carbon and nitrogen into the lattices of transition metals to produce a class of compounds with unique physical and chemical properties [1]. The outstanding features of these materials properties; as the hardness, the refractory and the covalent-metal feature, have varied the areas of their applications. TiN and

R. Gheriani (✉)
Laboratory of Radiation and Plasmas and Physics of Surfaces (LRPPS),
University of Ouargla, Ouargla, Algeria
e-mail: ragheriani@yahoo.fr

R. Mechiakh
University El-Hadj Lakhdar, 5000 Batna, Algeria
e-mail: raouf_mechiakh@yahoo.fr

E. K. Polychroniadis et al. (eds.), *International Multidisciplinary Microscopy Congress*,
Springer Proceedings in Physics 154, DOI: 10.1007/978-3-319-04639-6_5,

TiC films have many useful physical and chemical properties allowing their use in various technological applications; in corrosion and wear resistance, electronic and optical components, etc. They have been the subject of several academic and industrial researches [2–5]. In present work; thin titanium films were deposited on steel substrates by magnetron sputtering method; after the samples were subjected to vacuum annealing in order to activate the reaction between the thin films and substrates. The samples were characterised with scanning electron microscopy for a morphological analyze and with x-ray microanalysis in order to determine the chemical elements and their kinetics as a function of annealing temperature.

5.2 Experimental Procedure: Thin Films Deposition and Characterization

Before deposition, the chamber was evacuated to $\sim 10^{-6}$ mbar using a diffusion pump, the sputtering gas continuously pumped during the deposition. Before the introduction of substrates into the deposition system; they are mechanically polished; then cleaned chemically in an ultrasonic bath with trichloroethylene, distilled water and ethanol successively. For obtaining a pure film with good adhesion, the titanium target and substrates were cleaned by Ar^+ sputtering for about 20 min for each of them successively. In order to activate the atomic diffusion, the samples were subjected to vacuum heat treatment in the temperature range of 400 to 1000 °C.

5.3 Experimental Results

5.3.1 X-Ray Microanalysis Results (EDS)

The EDS patterns of the non-treated samples are shown in (Fig. 5.1). The spectra exhibit four EDS energetic lines from Ti and from substrates elements, Also, we note the apparition of EDS line from oxygen which was as an contamination element. In the case of the other samples, the pattern shows just the Ti, Fe and carbon lines, (Fig. 5.2). Other Spectra obtained for an increasing heat treatment temperatures, exhibit EDS lines from Ti, Fe, C, Mn, Cr, Si. The intensities increase respectively with the temperature and the concentration of the element in the samples, In the case of treated samples at 1000 °C for 30 min., group 2, we note the total absence of titanium lines and the apparition of high energetic line of oxygen, after his diffusion towards external layers from the internal layers of substrates under the effect of heat treatments. We think that the oxygen was absorbed into the substrates during the storage before the deposition, (Fig. 5.3).

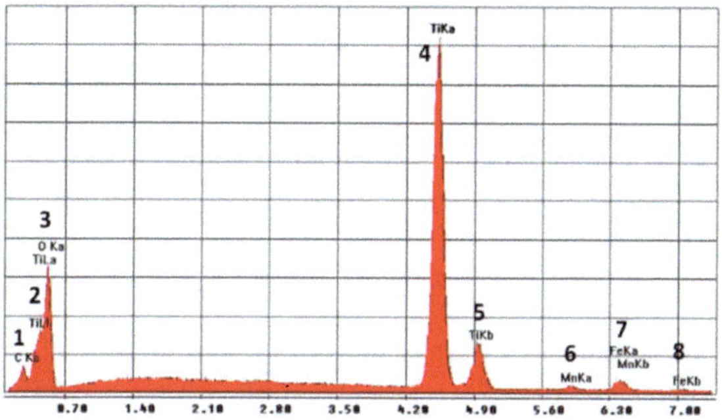

1	2	3	4	5	6	7	8
Ckα	TiLβ	Okα,TiLα	Tikα	Tikβ	Mnkα	Fekα,Mnkβ	Fekβ

Fig. 5.1 *X-ray* microanalysis pattern of non annealined and coated samples. (1)

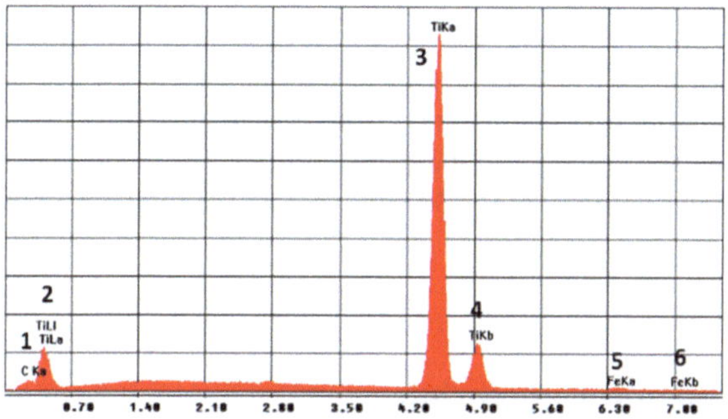

1	2	3	4	5	6
Ckα	TiLβ,TiLα	Tikα	Tikβ	Fekα	Fekβ

Fig. 5.2 *X-ray* microanalysis pattern of coated and non treated samples. (1)

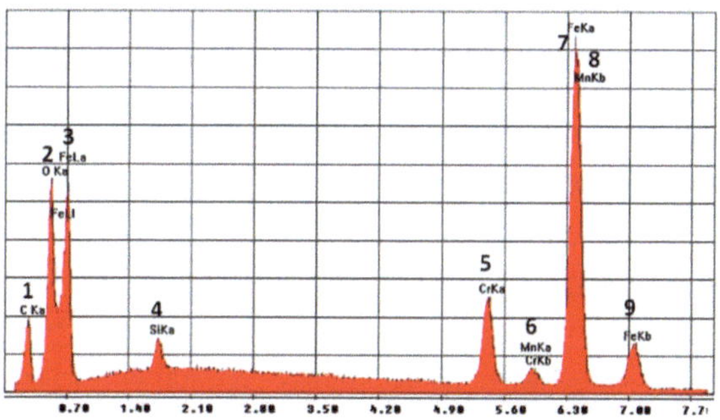

1	2	3	4	5	6	7	8	9
Ckα	Okα	FeLα	Sikα	Crkα	Mnkα,Crkβ	Fekα	Mnkβ	Fekβ

Fig. 5.3 *X-ray* microanalysis pattern of coated and treated samples. at 1000 °C, for 30 min

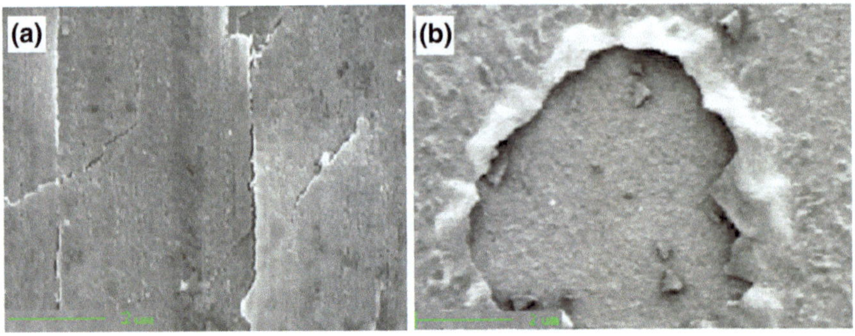

Fig. 5.4 Scanning electron micrograph obtained from coated and annealined samples for 30 min: **a** at 800 °C; **b** at 1000 °C

5.3.2 SEM Results

Scanning electron micrograph obtained from coated and treated samples at 800 °C, for 30 min, shows the cracking of the deposed films, Fig. 5.4a, in the case of coated and treated samples at 1000 °C for 30 min we note the total detachment of titanium films, Fig. 5.4b.

5.4 Discussion and Conclusion

1. The heat treatment of samples led to the reactions between the substrates and deposited thin films confirmed by the variations in the signal of chemical elements in the concentration and number.
2. The intense diffusion of the substrates chemical elements to the surface layers causes the cracking and total detachment of the coatings at high annealing temperatures.
3. The amount of oxygen present in the treated and non treated samples, can be as a result of operating conditions (vacuum).

References

1. M. Detroye, F. Reniers, C. Buess-Herman, J. Vereeckers, Appl. Surf. Sci. **120**, 85–93 (1997)
2. M. Kobayashi, Y. Dor, Thin Solid Film **54**, 67–74 (1978)
3. I. Dahan, U. Admon et al., Surf. Coat. Technol. **137**, 111–115(2001)
4. H. D. Bhat et al., Thin Solid Film **342**, 214–220 (1999)
5. Shouchi Ejiri, Toshihiko Sasaki, Yukio Hirose, Thin Solid Film **307**, 178–188 (1997)

Chapter 6
A Novel Three-Dimensional Visual Inspection Scheme for Defect Detection of Transparent Materials Based on the Digital Holographic Microscopy

Kwang-Beom Seo, Byung-Mok Kim, Jung-Sik Koo and Eun-Soo Kim

Abstract In the conventional visual inspection system with naked eyes, a camera has been mostly employed for detection of scratches or glass dust, so that it may be difficult to discriminate between the high-grade and the low-grade mass components. Therefore, here we propose a novel three-dimensional (3-D) visual inspection scheme for effective detection of defects of the transparent materials based on a digital holographic microscopy, which is optically implemented as a form of transmission-typed shearing interferometer. In the proposed method, quantitative data are obtained by converting the phase information of the transparent touch panel or the substrate glass into its depth or thickness information. Then, by using 3-D data extracted from the depth or thickness information of transparent touch panel or substrate glass, we can accurately discriminate between the high-grade and the low-grade mass components. Experiments with test objects confirm the feasibility of the proposed method in the practical applications. Based on these experimental results, the proposed method expects to improve the production yield rate of the original mother glass of the transparent touch panel or the substrate glass used in the LCM.

6.1 Introduction

Recently, in the mass production process related to the smart phones or display devices, a UV (ultra-violet) bonding technique with the resin has been employed for bonding between the LCM (liquid-crystal module) and the transparent touch glass. Before the UV bonding process, however, defects of scratch or glass dust which

K.-B. Seo · B.-M. Kim · J.-S. Koo · E.-S. Kim (✉)
HoloDigilog Human Media Research Center (Holodigilog), 3D Display Research Center (3DRC), Department of Electronic Engineering, Kwangwoon University, 20 Kwangwoon-ro, Nowon-gu, Seoul, Korea
e-mail: eskim@kw.ac.kr
URL: http://www.holodigilog.org, http://www.3drc.org

E. K. Polychroniadis et al. (eds.), *International Multidisciplinary Microscopy Congress*, Springer Proceedings in Physics 154, DOI: 10.1007/978-3-319-04639-6_6,

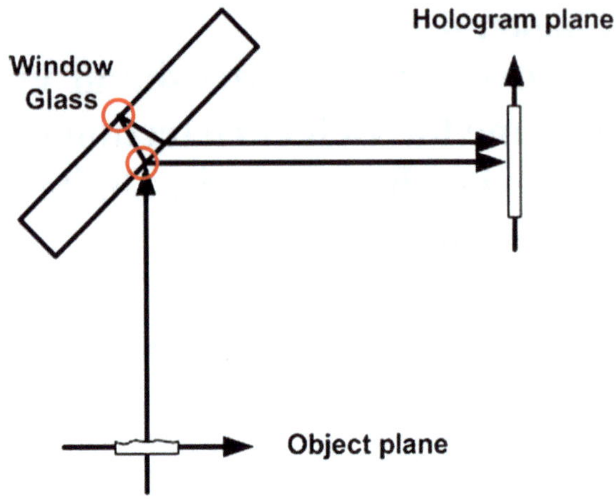

Fig. 6.1 Simple optical scheme for hologram formation

happened to be occurred in the process of manufacturing the original mother glass of the transparent touch panel or the substrate glass used in the LCM may make enormous errors of mass components.

In most conventional visual inspection systems, these errors can be detected by the examiner's naked eye or by analyzing the two-dimensional (2-D) high-definition image of the transparent touch glasses by using the SEM [1]. However, in this method, depth or thickness information about the transparent glasses may have been very limited.

For this reason, we propose a novel three-dimensional visual inspection scheme for effective detection of the defects of the transparent materials based on the transmission-typed digital holographic microscopy (DHM).

6.2 Theory

Digital holographic microscopy (DHM) is a very effective tool for 3-D imaging of micro-scale transparent materials since it can obtain the phase information of the materials as the interference patterns [2]. Here, it must be noted that phase information obtained from the interference patterns picked up with the DHM can be converted into the depth or the thickness information of the transparent materials. Also, the angular spectrum propagation (ASP) approach to the scalar diffraction theory can be used for numerical reconstruction of the picked-up hologram patterns from the DHM.

A. Hologram Formation in the DHM

Figure 6.1 shows a simple optical scheme for hologram formation. As shown in Fig. 6.1, a shearing interferometer seems to be very useful for direct holographic imaging of the transparent objects. The shear is created by reflections from the front and back surfaces of a glass plate. Accordingly, a holographic pattern can be obtained by a process of interference between the front and back of the surface of the glass at the detector.

B. Digital Reconstruction of Hologram Patterns of the Object

In the digital holography, reconstruction of the object field can be done by numerically simulating the diffraction process happened on the recorded holograms when they are illuminated by the reference wave by using the angular spectrum approach to the diffraction theory. Angular spectrum approach is helpful since it can describe the wavefront propagation over a short distance from the objects [3, 4].

C. Reconstruction by Using the Angular Spectrum Propagation Method

The complex wave field U(x, y, 0) at the hologram plane (x, y, 0) can be obtained by illuminating the hologram by the reference wave. That is, assuming h(x, y) and R(x, y) are the hologram transfer function and the complex reference wave, the complex wave filed is described as follows.

$$U(x, y; 0) = h(x, y)R(x, y) \tag{6.1}$$

Its angular spectrum is also given by the process of Fourier transform [5]

$$\hat{U}(f_x, f_y, 0) = \frac{1}{2\pi} \int\int_{-\infty}^{\infty} dxdyU(x, y, 0) \exp[-i(k_x x + k_y y)] \tag{6.2}$$

By applying a process of spatial filtering to the spectrum, the un-diffracted reference beam and the conjugate real image are removed and only the object spectrum can be selected by this filtering process as follows.

$$\overline{U}(x, y, 0) = \frac{1}{2\pi} \int\int_{-\infty}^{\infty} df_x df_y \, \mathit{filter} \, [\hat{U}(f_x, f_y, 0)] \exp[i(k_x x + k_y y)] \tag{6.3}$$

In (6.3), inverse-Fourier transformation of the filtered angular spectrum provides the complex amplitude at (x, y, 0) containing information about the object. Complex amplitude at a plane parallel to (x, y, 0) plane situated at a distance d from the filtered angular spectrum is given by

$$U(x, y, d) = \frac{1}{2\pi} \int\int_{-\infty}^{\infty} df_x df_y \, \mathit{filter} \, [\hat{U}(f_x, f_y, 0)] \exp[i\sqrt{k^2 - k_x^2 - k_y^2}d] \exp[i(k_x x + k_y y)] \tag{6.4}$$

D. Computation of the Phase and OPL Changes

The phase difference can provide the 3-D information of the object. Now, the phase of the hologram can be written as

$$\phi(x, y) = \arctan \frac{\mathrm{Im}[U(x, y, d)]}{\mathrm{Re}[U(x, y, d)]} \tag{6.5}$$

The phases of the wave fronts with and without the object are calculated from the complex amplitudes of two reconstructed holograms $U_{with\ object}(x, y, d)$ and $U_{without\ object}(x, y, d)$, respectively.

$$\phi_{with\ object}(x, y) = \arctan \frac{\mathrm{Im}[U_{with\ object}(x, y, d)]}{\mathrm{Re}[U_{with\ object}(x, y, d)]} \tag{6.6}$$

$$\phi_{without\ object}(x, y) = \arctan \frac{\mathrm{Im}[U_{without\ object}(x, y, d)]}{\mathrm{Re}[U_{without\ object}(x, y, d)]} \tag{6.7}$$

The phase difference is directly calculated by subtraction.

$$\Delta\phi(x, y) = \phi_{with\ object}(x, y) - \phi_{without\ object}(x, y) \tag{6.8}$$

In (6.9), this phase difference can be converted to the change of the optical path length of the object

$$\Delta\phi(x, y) = \frac{2\pi}{\lambda}\Delta n(x, y)\Delta L(x, y) \tag{6.9}$$

where $\Delta n(x, y)$ and $\Delta L(x, y)$ represent the changes of the refractive index and the thickness respectively. These results mean that a change of the phase information can provide the corresponding change of the physical thickness of the object.

6.3 Experimental Setup

Figure 6.2 shows the experimental setup for implementation of the transmission-typed shearing interferometer. Here, a randomly linearly-polarized He-Ne laser providing a CW output at the wavelength of 632.8 nm and the maximum output power of 2 mW is employed as the laser source. The transparent object can be magnified with microscopic objectives with the magnification of 40X, NA $= 0.65$ and 60X, NA $= 0.85$. The interference pattern is recorded with a charge-coupled device (CCD) camera (AVT Marlin F-145C2) with 8-bit dynamic range and a pixel size of $4.65\ \mu\mathrm{m} \times 4.65\ \mu\mathrm{m}$.

In order to detect the defects of a transparent object using the optical shearing interferometer, an original mother glass sorted out among the inferior goods is selected as the test object. The microscopic defects of this glass cannot be detected with a naked

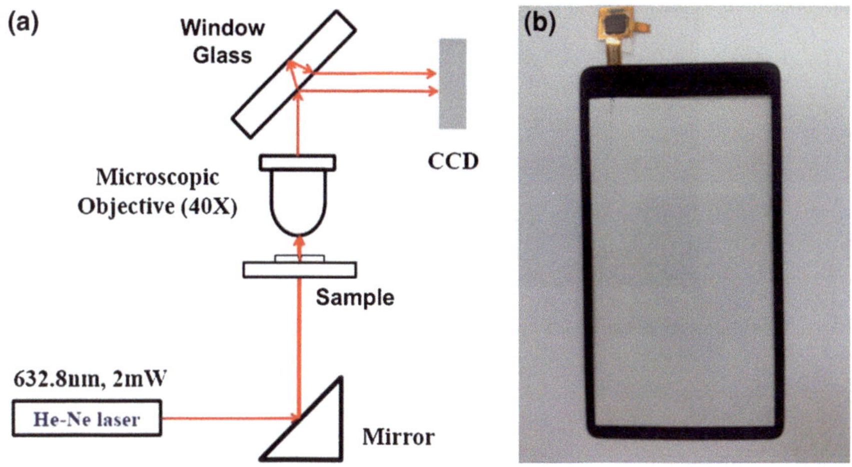

Fig. 6.2 (**a**) Experimental setup for the transmission-typed shearing interferometer, (**b**) original mother glass for the smart phone as a test object

eye. However, by using the proposed transmission-type optical shearing interferometer, 3-D phase information of the mother glass can be recorded on the CCD camera as the form of hologram patterns, then from them, depth or thickness information of the mother glass can be numerically obtained.

6.4 Experimental Results

Figure 6.3a shows the intensity image and the phase contrast image by using the 40X objective lens. Here we can't detect the scratches of the original mother glass with the naked eye as shown in Fig. 6.3a. On the other hand, a holographic image may show phase information of the mother glass by using the angular spectrum propagation method.

Figure 6.4 shows the numerically reconstructed images of the original mother glass by using the phase difference information. These images have been reconstructed by computer simulation with the MATLAB. The depth information of this transparent object can be quantitatively obtained as shown in Fig. 6.4c. Moreover, Fig. 6.4a shows 1-D information of the thickness by using an arrow in Fig. 6.4b. As a result in Fig. 6.4a, scratches on the original mother glass can be accurately detected in the range of less than 1 μm..

Now, here the objective lens with 40X has been changed into that of 60X. Figure 6.5a shows the thickness information with an arrow in Fig. 6.5b. Figure 6.5c shows the 3-D image of the thickness information of a transparent mother glass. Unlike the case with the 40X objective lens, the detection range of defects per unit

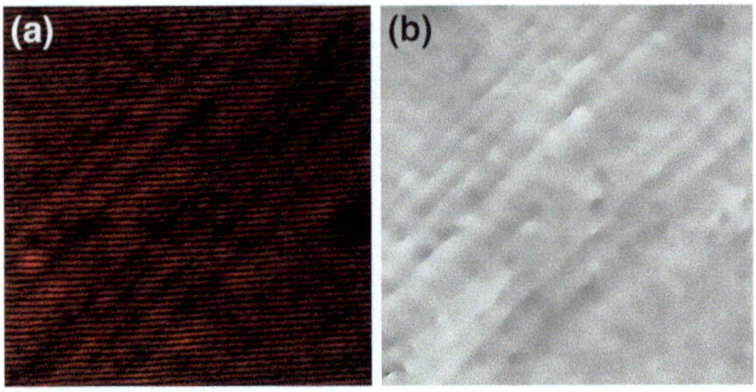

Fig. 6.3 (**a**) Intensity image and (**b**) phase contrast image for the original mother glass

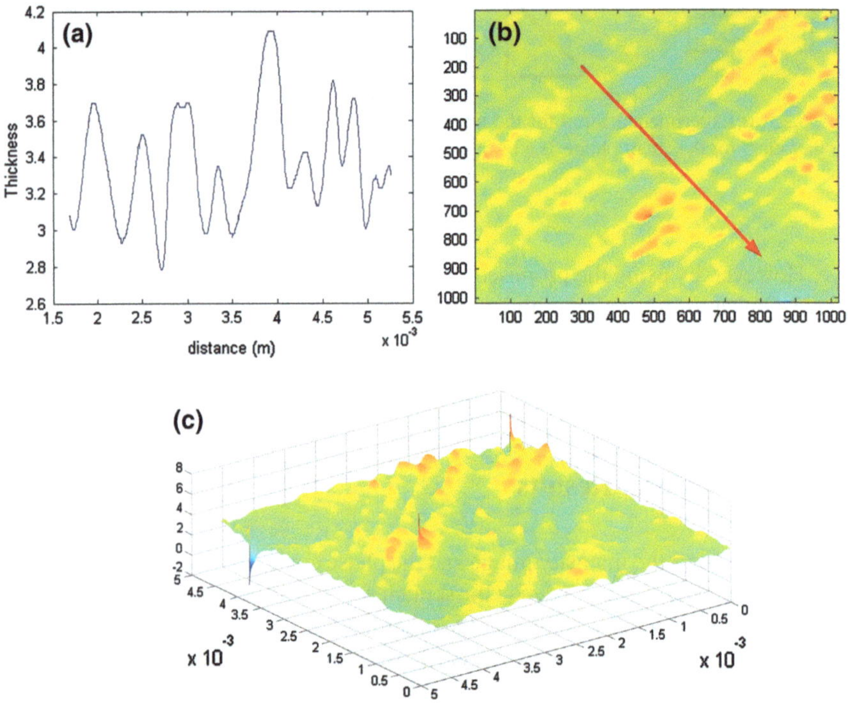

Fig. 6.4 Reconstructed images with a 40X microscopic objective. (a) 1-D, (b) 2-D and (c) 3-D images for the scratches on the transparent original mother glass

area may reduce compared to that with of the 60X objective lens. However, it is possible to precisely detect the thickness information of the object in the expanded range.

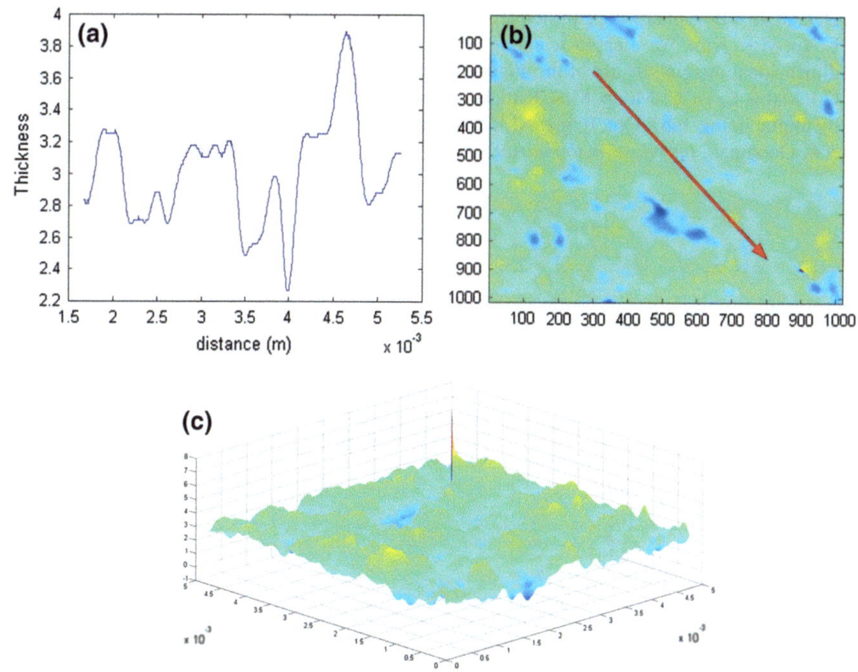

Fig. 6.5 Reconstructed images with a 60X microscopic objective. (a) 1-D, (b) 2-D and (c) 3-D images for the scratches on the transparent original mother glass

6.5 Conclusion

In this paper, a novel 3-D visual inspection scheme based on the transmission-type optical shearing interferometer has been proposed for effective detection of the defects occurring on the transparent original mother glass, which are mostly used in the smart phones or display devices. Unlike the conventional method, the proposed method can detect the depth or thickness information of the transparent glass in the range of less than 1 μm from the phase information recorded on the hologram patterns. From the experimental results, the feasibility of the proposed method in the practical application has been confirmed.

Acknowledgments This work was supported by the National Research Foundation of Korea (NRF) grant funded by the Korea government (MEST) (No. 2013-067321).

References

1. H. Okuda, T. Hiroi, Robust defect detection method using reference image averaging for high-throughput SEM wafer pattern inspection system. Metrol. Inspection Process Control Microlith. **XX**, 61524F (2006)

2. D. Gabor, A new microscope principle. Nature, **161**, 777 (1948)
3. L. Yu, M.K. Kim, Wavelength-scanning digital interference holography for tomographic three-dimensional imaging by use of the angular spectrum method. Opt. Lett. **30**(16), 2092 (2005)
4. C.J. Mann, L. Yu, C.-M. Lo, M.K. Kim, High-resolution quantitative phase-contrast microscopy by digital holography, Opt. Express **13**(22), 8693 (2005)
5. J.W. Goodman, *Introduction to Fourier Optics*, Chap. 3 (McGrawHill, NewYork, 1996)

Chapter 7
The Use of Scanning Electron Microscopy to Identify Zeolite Minerals

Rafal Panek, Magdalena Wdowin and Wojciech Franus

Abstract Zeolites are hydrated aluminosilicates of alkali elements and alkaline earth elements. The primary structural element is silicon-oxygen and aluminum-oxygen tetrahedrons, which are arranged in three dimension, form the spatial lattice about frame structure. The primary method of zeolite identification is the X-ray Diffraction. An alternative method for recognizing zeolites is a scanning electron microscopy (SEM) combined with chemical analysis in the microarea (SEM-EDS). Application of this method is justified because of the zeolites reach the crystal size in the range from a few microns to several millimeters about crystal habit: needle-like, cubic-like, plate-like. The one of needle-like zeolites is natrolite and carcinogenic erionite. Plate-like zeolites are represented by stilbite, heulandite, whether most commonly occurring clinoptilolite or synthetic zeolite type Na-P1. The zeolites with the cubic-like habits include philipsite, chabazite and harmotome and a number of synthetic zeolites such as Na-X and Linde-A. The use of SEM to observe the form of zeolite crystals confirmed by diffraction studies in full allows for the phase identification of these group minerals. Chemical analyzes in microarea additionally specify the nature of ion-exchange cations and the Si/Al ratio as well thermal stability and surface activity.

R. Panek (✉)
Division of Geotechnics, Lublin University of Technology, Nadbystrzycka 40, 20-618 Lublin, Poland
e-mail: r.panek@pollub.pl

M. Wdowin
Mineral and Energy Economy Research Institute of The Polish Academy of Sciences, Wybickiego 7, 31-261 Kraków, Poland
e-mail: wdowin@meeri.pl

W. Franus
Division of Geotechnics, Lublin University of Technology, Nadbystrzycka 40, 20-618 Lublin, Poland
e-mail: w.franus@pollub.pl

E. K. Polychroniadis et al. (eds.), *International Multidisciplinary Microscopy Congress*, Springer Proceedings in Physics 154, DOI: 10.1007/978-3-319-04639-6_7,

7.1 Introduction

Zeolites constitute microporous, hydrated aluminosilicates of alkali elements, alkaline earth metals or other cations, which have in their crystal structure, numerous channels and chambers of different sizes (in the order of several angstroms). A very distinctive feature of this group of minerals is the presence in their chemical composition the water molecules bound in a specific manner. This is so-called zeolite water. During heating of zeolite minerals to the temperature of 400 °C the water is continuous removed from the structure of mineral, without changing the shape of the crystal. However, during cooling the zeolite minerals in humid environment water molecules are also continuously re-absorbed in the crystal structure of the zeolites. Removed molecule water can be substituted by other ions or molecules, which gives zeolites the ion exchange properties commonly used in technologies of removal of environmental pollutants in the form of ammonium ions, ions of heavy metals, radionuclides, gas molecules [1–3]. Zeolites, because of the different sizes of channels and chambers are also used as molecular sieves that allow separation of gas mixtures. A special of such application of them is drainage and purifying of respiratory air as well as purifying oxygen in the cabins of spacecraft and breathe masks of astronauts.

Identification of mineral phases generally are carried out by means of X-ray powder diffraction that is the one of the most important techniques for determining the structures of unknown zeolites [4]. However, when the structures become too complex and a large portion of the reflections overlap in the powder diffraction pattern, structure determination from X-ray powder diffraction data is very difficult and should be supplemented by other available techniques [5].

Electron beam techniques namely Scanning Electron Microscopy (SEM) have ample applications on the characterization of various types of zeolites. They can provide morphological, structural and chemical information down to the atomic scale.

The traditional SEM, including its improvements for better spatial resolution like the warm or cold field emission gun (FEG) SEM, is used either in the secondary electron (SE) mode for crystal size and morphology study or the back-scattered electron (BSE) mode for study of the texture of zeolite-bearing rocks. In the case of various zeolites different species have different particle shapes. Fore instance mordenite forms acicular-fibrous crystals, heulandite-type zeolites (heulandite-clinoptilolite) form coffin-type crystals and faujasite forms well formed octahedral crystals. A significant contribution of SE images to the study of alkali zeolites is the determination of the mode of their origin. Secondary electrons reveal also the textural relationships of other authigenic minerals often associated with zeolites.

The use of BSE images are useful for microanalyses because they assist to avoid contamination by Ti and Fe oxides and sulfides and to distinguish coarsegrained minerals (quartz, feldspars) from zeolites. BSE images assist to identify accessory or trace phases, which are not detected with other conventional mineralogical techniques, such XRD [6].

7.2 Materials

Research material constituted natural and synthetic zeolites. Beyond the naturally occurring in the environment zeolites i.e. natrolite, erionite, stilbite, heulandite and clinoptilolite also synthetic zeolites were investigated.

Most of natural accumulation of zeolites are well recognize therefore investigation of their structure is not so troublesome as in described synthetic zeolites, that were derived in hydrothermal synthesis of F class fly ash. Fly ash come from Kozienice, Rybnik and Stalowa Wola Power Plants.

In series of chemical reactions of fly ash with sodium hydroxide it was possible to obtain a variety of zeolite materials from fly ash according to the following equation:

$$fly\ ash + x\ mol\ dm^{-3} NaOH \xrightarrow{time,\ temperature} Zeolite + residium \tag{7.1}$$

Depending on the synthesis conditions a number of zeolite materials were derived: Na-X, Na-P1, Linde-A, analcime and sodalite [7, 8].

After the synthesis reaction, it is important to identify whether the obtained zeolite material is suitable. In case of mixtures of synthetic zeolites (sometimes in synthesis reaction a few type of zeolite were derived) it is important to distinguish from each other obtained zeolite phases. Therefore very often the basic XRD diffraction method is not sufficient.

7.3 Methods

In order to identified zeolites morphological form, crystal structure, and chemical composition they investigated by scanning electron microscopy using an FEI Quanta 250 FEG SEM equipped with a chemical composition analysis system based on energy dispersion scattering, energy dispersive spectrometry (EDS EDAX).

7.4 Results

Observations of the morphology of zeolite crystals and their habit (description of mineral visible external shape) made by scanning electron microscopy allow the identification of individual representatives of this group of minerals. Observation of crystals (crystal habits) are useful in communicating what specimens of a particular mineral often look like, what allow to distinguish a large number of minerals.

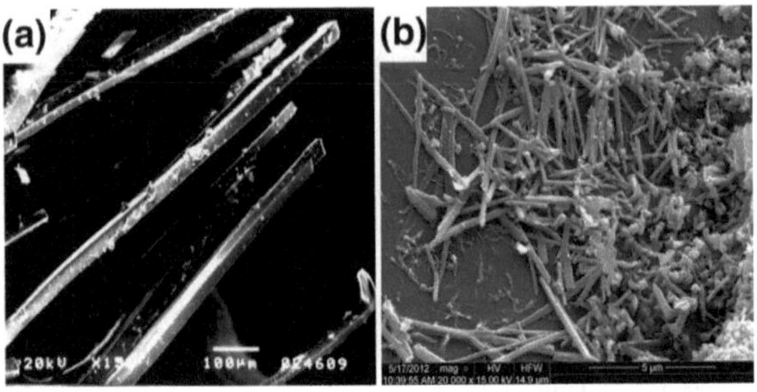

Fig. 7.1 SEM microphotographs of zeolites: **a** natrolite, **b** erionite

Depending on the zeolite type a various of crystals shape were obtained. It is cause by fact that crystallization is mainly a phase transition through which matter is transformed from a state of high free energy in a solvated state to one of low free energy in the crystal lattice [9].

Microphotographs of zeolites about needle-like habit (with one of predominant dimension of growth) are represent by a group of natural zeolites such as natrolite and carcinogenic erionite (in Fig. 7.1a, b). Such crystal habit compose a radiating mass of slender, acicular crystals, where minerals with this crystals tend to be fragile, and undamaged specimens can be uncommon. Crystals of these minerals are confined walls of the column and/or pyramid with a very small angle of the walls.

Zeolites of plate-like/coffin-type crystals (with two of predominant dimensions of growth) are represented by among others stilbite (first described in the nature of zeolites [10]), philipsite, heulandite, and the most popular in nature zeolite—clinoptilolite (Fig. 7.2a–c). In the group of synthetic zeolites plate-like habit have for example zeolite from the group of gismondite—Na-P1 (Fig. 7.2d). Plate-like habit composed flattened and thin crystals (like plates) but wider than bladed and thinner than tabular.

Zeolites with cubic-like and octahedral crystals (growth of crystals parallel in three dimensions) include natural analcime, chabazite and philipsite and their synthetic analogies zeolites types Na-X and Linde-A (Fig. 7.2e–h). In cubic crystals it is eight points, six faces and twelve edges that are perpendicular to each other forming 90° angles and square cross-sections. However octahedral crystals form a three-dimensional shape bounded by eight triangular faces and having six corners.

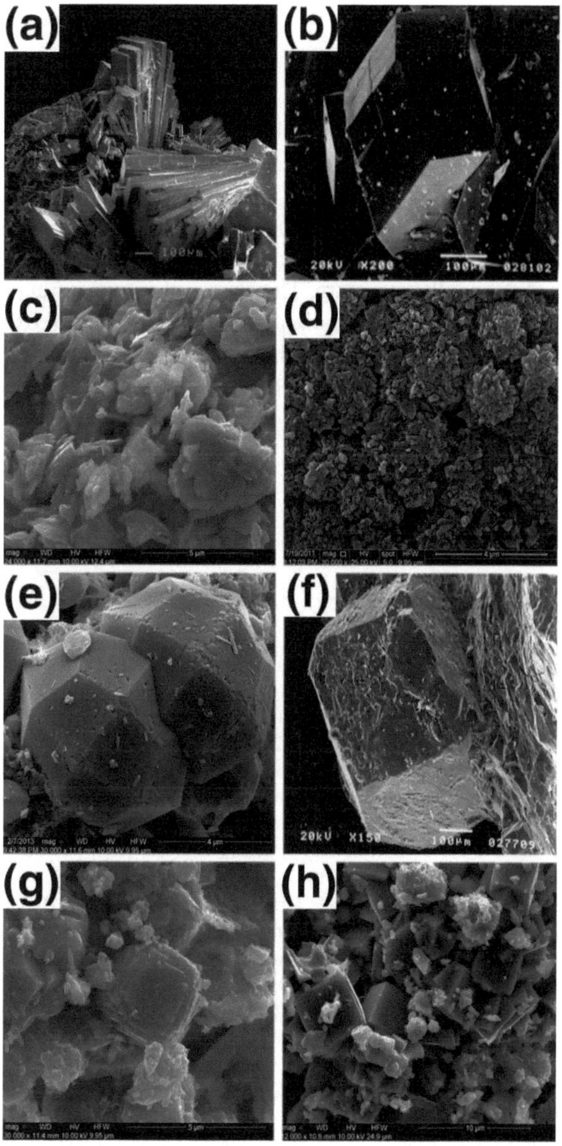

Fig. 7.2 SEM microphotographs of zeolites: **a** stilbite [11], **b** philipsite, **c** clinoptilolite, **d** Na-P1, **e** analcime, **f** chabazite, **g** Na-X, **h** Na-A

7.5 Conclusion

The use of scanning electron microscopy to observation of the form of zeolite crystals confirmed by diffraction studies in full allows for the identification phase of the group minerals. Additionally scanning microscope equipped with Energy-dispersive X-ray system (EDS) provides an analysis of the chemical composition of zeolites in the microarea. The proportions between the main chemical components of the zeolite framework (the ratio of Si^{4+}/Al^{3+}) provide information on the thermal stability and surface activity of the zeolite structure. The chemical composition of ions located in exchangeable positions allows the indirect determination of ion exchange capacity of the zeolite structure. Finally the scanning electron microscopy method in combination with XRD analysis can ultimately diagnose which zeolite phase occurring in the tested material.

Acknowledgments This researches are financed by NCBiR within Project no PSB1/A2/7/2012.

References

1. W. Franus, M. Wdowin, Removal of ammonium ions by selected natural and synthetic zeolites. Gosp. Sur. Miner.-Miner.Resour. Manag.**26**(4), 133–148 (2010)
2. S. Chalupnik, W. Franus, M. Wysocka, G. Gzyl, Application of zeolites for radium removal from mine water. Environ. Sci. Pollut. Res. **20**(11) 7900–7906 (2013)
3. M. Wdowin, W. Franus, R. Panek, Preliminary results of usage possibilities of carbonate and zeolitic sorbents in CO_2 capture. Fresen. Environ. Bull. **21**(12), 3726–3734 (2012)
4. T. Wessels, Ch. Baerlocher, L.B. McCusker, E.J. Creyghton, An ordered form of the extra-large-pore zeolite UTD-1: synthesis and structure analysis from powder diffraction data. J. Am. Chem. Soc. **121**, 6242–6247 (1999)
5. J. Sun, Z. He, S. Hovmoller, X. Zou, F. Gramm, Ch. Baerlocher, L.B. McCusker, Structure determination of the zeolite IM-5 using electron crystallography. Kristallogr **225**, 77–85 (2010)
6. G.E. Christidis, Application of electron microscopy to the study of smectites and zeolites. Revista de la sociedad española de mineralogía no **11**, 9–10 (2009)
7. A. Derkowski, W. Franus, E. Beran, A. Czímerová, Properties and potential applications of zeolitic materials produced from fly ash using simple methods of synthesis. Powder Technol. **166**(1), 47–54 (2006)
8. W. Franus, Characterization of X-type zeolite prepared from coal fly ash. Pol. J. Environ. Stud. **21**(2), 337–343 (2012)
9. J.J. De Yoreo, P. Vekilov, Principles of Crystal Nucleation and Growth Biomineralization, ed. by P.M. Dove, J.J. De Yoreo, S. Weiner. Mineralogical Society of America, (2003), p. 54
10. A.F. Cronstedt, Observation and description of an unknown kind of rock to be named zeolites. Kongl Vetenskaps Acad Handl Stockh **17**, 120–123 (1756)
11. http://www.mineral-paradise.net/mineralphotosq-s.html#anchor_42

Chapter 8
Fractal Analysis of the Pore Space in Sandstones as Derived from Mercury Porosimetry and Image Analysis

Grazyna Stanczak

Abstract The pore space fractal dimension was measured using a model of a porous body based on the Menger sponge and mercury porosimetry data for selected samples of the reservoir sandstones of the Weglowka oil field (SE Poland) in a pore-throat diameter range between 91 and 0.0090 μm. Based on the digital analyses of the two-dimensional images of thin sections impregnated with blue-dyed epoxy and taken under an optical microscope as well as the images of thin sections taken under a cold field emission gun scanning electron microscope (FEGSM) in the backscattered electron image mode, the current paper tries to quantify the pore space of sandstones by using the box counting method. The results derived from analysis of the pore-throat diameter distribution by mercury porosimetry revealed the multifractal structure of the pore space of sandstone in two separated ranges of the pore-throat size considerably lesser than the pore-throat diameters (10–50 μm) corresponding to threshold pressures. This means that only the pore throats connecting wider parts of the pore network (pores) exhibit the fractal structure. The assumption that the fractal dimension monitoring the distribution of the pore-space volume within the smallest pore-throat diameters characterizes the overall pore-throat network in the rock sample provides a device to set apart the distribution of the pore-throat volume from the distribution of the pore volume. On the other hand, the fractal dimensions derived from the image analysis of thin sections describe the pore space as a whole.

8.1 Introduction

The pore space of a sandstone is known as an extremely complicated, irregular system which can be considered as a chaotic structure with a high degree of heterogeneity. Owing to this complexity, it is very difficult to acquire a complete description of the

---

G. Stanczak (✉)
Faculty of Geology, Geophysics and Environmental Protection, AGH University of Science and Technology, al. Mickiewicza 30, 30-059, Krakow, Poland
e-mail: gstanczak@geol.agh.edu.pl

E. K. Polychroniadis et al. (eds.), *International Multidisciplinary Microscopy Congress*, Springer Proceedings in Physics 154, DOI: 10.1007/978-3-319-04639-6_8,

pore geometry. In order to obtain a quantitative characterization of this pore network, a number of approaches such as the mercury intrusion porosimetry and the image analysis of thin sections have been used for measuring structural parameters [2, 5, 16, 22, 27]. The application of the fractal geometry to describe properties of highly complex structures makes it possible to obtain extra information on the pore network using the fractal dimension, being a global measure of the roughness of the features [15] or of the surface irregularity, describing a heterogeneous surface with single numerical value [20]. The fractal nature of the pore system in sedimentary rocks such as sandstones has been reported by a number of authors [6, 8, 12–14, 24].

The author presents here the measurement of the pore space fractal dimensions, using two methodologies: the mercury intrusion porosimetry and the image analysis of thin sections. Samples included sandstones from outcrops of the Lower Cretaceous Lgota Beds located within the Weglowka fold, which is a structural trap for oil accumulations in the Weglowka oil field in the SE Poland (the Outer Carpathians).

The fractal geometry is used to describe the irregular and fragmented patterns which are characterized by their invariability at any scale of scrutiny (i.e. self-similarity) [15]. Its crucial purpose is to provide means by which seemingly chaotic and random systems can be quantified by fractal dimensions. The fractal dimensions in a multi-fractal pattern change at different levels of scrutiny, since different aspects of the pattern either become apparent or become no longer noticeable [7].

Complex structures observed in the nature reveal fractal properties (scale invariance) only in (i) a statistical sense and (ii) within a certain range of scale. It means that these fractal structures may exhibit a statistical distribution of objects which are more or less at random and if the number of objects N with a characteristic linear dimension greater than r satisfies the relation (i.e. a cumulative form of fractal relationship):

$$N = \frac{C}{r^D} \tag{8.1}$$

where C is a proportionality constant, r is a characteristic linear dimension, and D is the fractal dimension. Additionally, there is the range of scales, determined by a lower and upper limits, over which the scale invariance extends. These limits, called inner and outer cut-offs of the fractal regime, are not less important parameters than the fractal dimension (D) itself [21, 25, 26].

The Menger sponge used as a model for a porous medium is constructed from cubes of density ρ_0 and size r_0. It is assumed that each cube is solid and has no porosity [25, 26]. The first-order Menger sponge constructed from the zero-order cubes has the size $r_1 = 3^1 r_0$ and consists of 20 such cubes ($N_1 = 20^1$) so that the first-order density is $\rho_1 = 20\rho_0/27$. At the next smaller scale the size of the second-order sponge is $r_2 = 3^2 r_0$ and there are 400 ($N_2 = 20^2$) solid cubes of size r_0 with density ρ_0. Thus the density of the second-order Menger sponge is $\rho_2 = 400\rho_0/729$. The density of the nth-order Menger sponge is

$$\frac{\rho_n}{\rho_0} = \left(\frac{r_0}{r_n}\right)^{3-\frac{\ln 20}{\ln 3}} \tag{8.2}$$

This is a power-law relation, where the fractal dimension of the Menger sponge is $D = ln\ 20/\ln 3 = 2.727$. Thus for a fractal distribution of pore volume, the density of a rock increases with its size *r* according to

$$\frac{\rho}{\rho_0} = \left(\frac{r_0}{r}\right)^{3-D} \tag{8.3}$$

The density of a fractal solid systematically decreases with the increasing size of the sample considered [3, 25, 26].

8.2 Material and Methods

Samples and thin-section images. Four sandstone samples were selected for this study. They have a narrow range of densities (2.66–2.70 g/cm^3) determined with a helium pycnometer (Micromeritics Accypyc 1330). Porosities calculated using the densities and the data from mercury porosimeter vary widely from 9.55 to 28.42 % (Table 8.1). One polished thin section impregnated with blue-dyed epoxy was made per sample. All thin sections were examined with a standard optical microscope. Moreover, all of the samples were examined by scanning electron microscopy, using a Hitachi S-4700 FEGSM (a cold field emission gun scanning electron microscope) on other thin sections coated with coal. The field emission SEM was operated at an accelerating voltage of 20 kV and a working distance of 12.9–14.4 mm.

Image analysis. Images of the analyzed thin-sections were acquired with both the optical microscope (OM) by using the DS-Fi1 color camera (Nikon) and the FEGSM at backscatter imaging with an Autrata modified YAG crystal. The OM images were digitized as 1280 × 1024 24-bit color images covering an area of approximately 550 × 440 μm for 10 ×-objective magnification. The FEGSM micrographs digitized as 2560 × 1920 8-bit gray-scale images cover an area of approximately 1150 × 860 μm for 110 ×-magnification (length of scale 500 μm) or 976 × 730 μm for 130 ×-objective magnification (length of scale 400 μm).

The original images acquired in this study were binarized into a pore phase and a solid phase using the Nikon's imaging software NIS-Elements BR (version 3.10). A simple *auto detect filled tool* using threshold technique allows a clear distinction between these two phases because of the high quality of the original images. The OM images were binarized by assigning pore phase to all pixels containing blue color tones. The FEGSM images have sufficient contrast to apply this tool to separate between dark pore space and light solid phase. The edges of the detected pore phase were smoothed by applying *a close transformation* with the 8-connectivity model. This procedure is performed as a dilation followed by erosion, resulting in smoothing of contours, filling of small holes and connecting very close objects (all pixels neighboring by the corner belong to one object). Continuously comparing the binarized image with the original thin-section image ensures the accuracy of the

Table 8.1 Density and effective porosity measured with proper device (pycnometer and porosimeter, respectively), and total porosity calculated for each sample. The box-counting fractal dimensions of the total optical porosity (TOP)

	Sample unit	1	2	3	4	Min	Max
Density	g/cm^3	2.70	2.70	2.67	2.66	2.66	2.70
Effective porosity	%	25.13	18.38	13.42	9.06	9.06	25.13
Total porosity	%	28.42	20.36	14.35	9.55	9.55	28.42
OM images							
D_{OM}		1.522	1.486	1.320	1.538	1.320	1.538
R^2		0.9953	0.9973	0.9968	0.9987	0.9953	0.9987
Total optical porosity	%	28.64	26.28	18.72	24.02	18.72	28.64
	μm^2	68,733.6	63,060.0	44,918.4	57,652.8	44,918.4	68,733.6
FEGSM images							
D_{FEGSM}		1.561	1.448	1.563	1.564	1.448	1.564
R^2		0.9991	0.9999	0.9996	0.9997	0.9991	0.9999
Total optical porosity	%	28.17	11.83	16.32	7.47	7.47	28.17
	μm^2	280,894.8	117,905.2	116,673.7	53,417.7	53,417.7	280,894.8

D_{OM}, D_{FEGSM}—the box-counting fractal dimension for the OM image and the FEGSM image, respectively; R^2—coefficient of determination

binarization process. For every studied sample two representative micrographs, the OM image and the FEGSM image, were chosen to image analysis.

The box-counting fractal dimension. Fractal dimension of pore space was determined with the box counting method performed on binary images. This method involve superimposing boxes with the box size δ onto the image and recording the number of boxes $N(\delta)$ covering the pore spaces. This process is repeated for different box sizes. If the distribution of the pore system within the two-dimensional images is fractal, then the $N(\delta)$ and the δ satisfy the relation:

$$N(\delta) \propto \delta^{-D} \tag{8.4}$$

where D is the box counting dimension. Thus, for fractal objects a double-logarithmic plot of $N(\delta)$ against δ yields a straight line of a slope equal to $-D$ [18]. For every calibrated image a sequence of five grids of boxes was used where the box size δ was reduced by a factor 1/2 (δ = 100, 50, 25, 12, 6 μm).

Mercury porosimetry and fractal dimension. Mercury porosimetry used in petrophysics to characterize the pore volume distribution of rocks is based on the gradual injection of mercury into an evacuated pore network as an external pressure is applied. Assuming that the pores are bundles of cylindrical capillary tubes, the Washburn equation relates capillary pressure and capillary diameters:

$$P_c = -\frac{4Y\cos\theta}{d_c} \tag{8.5}$$

where P_c is the capillary pressure required to force mercury into the evacuated pore space, Y is the surface tension of mercury, θ is the contact angle between mercury and the capillary tube, and d_c is the diameter of a cylindrical capillary tube, which models the diameter of a tubular pore throat. The increasing capillary pressure is plotted against the mercury saturation in terms of the percentage of rock pore volume saturated by mercury, resulting in the injection curve. Being the approximation of the distribution of pore volume accessible by throats of given effective size [27]. Thus, (8.5) becomes a relation between the pore space volume (V_p) and the injection capillary pressure (P_c):

$$V_p \propto P_c \tag{8.6}$$

We can consider the pore system of the rock as the Menger sponge fractal model where the pore diameter is assigned to the size of solid cube (r_0) of which the fractal model is constructed from. If the Menger sponge is injected with mercury as the injection pressure is applied then with increasing pressure smaller and smaller pores are filled so the volume of injected mercury increase and the density of the sponge increase consequently. Therefore, (8.3) and the relation (8.6) can be integrated to give the following expression for pore space volume (V_p) in function of pore fractal dimension (D_p) and capillary pressure (P_c):

$$V_p \propto P_c^{3-D_p} \tag{8.7}$$

The fractal dimension of pore volume (D_p) can therefore be deduced from the slope of the linear log-log plot of pore volume V_p against P_c [1, 3].

A Micromeritics Autopore 9500 mercury porosimeter was used to measure pore volume distribution as a function of pressure from 3.03×10^{-3} to 138 MPa (0.44 to 20×10^3 psia) or pore-throat diameters from 0.0090 to 413 μm on air-dried sandstone samples of the range of weight from 4.34 to 7.11 g. Values for the surface tension of mercury of 0.485 N/m and a contact angle on rock of 130° were used with the Washburn equation (8.5), assuming system of cylindrical capillary tubes in the calculation. The mercury-injection curve was describe by 32 pairs of applied pressure and intruded volume values but the initial portion of the intrusion curve associated with surface defects in samples was not taken into account in calculations of fractal dimensions [9].

8.3 Results

Results of image analysis of pore system. Each sample is represented by a pair of images, the OM image and the FEGSM image. All of the images have approximately similar resolutions, from 0.45 μm/pixel for two of the FEGSM images and 0.43 μm/pixel for all OM images to 0.38 μm/pixel for another two FEGSM images. The former FEGSM images cover an area of approximately 997,000 μm^2 while the latter ones include an area nearly of 715,000 μm^2 whereas the OM images span an area close to 240,000 μm^2.

For binary images created by identifying the pixels associated with the pore phase including the overall porosity (total porosity), the ratio of the pore phase area to the total image area was calculated. This ratio is termed *the total optical porosity* or TOP [5]. The values of the TOP estimated for the OM images are usually higher than the physically measured total porosities of the samples (Table 8.1). Merely one value of the TOP (28.64 %) is consistent with the total porosity of the sample (28.42 %). However, the TOP-s calculated for the FEGSM images mostly correspond to the measured total porosities. There is one value of the TOP (11.83 %) lower than the total porosity (20.36 %).

Generally, the total porosity (TOP) defined by the image analysis is dependent on the thin-section selection and the field of view under the microscope. In addition, the porosity values estimated for the images taken from the optical microscopy (OM) are determined by the penetrating power of epoxy impregnation and the overlapping phenomena, being a consequence of that the transmission light penetrating the entire 20–30 μm of the thin section displays blurred pore outlines in contact to transparent or translucent minerals [2, 4].

The calculations of the fractal dimensions using the box counting method have been performed for the total optical porosity (TOP). The linear log-log diagrams of the number of boxes $N(\delta)$ against the box sizes (δ) obtained for all the images yield the straight regression lines fitting to the data. Thus, the power-law relation is proved for the pore phases revealed within two-dimensional images of thin sections. The box-counting fractal dimensions for the OM images range from 1.320 to 1.538, whereas for the FEGSM images vary from 1.448 to 1.564. The range of the coefficients of determination is wider for the OM images (0.9953–0.9987) while for the FEGSM images is narrow from 0.9991 to 0.9999 (Table 8.1).

Fractal analysis of mercury porosimetry data. The calculations of the fractal dimensions of the pore volume basing on the mercury porosimetry data and using the Menger sponge fractal model have been carried out for the pore-throat diameters range from 91 to 0.0090 μm. The results of the fractal analysis for all the samples studied are summarized in Table 8.2.

For each sample the double-logarithmic injection curve of the cumulated pore volume against the cumulated injection pressure displays two linear components which can be fitted with piece-wise regression lines to estimate two different fractal dimensions [cf [17]]. For each linear element the regression line was fitted for as narrow as possible the range of the pore-throat diameters (cumulative pressures) and for as high as possible the coefficient of determination (R^2).

These two linear components representing two separate fractal elements appear in two distinct ranges of the pore-throat diameters (or cumulated injection pressures) considerably lesser than *the threshold pore-throat diameters* (11.32–45.37 μm) corresponding to *the threshold pressures* (64.81–525.66 kPa). The inflection point of the rapidly rising portion of the injection curve indicates *the threshold pressure* as the pressure at which mercury forms a connected pathway across the sample (*percolation backbone*), [1, 9, 10].

Table 8.2 The results of fractal analysis of pore volume based on mercury porosimetry data

		Sample	1	2	3	4	Min	Max
The first fractal element								
Fractal dimension D_1			2.956	2.981	2.950	2.974	2.950	2.981
Pore-throat diameter	min	μm	0.0090	0.0090	0.0090	0.0090	0.0090	0.0090
	max	μm	0.2984	0.0905	0.2987	0.0905	0.0905	0.2987
R^2			0.9153	0.9093	0.9183	0.9819	0.9093	0.9819
The second fractal element								
Fractal dimension D_2			2.777	2.900	2.419	2.201	2.201	2.900
Pore-throat diameter	min	μm	0.9024	0.1800	1.2061	1.2057	0.1800	1.2061
	max	μm	4.5367	1.7661	3.6412	3.6387	1.7661	4.5367
R^2			0.9925	0.9794	0.9915	0.9961	0.9794	0.9961
Threshold pore-throat diameter		μm	45.37	30.23	11.34	11.32	11.32	45.37
Threshold pressure		kPa	64.81	143.89	525.52	525.66	64.81	525.66

R^2—coefficient of determination

The first fractal element tied to the range of the smallest pore-throat diameters (0.0090–0.2987 μm) is characterized by the fractal dimensions (D_1) varying from 2.950 to 2.981, whereas the fractal dimensions (D_2) for the second fractal element linked to the range of the greater pore-throat diameters (0.1800–4.5367 μm) extend from 2.201 to 2.900. The values of the coefficients of determination (R^2) are different for each linear unit 0.9093–0.9819 and 0.9794–0.9961, respectively (Table 8.2).

The two-section fractal distribution of pore-space volume identified by two different fractal dimensions within two separate ranges of the pore-throat diameters implies that the pore space is the multifractal pattern at two levels of scrutiny (ranges of scale or ranges of the pore-throat diameters). The first fractal exhibited at the highest values of the injection pressures (the smallest diameters) is referred to as *the textural fractal*, whereas the second fractal manifested at the lower values of the injection pressures (the greater diameters) is labeled as *the structural fractal* [11].

The following inference is based on the results of the research of Krohn and Thompson [14] suggesting that the microvolumes of the pore space of sandstones exhibit the fractal structures. Therefore the textural fractal dimension (D_1) can be considered as controlling the distribution of volume of the smallest pore-throat (necks). The assumption that the textural fractal dimension (D_1) characterizes the overall pore-throat network in studied sample provides a device to set apart the distribution of the pore-throat volume (only necks) from the distribution of the pore-space volume (pores and necks). The extrapolation of the regression line fitted to the included data points into all known cumulative injection pressures gives the cumulative distribution of the pore-throat volume which is consecutively normalized by assigning the zero value of the volume to the threshold pressure (threshold diameter). The difference between the cumulative distribution of the pore-space volume and the pore-throat volume gives the cumulative distribution of pore volumes. This method of the partition of the distribution of the pore-space volume was introduced by Such and Lesniak [23].

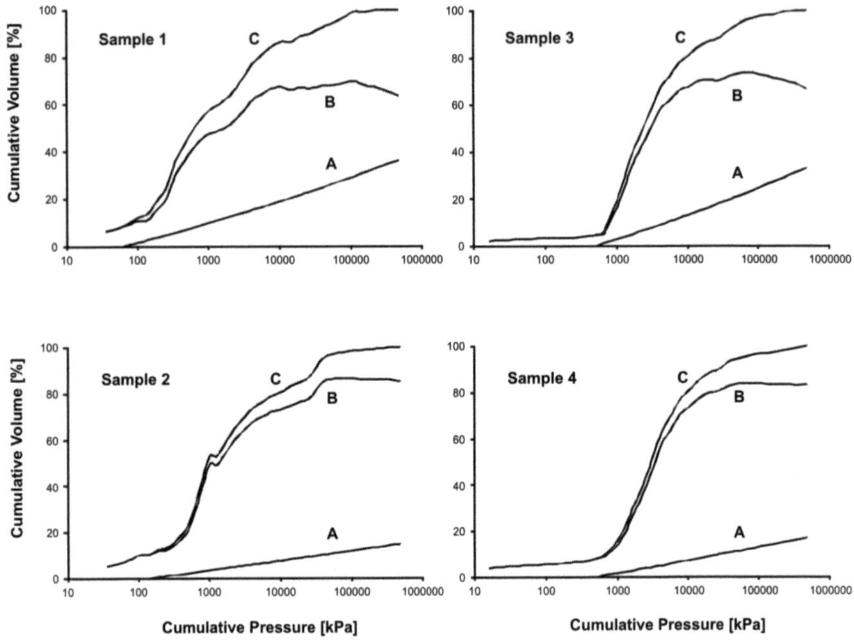

Fig. 8.1 The cumulative distributions of the pore-throat volume (A), the pore volume (B), and the pore-space volume (C) for studied samples

The distributions of the pore-throat volumes depend on the value of the textural fractal dimensions (D_1). The highest values of the fractal dimensions D_1 (2.974 and 2.981), typical of the narrow ranges of the pore-throat diameters (0.0090–0.0905 μm), result in the gradual distributions of the pore-throat volume in contrast to the steeper distributions determined by the lower values of the fractal dimensions D_1 (2.950 and 2.956) which characterize the wider range of the pore-throats (0.0090–0.2987 μm) (Figure 8.1).

8.4 Conclusions

The application of fractal geometry to studying the pore space in sandstones has been demonstrated for two different data sets (generated by mercury porosimetry and binary images) using two different methods.

The measurements of the fractal dimensions applying mercury porosimetry are based on the Menger sponge fractal model and indicate that the pore space is a bifractal pattern within two distinct ranges of the pore-throat diameters (0.0090–0.2987 μm) and (0.1800–4.5367 μm). The assumption that the textural fractal dimensions identified for the ranges of the smaller pore-throat diameters (0.0090–0.2987 μm) characterize the overall pore-throat network in the sandstones yields a device making possible partitioning the cumulative distribution of the pore-space volume into the pore-throat volume and pore volume cumulative distributions (Figure 8.1).

The measurements of the fractal dimensions using the box counting method have been carried out for the two-dimensional images of thin sections of the sandstones which had been processed to extract the pore phase. The box-counting fractal dimensions (1.320–1.564) evaluated for the total optical porosity (45,000–281,000 μm^2) are different from those obtained for the mercury porosimetry data (2.201–2.981). Assuming that the sandstones studied are isotropic structures, the box-counting dimensions may be increased by a unity to obtain the corresponding fractal dimension of the pore volume in 3D [15]. The new values of the fractal dimensions (2.320–2.564) calculated in this way are included within the range of the fractal dimensions found for mercury porosimetry data. All values of the fractal dimensions computed in this study for the pore volume correspond to those found by Krohn and Thompson [14], Hansen and Skjetorp [6] and Pérez Bernal and Bello López [19]. The two methods applied indicate that the pore space of the sandstones studied is fractal, and the more the fractal dimensions approach a value of 3 the more complex is the pore volume.

Acknowledgments The FEGSM examinations were conducted at the Laboratory of Field Emission Scanning Electron Microscopy and Microanalysis of the Institute of Geological Sciences of the Jagiellonian University. This work was financed by the AGH-UST (statutory grant No. 11.11.140.320).

References

1. R.F. Angulo, V. Alvarado, H. Gonzalez, Fractal dimensions from mercury intrusion capillary tests. SPE **23695**, 255–263 (1992)
2. F.S. Anselmetti, S. Luthi, G.P. Eberli, Quantitative characterization of carbonate pore systems by digital image analysis. AAPG Bull. **82**(10), 1815–1836 (1998)
3. F. Bartoli, N.R.A. Bird, V. Gomendy, H. Vivier, S. Niquet, The relation between silty soil structures and their mercury porosimetry curve counterparts: fractals and percolation. Eur. J. Soil. Sci. **50**, 9–22 (1999)
4. R. Ehrlich, S.J. Crabtree, S.K. Kennedy, R.L. Cannon, Petrographic image analysis I, analysis of reservoir pore complexes. J. Sediment. Petrology **54**(4), 1365–1378 (1984)
5. R. Ehrlich, S.J. Crabtree, K.O. Horkowitz, J.P. Horkowitz, Petrography and reservoir physics I: objective classification of reservoir porosity. AAPG Bull. **75**(10), 1547–1562 (1991)
6. J.P. Hansen, A.T. Skjeltorp, Fractal pore space and rock permeability implications. Phys. Rev. B **38**(4), 2635–2638 (1988)
7. J.P. Hyslip, L.E. Vallejo, Fractal analysis of the roughness and size distribution of granular materials. Eng. Geol. **48**, 231–244 (1997)
8. A.J. Katz, A.H. Thompson, Fractal sandstone pores: implications for conductivity and pore formation. Phys. Rev. Lett. **54**(12), 1325–1328 (1985)
9. A.J. Katz, A.H. Thompson, Quantitative prediction of permeability in porous rock. Phys. Rev. B **34**(11), 8179–8181 (1986)
10. A.J. Katz, A.H. Thompson, Prediction of rock electrical conductivity from mercury injection measurements. J. Geophys. Res. **92**, 599–607 (1987)

11. B.H. Kaye, Image analysis techniques for characterizing fractal structures, in *The Fractal Approach to Heterogeneous Chemistry*, ed. by D. Avnir (Wiley, Chichester, 1989), pp. 55–66
12. C.E. Krohn, Sandstone fractal and Euclidean pore volume distributions. J. Geophys. Res. **93**(B4), 3286–3296 (1988)
13. C.E. Krohn, Fractal measurements of sandstones, shales, and carbonates. J. Geophys. Res. **93**(B4), 3297–3305 (1988)
14. C.E. Krohn, A.H. Thompson, Fractal sandstone pores: automated measurements using scanning-electron-microscope images. Phys. Rev. B **33**(9), 6366–6374 (1986)
15. B.B. Mandelbrot, *The Fractal Geometry of Nature* (Freeman, New York, 1983)
16. C.A. McCreesh, R. Ehrlich, S.J. Crabtree, Petrography and reservoir physics II: relating thin section porosity to capillary pressure, the association between pore types and throat size. AAPG Bull. **75**(10), 1563–1578 (1991)
17. J.D. Orford, W.B. Whalley, The use of the fractal dimension to quantify the morphology of irregular-shaped particles. Sedimentology **30**, 655–668 (1983)
18. H.O. Peitgen, H. Jürgens, D. Saupe, *Chaos and Fractals: New Frontiers of Science* (Springer, New York, 1992)
19. J.L. Pérez Bernal, M.A. Bello López, The fractal dimension of stone pore surface as weathering descriptor. Appl. Surface Sci. **161**, 47–53 (2000)
20. P. Pfeifer, D. Avnir, Chemistry in noninteger dimensions between two and three. I. Fractal theory of heterogeneous surfaces. Jour. Chem. Phys. **79**(7), 3558–3565 (1983)
21. P. Pfeifer, M. Obert, Fractals: basic concepts and terminology, in *The Fractal Approach to Heterogeneous Chemistry*, ed. by D. Avnir (Wiley, Chichester, 1989), pp. 11–43
22. E.D. Pittman, Relationship of porosity and permeability to various parameters derived from mercury injection-capillary pressure curves for sandstone. AAPG Bull. **76**(2), 191–198 (1992)
23. P. Such, G. Lesniak, Study of pore space parameters of rocks. Prace Instytutu Gornictwa Naftowego i Gazownictwa **119**, 3–63 (2003) (summary in English)
24. A.H. Thompson, A.J. Katz, C.E. Krohn, The microgeometry and transport properties of sedimentary rock. Adv. Phys. **36**(5), 625–694 (1987)
25. D.L. Turcotte, *Fractals and Chaos in Geology and Geophysics* (Cambridge University Press, Cambridge, 1992)
26. D.L. Turcotte, J. Huang, Fractal distribution in geology, scale invariance, and deterministic chaos, in *Fractals in the Earth Sciences*, ed. by C.C. Barton, P.R. La Pointe (Plenum Press, New York, 1995), pp. 1–40
27. C.L. Vavra, J.G. Kaldi, R.M. Sneider, Geological applications of capillary pressure: a review. AAPG Bull. **76**(6), 840–850 (1992)

Chapter 9
Single Asperity Scratch Behaviour of Cast Stellite 6 Alloy

M. A. Ashraf, M. El-Ameen, R. Ahmed, N. H. Faisal, A. M. El-Sherik and M. F. A. Goosen

Abstract The aim of this paper is to investigate the nano-scale sliding wear behaviour of cast Stellite 6 (Co-28Cr-4.5W-1C) with a view to comprehend single asperity deformation. A nanoindentation system (NanoTest^{TM}—Micro Materials Limited, UK) equipped with wear testing module was used to simulate single asperity deformation behaviour using a sphero-conical indenter of 10 μm tip radius, 60° included angle for a sliding distance of 60 μm under 50 and 100 mN load, with sliding velocity of 2 μm/s. The test load was increased linearly over the sliding distance. Post-test evaluations included X-ray diffraction (XRD), scanning electron microscopy (SEM) and atomic force microscopy (AFM) measurements to determine the failure mode and wear volume. An elastic-plastic finite element model (FEM) was used to compare the displaced volume of alloy with the experimental data. There is limited work available to understand the nanotribological properties of Stellite alloys in published literature. In the current investigation results are discussed in terms of microstructural and tribo-mechanical evaluations to provide an understanding of the structure-property relationships. Results indicate that the wear behaviour at the nano-scale was dominated by the interaction of interdentritic carbides and metal matrix. It can be

---

M. A. Ashraf · M. El-Ameen · R. Ahmed · N. H. Faisal
College of Engineering, Alfaisal University, P.O. Box 50927, Riyadh 11533, Saudi Arabia

R. Ahmed (✉)
School of Engineering and Physical Sciences, Heriot-Watt University, Edinburgh EH14 4AS, UK
e-mail: rahmed@alfaisal.edu

N. H. Faisal
School of Engineering, Robert Gordon University, Garthdee Road, Aberdeen AB10 7GJ, UK

A. M. El-Sherik
Research and Development Centre, Saudi Aramco, Dhahran 31311, Saudi Arabia

M. F. A. Goosen
Office of Research and Graduate Studies, Alfaisal University, P.O. Box 50927,
Riyadh 11533, Saudi Arabia

E. K. Polychroniadis et al. (eds.), *International Multidisciplinary Microscopy Congress*,
Springer Proceedings in Physics 154, DOI: 10.1007/978-3-319-04639-6_9,

concluded that nano-scale sliding wear behaviour of cast Stellite 6 alloy is controlled by single asperity deformation. The predictions of nano-scale elastic-plastic FEM to calculate the volume displaced by the scratch are also discussed.

9.1 Introduction

Stellite 6 which has a nominal wt.% composition of Co-28Cr-4.5W-1C, is one of the most widely used cobalt based alloys for industrial applications. This cobalt-based alloy is generally used in power generation, marine, automotive, aerospace, and oil and gas industries to provide wear resistant components particularly in lubrication-starved, high temperature, or corrosive environments. Stellite® cobalt base alloys[1] consist of mixed carbides in a CoCr-based solid solution strengthened alloyed matrix [1]. The tribomechanical behaviour, especially the sliding wear properties of alloy components are significantly improved by introducing one or more hard phases. As a result, the wear resistance of Stellite alloys is due to the unique inherent wear characteristics properties imparted by a hard carbide phase dispersed in a cobalt alloy matrix.

Due to the variations in load during sliding wear conditions, a high toughness of the bulk material is of high importance. As a result, cast Stellite 6 alloys are often used because of their higher toughness compared to their individual hard carbide phases in the material. However, a high toughness is usually accompanied by a lower hardness that in turn may lower the wear resistance, especially in the dual or multi-phase alloy materials. Deformation in softer matrix and cracking in the harder phases can appear in material due to in service loading due to asperity interactions, single or multiple scratches. In order to obtain a more detailed understanding of materials failure, knowledge of the instrumented scratch depth-distance profile characteristics is essential. Over the past decade, a number of investigations have considered nanoscratch test for a number of industrial applications. These investigations take advantage of the typical features in the depth-distance profile recorded during single or multi-pass scratch test in bulk materials and thin films. However, there is no work available to comprehend the nanoscratch test results of Stellite alloys. Previously, the authors investigated the influence of two manufacturing routes, namely casting, HIP-consolidation (Hot Isostatic Pressing or HIPing), and re-HIPing on the properties of Stellite 4, 6, and 20 alloys [2, 3]. The results indicated that the three manufacturing processes result in distinct microstructural features; especially in terms of the content, shape and size of carbides, which has a significant influence on their impact and fatigue resistance.

The aim of this paper was to provide microstructural and tribo-mechanical evaluation of cast Stellite 6 alloys via Scanning Electron Microscopy (SEM), Atomic Force Microscopy (AFM), X-ray diffraction (XRD), EDS (Energy Dispersive

---

[1] Stellite is a registered trade name of Deloro Stellite Company Inc.

Table 9.1 The Chemical compositions of cast Stellite 6 alloy (wt.%).

Co	Cr	W	C	Mo	Fe	Ni	Mn	Si
Bal.	27.10	4.95	0.95	0.30	1.10	0.60	0.90	1.24

X-ray Spectroscopy) and single asperity nanoscratch behaviour. Comparison was also made with FEM (Finite Element Modelling) analysis. Variations in measured scratch profile with the sphero-conical diamond probe for this cobalt based cast alloy were investigated to provide an understanding of the structure-property relationship at nano-scale.

9.2 Experimental Test Procedures

9.2.1 Materials and Microstructure

The cast alloy samples were produced via sand casting technique. The chemical composition of different phases developed in the powders and alloys were determined using EDS and XRD with Cu-K_α radiation (wavelength = 1.5406 Å). Image analysis was also conducted to ascertain the area fractions of individual phases. Table 9.1 summarises the chemical composition of cast Stellite 6 alloy.

9.2.2 Nanoindentation Tests

Nanoindentation testing which included nanohardness and elastic modulus measurements were performed using a calibrated nanoindentation system (NanoTest^{TM}—Micro Materials Limited, UK) equipped with a standard Berkovich nanoindenter tip. Measurements were performed at room temperature (~23 °C) in load control mode at a load of 50 mN. The indentation procedures were programmed as three segments of trapezoidal shape with 10 s loading, 5 s hold and 10 s unloading segments. Eighteen equally spaced measurements were performed on the sample. The *P-h* profiles were analysed using standard methods with the area function for the Berkovich indenter which was determined by indentations into fused silica with elastic modulus of 69.9 GPa. The raw data (*P-h* profile) were used to evaluate hardness (*H*) and reduced elastic modulus (E_r) using Oliver and Pharr method [4]. The elastic modulus (E_i) and Poisson ratio (ν_i) of the diamond indenter was taken as 1140 GPa and 0.07, respectively. To calculate the elastic modulus (E_s) of the specimen (cast Stellite alloy), the Poisson ratio for cast Stellite alloy (ν_s) was taken as 0.3. The indentation instrument also consisted of an optical microscope (OM) and an integrated atomic force microscope (AFM), directly linked by an automated positioning system.

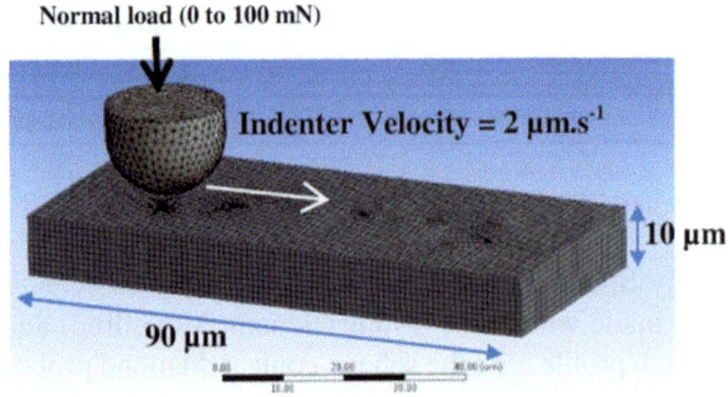

Fig. 9.1 FEA geometry of scratch

9.2.3 Nanoscratch Sliding Wear Tests

Nanoscratch (sliding wear) tests were performed with a sphero-conical diamond indenter of 10 µm tip radius and 60° included angle. The scratch tests were performed as multi-pass tests at 2 µm.s^{-1} over a 60 µm track with (a) pre-scratch topographic scan, (b) a levelling distance followed by a ramped scratch, and (c) a post-scratch low load scan. In the pre- scratch and post-scratch scan the applied load was sufficiently low (10 µN) so that no wear occurs. In the 'on-load' scratch scan the load was ramped after 20 µm travel at 2.5 mNs^{-1} to reach 50 mN and at 5 mN.s^{-1} to reach 100 mN. Six repeat tests were performed to test the reproducibility of scratch behaviour. The individual scratch was spaced at 25 µm and it is expected that the probe is not deflecting the path when it encounters a large carbide phase in the soft matrix. Post-test imaging using AFM and SEM has also been used to investigate the residual surface features after scratch testing. Wear volume was calculated by multiplying the projected area of the scratch by average scratch width. Scratch projected area was imported in to MATLAB® as topography map and volume was calculated using trapezoidal rule function.

9.2.4 Finite Element Modelling of Scratch Test

The nanoscratch scenario was modeled as elastic-plastic contact model implemented in commercial FEM platform of ANSYS (R14). In this study, a spherical indenter (diamond) comes in contact and slides with a ramping force on the surface of Stellite 6 alloy, which is assumed to display an elasto-plastic behavior [5]. The indenter was modeled as rigid with a tip radius of approximately 10 µm. Figure 9.1 shows the model with a grid in 3D. The mesh follows an element size of 1 µm to keep balance of accuracy and computation time. The model constituted of 49665 elements of the CONTA171 and TARGEI69 type. The scratch and loading dimensions were in accordance with the experimental set-up for a load of 100 mN.

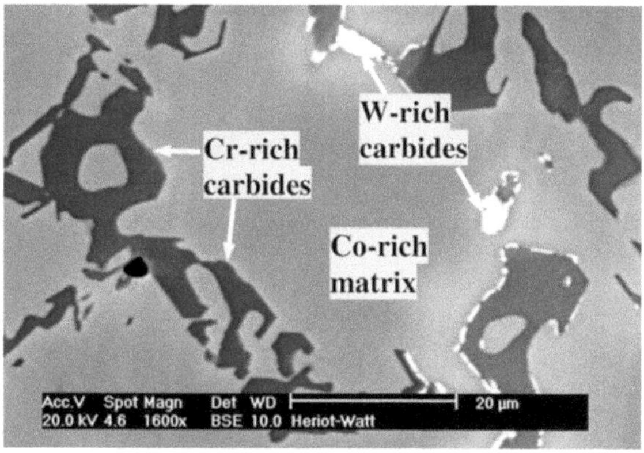

Fig. 9.2 SEM image showing microstructure of cast Stellite 6 alloy

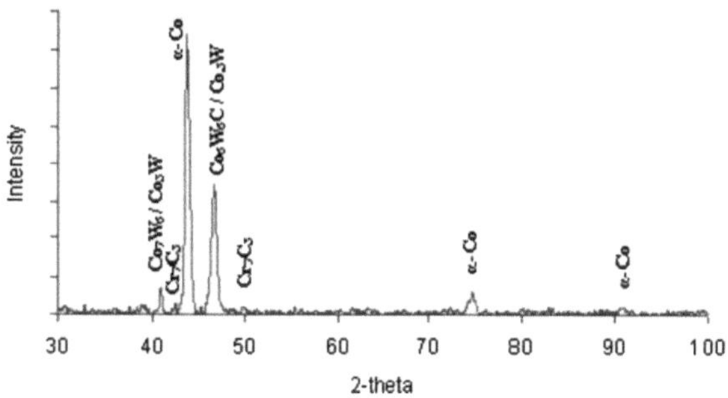

Fig. 9.3 XRD pattern of Cast Stellite 6

9.3 Results

9.3.1 Microstructure and Phase Analysis

The SEM and XRD evaluations of the cast alloy were assessed. Figures 9.2 shows the hypoeutectic microstructure of the cast alloy, which consists of Cr-rich carbides (dark phase), W-rich carbides (bright phase), and the Co-rich dendritic matrix (grey region). The XRD analysis shown in Fig. 9.3 confirms the presence of Cr-rich and W-rich carbides. The image analysis results indicated that the average area fraction of these carbides was $14.5 \pm 6.6\,\%$ and $1.0 \pm 0.5\,\%$, respectively. The analysis was based on five measurements and high standard deviation is due to the different sizes of inter-dendritic carbides.

9.3.2 Nanohardness and Modulus

The mean nanohardness and modulus values of the cast Stellite 6 were measured to be 8.5 ± 2.8 GPa and 203 ± 14 GPa, respectively. The hardness and modulus of

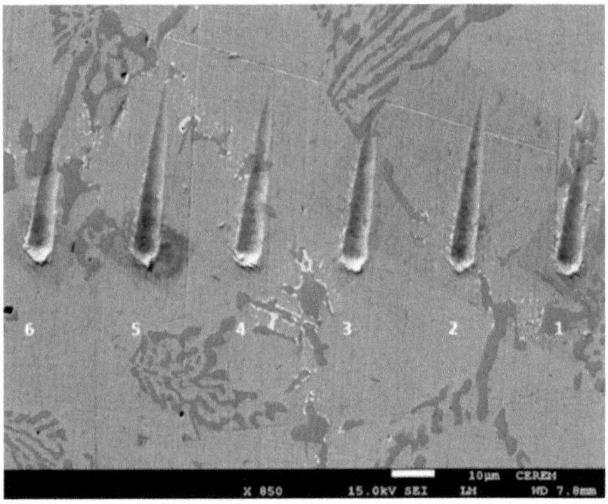

Fig. 9.4 SEM observation of wear scars after nanoscratch tests at 100 mN load

indentations which landed on the carbide phases were in the range of 16–9.5 and 243–204 GPa, respectively. Similarly, the hardness and modulus of indentations which landed on the metal matrix phases were in the range of 6.4–6.9 GPa and 187–193 GPa, respectively. However, these values are only indicative as the measurements on the matrix and carbide can be influenced by the neighbouring and underlying material.

9.3.3 Nanoscratch Testing

The AFM, SEM and topography imaging was conducted for all scratches. Figure 9.4 shows typical scratch observations for the 100 mN loading tests. The corresponding topography map of scratches 1–3 is shown in Fig. 9.5. The pre and post-scratch topography scan of scratch 2 at 100 mN load (Fig. 9.4) is shown in Fig. 9.6. High magnification SEM observation of this scratch is shown in Fig. 9.7. Figure 9.8 shows the pre and post-scratch topography for scratch 6 at 100 mN load. The corresponding high magnification image of the wear scar is shown in Fig. 9.9. The displaced average wear volume during scratch testing for the 50 and 100 mN load was computed as $5.1 \pm 1.7\,\mu m^3$ and $13.6 \pm 2.2\,\mu m^3$, respectively.

9.3.4 Finite Element Modelling

The model described earlier was solved in successive displacement iterations with load ramping of 100 mN. Figure 9.10, shows the actual and modelled scratches. To further verify the accuracy of the model, the volume displaced by both scratches is

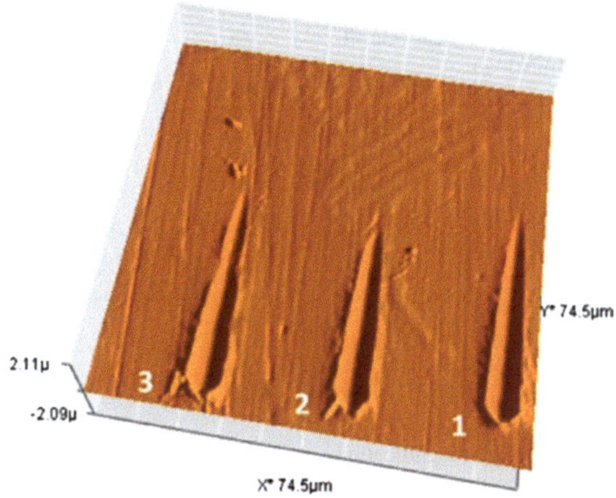

Fig. 9.5 Surface Topography of wear scars #1, 2, and 3 after scratch tests at 100 mN load

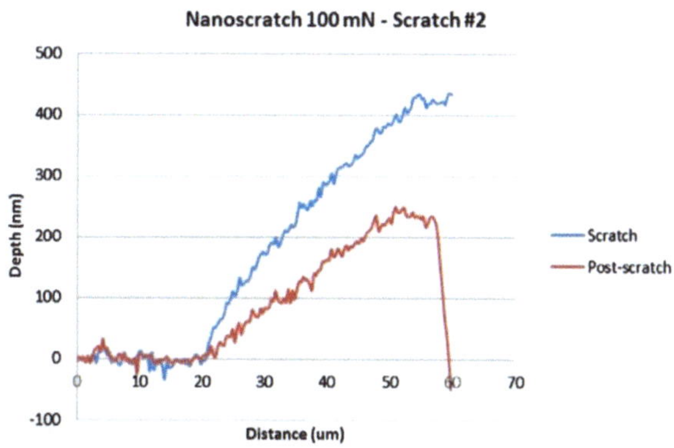

Fig. 9.6 Pre- and post-scan depth profile of wear scar #2 at 100 mN load

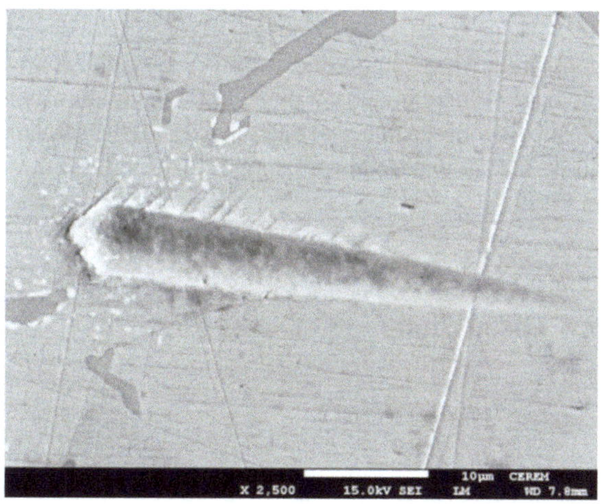

Fig. 9.7 High magnification SEM observation of scar #2 at 100 mN load

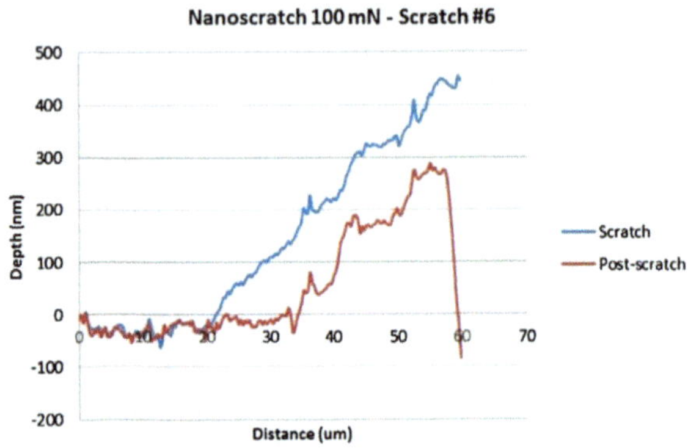

Fig. 9.8 Pre- and post-scan depth profile of wear scar #6 at 100 mN load

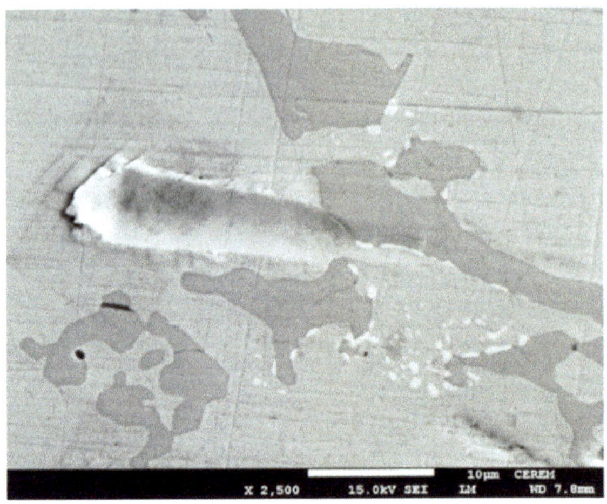

Fig. 9.9 High magnification SEM observation of scar #6 at 100 mN load

compared. The volume displaced in FEM analysis was calculated by difference of the base volume under x-z plane before and after the scratch. The indenter causes a plastic flow of the material on both sides of the scratch. The edge built up is therefore above the x-z plane and corresponds to the volume removed from the scratch depicted in Fig. 9.10.

9.4 Discussion

9.4.1 Microstructure

The cast CoCr28W alloy had a hypoeutectic microstructure (Fig. 9.2), which consists of Co-rich dendrites (grey region), set in lamellar eutectic Cr-rich (dark phase) and W-rich (bright phase) carbides.

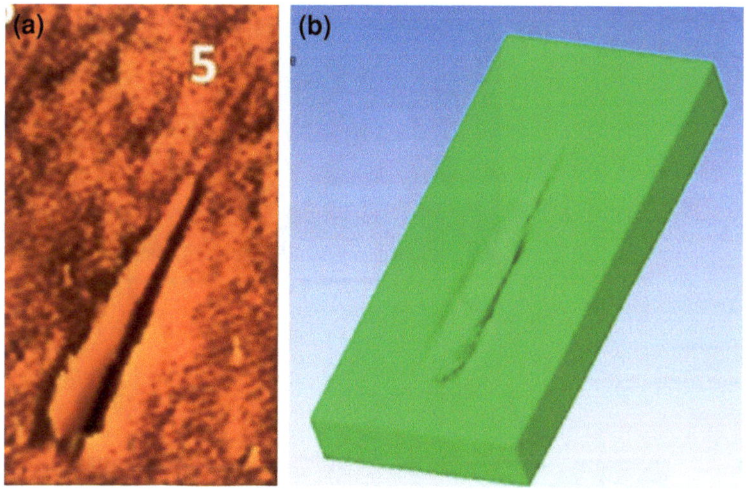

Fig. 9.10 Comparison of post nanoscratch test observations at 100 mN load **a** surface topography of scratch #5, and **b** FEA output from AutoDYN

The Cr-rich eutectic carbide had a composition of $(Cr_{0.71}Co_{0.25}W_{0.03}Fe_{0.005})_7C_3$, as approximated by the EDS analysis. The XRD analysis (Fig. 9.3) revealed that the carbides were Cr_7C_3 and Co_6W_6C, whilst α-Co (F.C.C.) was the primary phase in the solid solution, together with the inter-metallic compounds, Co_3W and Co_7W_6. This dendritic microstructure is typical of cast CoCr28W alloy in which carbide and grain size can be controlled by the rate of cooling. Within the family of the cast cobalt-based alloys, the relatively large carbide size seen in the cast microstructure indicated slow freezing during the casting process. The microstructure of cast satellite 6 alloys has been a topic of research for a number of investigations and further details of the influence of cooling rate on the grain size of cast cobalt-based alloys can be appreciated elsewhere [6]. The scope of the discussion here is therefore on understanding the structure-property relationships during tribo-mechanical performance at nano-scale.

9.4.2 Nanohardness and Modulus Analysis

The mean value of hardness and modulus provided an overall assessment of cast Stellite 6 alloy at nano-scale. The mean was dominated by the indentations on the softer matrix which comprised almost 85 % of the surface area of alloy as indicated by the image analysis. However, the difference in the hardness of carbides and matrix, where the hardness of carbides was in the range of 16–9.5 GPa in comparison to 6.4–6.9 GPa for the matrix, compensated for the difference in area fractions and hence

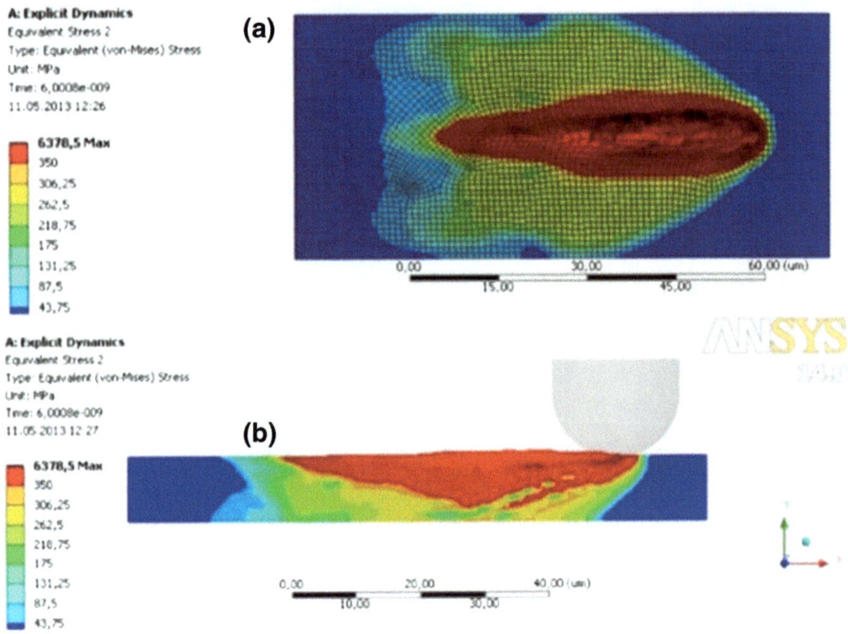

Fig. 9.11 Von-Mises stress distribution for 60 μm at 100 mN load **a** along thickness (y-axis) **b** laterally (z-x-axis)

represented an overall behaviour. The difference in modulus values was although not as substantial but reflected the changes in the bonding mechanism within the carbides and metal matrix.

9.4.3 Nanoscratch Tests

The nanoscratch behaviour indicated that cast Stellite predominantly deformed plastically with no fracture of carbides which is consistent with the high toughness of this alloy [2, 3]. Figures 9.4 and 9.5 indicate that the methodology of nanoscratch testing is able to evaluate the nano-scale differences in the wear behaviour of this alloy between the matrix and carbide phases. This is further highlighted in Figs. 9.6 and 9.8. The scratch behaviour was predominantly plastic in Fig. 9.6 which was further verified by the SEM observation on Fig. 9.7. However, observations shown in Figs. 9.8 and 9.9 highlight the differences when the indenter scratches over a combined matrix and carbide phase. The post-scratch profile between the 20 μm and approximately 35 μm sliding distance in Fig. 9.8 shows very little deformation. This part of the scratch relates to the period when the indenter scratched over the carbide phase as shown in the SEM observation in Fig. 9.9. Further flattening of depth curve in Fig. 9.8 was also observed between the scar length of 42–50 μm due to similar effect. These effects were further quantified in the displaced volume of $4.7 \pm 2\,\mu m^3$ and $14.5 \pm 2.8\,\mu m^3$, respectively for the 50 and 100 mN load and indicated nonlinear relationship between wear and loading at nano-scale, which is attributed to the differences in the elastic plastic ratio with load. In general it can be concluded that the differences in the wear response at nano-scale were well observed and quantified using the nanoscratch testing.

9.4.4 FEA Analysis

The FEM provided further insights in to the wear behaviour and indicated that the depth of plastic zone under the indenter was maximum up to a depth of $9\,\mu m$ (Fig. 9.11) for a plastically deformed wear scar depth of 300 nm (Figs. 9.6 and 9.8). The comparison of wear morphology was consistent with the experimental observations (Fig. 9.10), which was reflected in the quantitative analysis of wear volume and its comparison with the experimental analysis.

9.5 Conclusions

(a) Nanoscratch testing of cast Stellite 6 alloy indicated differences in the wear behaviour at nano-scale between the metal matrix and carbide phases.
(b) The FEM analysis provided further insights on the calculation of plastically deformed zone under the contact region and wear volume displaced.
(c) The nanohardness and modulus measurements indicated an overall response of the material where the differences in the area fractions of the carbide and metal matrix phases were compensated by the large difference in the nano-mechanical properties of these phases.

Acknowledgments The work presented in this paper has been supported by the project titled "Wear behavior characterization of Stellite Alloys using computational modeling", funded by Saudi Aramco (Contract number 6600028941).

References

1. P. Crook, Cobalt-base alloys resist wear, corrosion, and heat. Adv. Alloy. Prog. **145**, 27–30 (1994)
2. H. Yu, R. Ahmed, A comparison of the tribo-mechanical properties of wear-resistant cobalt based alloys produced by different manufacturing processes. ASME J. Tribology **129**, 586–594 (2007)
3. H. Yu, R. Ahmed, S. Davies, Influence of manufacturing process and alloying element content on the tribomechanical properties of cobalt-based alloys. ASME J. Tribology **131**, 11601–11613 (2009)

4. W.C. Oliver, G.M. Pharr, An improved technique for determining hardness and elastic modulus using load and displacement sensing indentation experiments. J. Mater. Res. **7**, 1564–1583 (1992)
5. M.A. Ashraf, B. Sobhi-Najafabadi, Ö. Göl, D. Sugumar, Numerical simulation of sliding wear for a polymer-polymer sliding contact in an automotive application. Int. J. Adv. Manuf. Technol. **41**, 1118–1129 (2007)
6. W.L. Silence, Effect of structure on wear resistance of Co-, Fe-, and Ni- base alloys. J. Lubr. Technol. Trans. ASME **100**, 428 (1978)

Chapter 10
A Study of Ink Film Tribology

Larisa G. Varepo, Alina S. Borisova and Evgeny V. Trapeznikov

Abstract Abrasion strength of ink film onto surface of different packaging materials is significant working parameter during transport to customers and handling. The process of surface damage were investigated by evaluation of the tribological parameters of ink films onto papers, boards and aluminum foil in different ways using "High Temperature Tribometer" firm CSM Instruments (adopted method). We have obtained a graph illustrating of the ink film destruction process. It is shown abrasion strength of the ink film is determined by directly nature and microgeometry of the printing material surface layer. Distribution of ink film inside pores of papers were investigated by SEM using JSM7500F (Jeol) and EDS X-max80 (Oxford). The mathematical model based on the strength of the ink film Ra roughness and porosity of the material has been developed. The adventage of the model is easy strength simulation of different systems ink-substrate.

10.1 Introduction

The evaluation of "ink-substrate" system strength by methods, allowing to exactly fix force of destruction and elaboration of adhesion strength model of the system are actual tasks.

L. G. Varepo (✉) · A. S. Borisova
Doctor of Engineering Science, Omsk State Technical University, Omsk, Russia
e-mail: larisavarepo@yandex.ru

A. S. Borisova
e-mail: lianilla@mail.ru

E. V. Trapeznikov
Postgraduate Student, Omsk State Technical University, Omsk, Russia
e-mail: evtrapeznikov@yandex.ru

E. K. Polychroniadis et al. (eds.), *International Multidisciplinary Microscopy Congress*, Springer Proceedings in Physics 154, DOI: 10.1007/978-3-319-04639-6_10,

Table 10.1 Results of tribological tests

Sample's number	Material	Ra, μm	Strength values			
			Friction factor	Tracks quantity	Track length, until damage, M	Time, s
1	Cellulose cardboard with double coated layer	0.411	0.39	54	1.53	7.5
2	Cellulose cardboard from virgin fiber with double coated layer	0.599	0.23	46	1.3	6.5
3	Combination material	0.709	0.24	135	3.8	34
4	Combination material	0.474	0.23	253	7.2	64

Therefore, the study of adhesion strength of ink films onto porous and non-absorbant materials on base tribological tests and evaluation of real contact area for (and) modeling is the subject-matter of this paper.

The process of surface damage were investigated by evaluation of the tribological parameters of ink films onto papers, cardboards and aluminum foil in different ways using "High Temperature Tribometer" firm CSM Instruments (adopted method).

Distribution of ink film inside pores of papers were investigated by SEM using JSM7500F (Jeol) and EDS X-max80 (Oxford) in secondary electron detection mode over the range 0.5–0.8 kV acceleration voltage.

10.2 Results and Their Discussion

Abrasion strength of ink film onto sample surface is estimated by next parameters: quantity of tracks, length of tracks, time (Table 10.1). The tribological test graph of "ink-substrate" system for one of a test sample is represented on Fig. 10.1.

Tests show that ink film abrasion strength is defined by ink distribution onto substrate and roughness of a substrate.

Spectrums of element distributions in different points of ink—coated layer interface on cross section of cellulose cardboard with double coated layer are represented on Fig. 10.2.

They show that carbon, oxygen, natrium are presented in composition of cardboard (Fig. 10.2, paper). Elementary composition of upper coated layer is characterized by presence C, O, Ca, Al, Si (Fig. 10.2, paint). Similar composition is observed for second coated layer but quantity of Al and Si are less (Fig. 10.2, interface). Separated particles of coated layer *(particle in paper)* are Ca, probably. Carbon is presented as expected in composition of cardboard (paper) and black ink film (upper layer), because main their component are cellulose and carbon black, respectively.

More homogeneous substrate surface than more strength of ink layer.

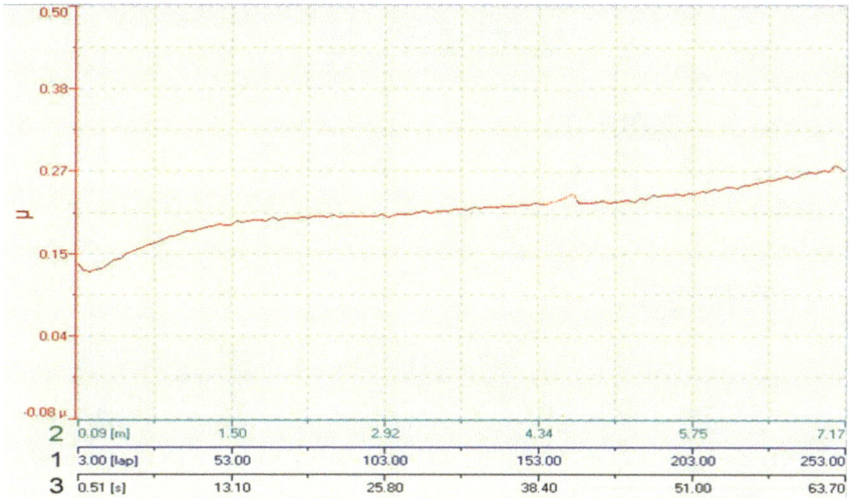

Fig. 10.1 Tribological test of "ink-substrate" system, where *1*—tracks quantity, *2*—track length, *3*—time

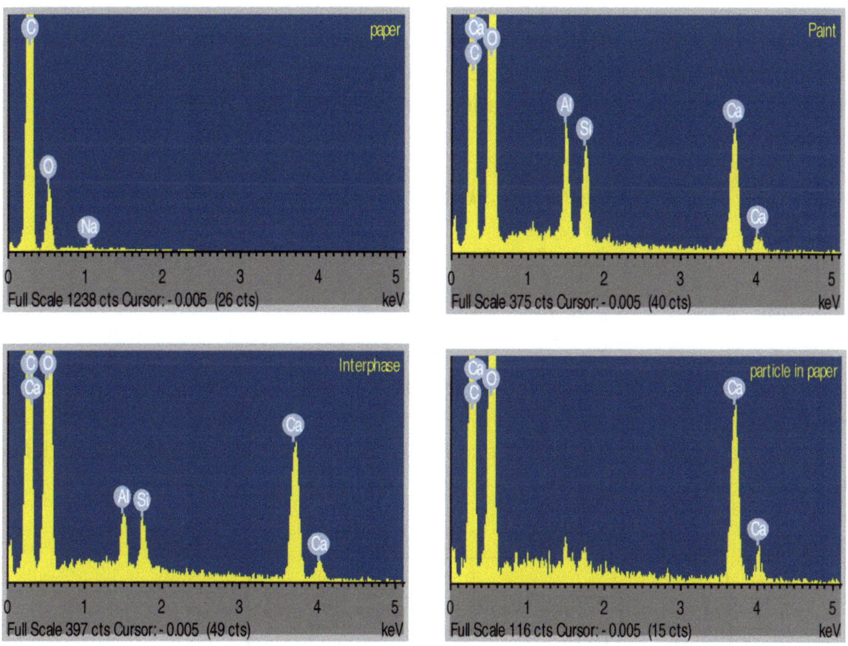

Fig. 10.2 EDS—spectrum of cross section of cellulose cardboard with ink layer here *paper*—base of cardboard; *interphase*—second coated layer, *particle in paper*—separated particles of coated layer

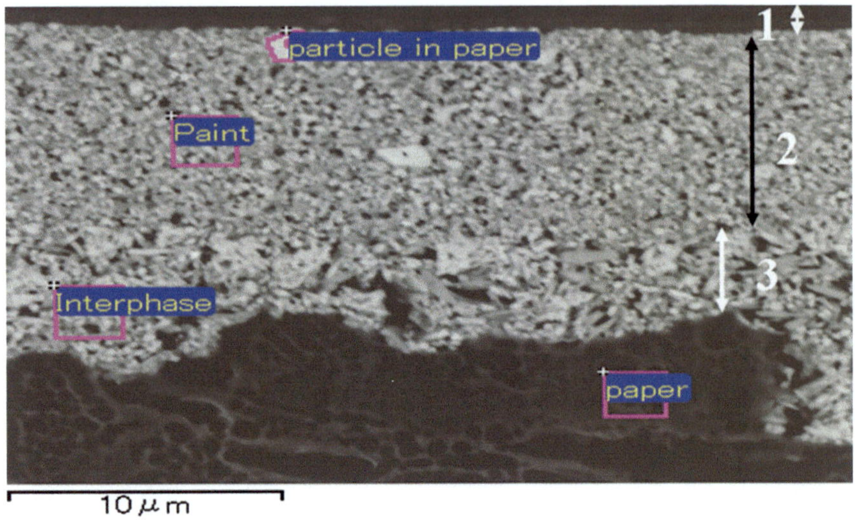

Fig. 10.3 SEM micrograph of cross section of cellulose cardboard with ink layer (*1*), where *2*—first coated layer, *3*—second coated layer

As rule adhesion strength of "ink-substrate" system is defined by value of real surface. In this study we have offered to use equation of surface profile as Fourie polynomial for more precise estimation of real contact area [1, 2]:

$$f_i(x) = z_i = \sum_{k=1}^{n} b_{ki} \sin \frac{\pi k x}{a}; i = 1, 2, \ldots, m,$$

$$b_{ki} = \frac{2}{a} \int_0^a f_i(x) \sin \frac{\pi k x}{a} dx.$$

Main part of ink films is onto tested substrate surface, penetration in pores is low (Fig. 10.3).

In this case real contact area per square centimeter of substrate space can defied as

$$S = h \sum_{i=1}^{m} \left(\int_0^a \sqrt{1 + \left(\sum_{k=1}^{n} b_{ki} \frac{\pi k}{a} \cos \frac{\pi k x}{a} \right)^2} dx - \sum_{u=1}^{r_i} \Delta_u x \right)$$
$$+ \left(h \sum_{i=1}^{m} \left(\int_0^a \sqrt{1 + \left(\sum_{k=1}^{n} b_{ki} \frac{\pi k}{a} \cos \frac{\pi k x}{a} \right)^2} dx - \sum_{u=1}^{r_i} \Delta_u x \right) \right) \cdot \left(n \pi^2 d^2 \sqrt{\frac{Pt}{\eta}} \right).$$

As adhesion work is proportional to real contact area independently from nature of bonding, than

$$Ad = æh \sum_{i=1}^{m} \left(\int_0^a \sqrt{1 + \left(\sum_{k=1}^{n} b_{ki} \frac{\pi k}{a} \cos \frac{\pi k x}{a} \right)^2} dx - \sum_{u=1}^{r_i} \Delta_u x \right) \cdot \left(1 + n\pi^2 d^2 \sqrt{\frac{Pt}{\eta}} \right)$$

Ad—adhesive strength; æ—energy of a unit average adhesive bond, which is multiplied by the number of adhesive bonds per 1 cm^2 of the real contact area; *h*—partitioning step of the sample in the unit area ($h = 3\,\mu$); *a*—length of the sample; b_{ki}—Fourier factors; *x*—the variable of integration, ($0 \le x \le a$); *n*—the number of pores per 1 cm^2 of surface; *d*—pore diameter; *P*—the pressure under the action of which the adhesive flows into the micropores of the substrate; *t*—the contact time; η—viscosity of the adhesive; $h \sum_{u=1}^{r_i} \Delta_u x$—sum of horizontal sections areas corresponding empty cavities for each of the *m* cylindrical surfaces; r_i—quantity of unfilled cavities corresponding to the cross section $y = y_i$.

Practice realization of this formula has proven its value by Fisher factor with confidence probably 0.95.

10.3 Conclusions

Tribological tests of ink films onto different printing materials have showed that abrasion strength depends on roughness of substrate surface. The mathematical model for abrasion strength evaluating of "ink-substrate" system on base estimating of real contact area has been obtained. The model is different from their analogs due to possibility of records of a real substrate surface profile by its representation as Fourie polynomial.

References

1. L.G. Varepo, in *Certificate of the state registration of computer software*, ed. by L.G. Varepo, A.S. Borisova, O.A. Kolozova, E.V. Trapeznikov, The Program for the Analytic Representation of Surface Profile and Roughness Estimates of Print Materials, No. 1201161039 (FIPS, Moscow, 2011) 11 Jan 2011
2. L.G Varepo, in *Proceedings of the Higher Education Institutions. Problems Polygraphy and Publishing: scientific and Technical Journal*, ed. by L.G. Varepo, A.S. Borisova, O.A. Kolozova, Analytical Representation of the Estimate of the Print Materials Surface Profile, vol (MGUP, Moscow, 2010), pp. 16–23

Chapter 11
The Numerical Calculation of the a Viscous Incompressible Fluid Transfer onto Porous Surface Between Rotating Cylinders

Alekxey V. Panichkin and Larisa G. Varepo

Abstract The process of viscous incompressible fluid (ink) transfer in offset printing is observed. Ink flow modeling between rotary cylinders with following splitting down of ink film is carried out with the help of finite—difference methods. It is shown creation possibility of local ink liftoff areas from paper surface, when escaping the printing engagement zone. The numerical calculations for motion of free boundaries were investigated for various values of viscosity, with the help of moving boundary nodes on node lines on a fixed grid.

11.1 Introduction

The most important source of productive efficiency growth is constant increase of engineering performance standard and quality of output products. Reproduction quality of printed output originals is determined both by characteristics of materials taking part in printing process, in particular inks and materials—information carriers, and by parameters of technological printing process.

Irrespective of used printing process type, ink curing represents a set of complex processes the course of which is made conditional upon physical—chemical properties of interrelated systems. Complexity of phenomena taking place in printing zone is determined not only by the process of ink transfer onto paper but also by the escaping process of substrate from engagement zone which is accompanied by splitting down of ink film. This is connected with adhesive forces action causing

A. V. Panichkin (✉)
Doctor of Physics and Mathematics, Omsk branch of the Institute of mathematics of the SB RAS. S.L. Soboleva, Omsk, Russia
e-mail: panich@ofim.oscsbras.ru

L. G. Varepo
Doctor of Engineering Science, Omsk state technical university, Omsk, Russia
e-mail: larisavarepo@yandex.ru

E. K. Polychroniadis et al. (eds.), *International Multidisciplinary Microscopy Congress*, Springer Proceedings in Physics 154, DOI: 10.1007/978-3-319-04639-6_11,

paper adhesion to cylinder surface (image carrier) in the escaping moment from the engagement zone. As a result, ultimate departure of paper from cylinder surface takes place with some delay, that is, ink transfer from plate onto paper is the result of formation and destruction of adhesive compounding.

However, so far there is no full concept about the character of adhesive contact linking during ink transfer onto substrate and determination of ink film splitting down coordinate.

More simply, ink transfer can be described as repeating ink offsetting under pressure in the printing engagement zone with the following splitting down process. In most cases, it results in ink maldistribution on both surfaces which are formed as a result of ink film rupture.

Due to the great significance which is given to ink transfer process in printing, highly interesting experimental studies were started in this field.

According to analysis of literary sources, a great number of scientific works are devoted to questions of modeling and ink transfer calculation.

Thus, in most literary sources data, are cited on ink splitting with reductive assumption of half splitting with zero absorbency. It is accepted on condition that ink film is divided in half on splitting angle Ψ.

The most complete and comprehensive study of ink transfer mechanism must necessarily be taken into consideration when forming processes of adhesive engagement as well as phase structure of ink film. It predetermines its splitting coordinate depending on ink film thickness and a number of other factors which have direct effect on this process.

Recent studies came to appreciable deepening of scientific knowledge in the field of ink transfer in printing process. However, until now their complete mathematical formulation is not available.

Therefore, the study of ink transfer mechanism on the basis of modern means and research methods as well as the necessity of theory development, allowing to model ink flow in print engagement zone and taking into consideration the above-mentioned, are important even today. They are of great scientific and practical interest, and reflect the goal of this work.

11.2 Methods

Different polymeric materials on composition structure, including paper, are explored as substrates.

The equations by Naïve—Stocks are used for numerical modeling of viscous incompressible fluid flow with free boundaries between rotary cylinders; with calculation of incompressible fluid flows (averaged according to Reynolds). Algorithm with loading diagrams on regular nets for Navie-Stocks equations in attendant cylindrical co-ordinates is developed in two-dimensional position [1].

11.3 Results and Their Discussion

For modeling of tear-off flows of viscous incompressible fluid from the outlet zone between two rotary cylinders special attention was paid to adequate reproduction of flow near the body surface. Fluid flow procedure is characterized by Reynolds number (Re). When the value Re is less than some critical number Re_{cr}, laminar flow of fluid takes place. If $\mathrm{Re} > \mathrm{Re}_{cr}$, flow procedure can become turbulent. It is necessary to use finite-difference nets with corresponding resolving capacity in range of Reynolds critical numbers when buckling failure of laminar boundary ink film takes place and vortical formations arise inside boundary ink film. In this case, as ink is highly viscous fluid, so at stages for modeling of flow in working layer it is necessary to apply model of viscous fluid flow describing laminar flow at certain limits of cylinders speed.

The process of layer splitting and transfer of its part onto substrate is complicated by many unexplored phenomena. It is proposed to take into consideration the following factors: surface characteristics; displacement of ink particles inside a sector, its rheological properties; printing specific pressure; contact area; printing speed and geometrics of cylinders; amount of clearance between cylinders.

As in modeling it is observed movement of two cylinders between which the movement of fluid with initial applied layer on upper cylinder takes place, it was worthwhile to examine decision region in attendant co-ordinates connected with upper cylinder and center.

The obtained numerical solutions of flow characteristics showing the formation possibility of local zones of ink breaking off from substrate surface when the latter leaves the contact zone. The implementation of given approach is carried out with the help of developed software for numerical solution of Navie-Stocks equations for incompressible fluid in coordinates on finite-difference nets (11.1).

$$\left.\begin{aligned}
&\tfrac{\partial u}{\partial t} + \upsilon\tfrac{\partial u}{\partial y} + \tfrac{uR}{R-y}\tfrac{\partial u}{\partial x} - \tfrac{\upsilon u}{R-y} - 2\upsilon\omega \\
&\qquad = -\tfrac{1}{\rho}\tfrac{R}{R-y}\tfrac{\partial P}{\partial x} + \nu\left(\nabla^2 u - \tfrac{u}{(R-y)^2} - \tfrac{2R}{(R-yr)^2}\tfrac{\partial \upsilon}{\partial x}\right), \\
&\tfrac{\partial \upsilon}{\partial t} + \upsilon\tfrac{\partial \upsilon}{\partial y} + \tfrac{uR}{R-y}\tfrac{\partial \upsilon}{\partial x} + \tfrac{(u+\omega(R-y))^2}{R-y} \\
&\qquad = -\tfrac{1}{\rho}\tfrac{\partial p}{\partial y} + \nu\left(\nabla^2 \upsilon - \tfrac{\upsilon}{(R-y)^2} + \tfrac{2R}{(R-y)^2}\tfrac{\partial u}{\partial x}\right) \\
&\tfrac{\partial \upsilon}{\partial y} - \tfrac{\upsilon}{R-y} + \tfrac{R}{R-y}\tfrac{\partial u}{\partial x} = 0,
\end{aligned}\right\} \tag{11.1}$$

where

$$\nabla^2 U = -\frac{1}{R-y}\frac{\partial U}{\partial y} + \frac{\partial^2 U}{\partial y^2} + \frac{R^2}{(R-y)^2}\frac{\partial^2 U}{(\partial x)^2}.$$

where ν—the kinematic viscosity, ρ—density of the liquid(ink), P—pressure, R—radius cylinder, ω—angular velocity of rotation.

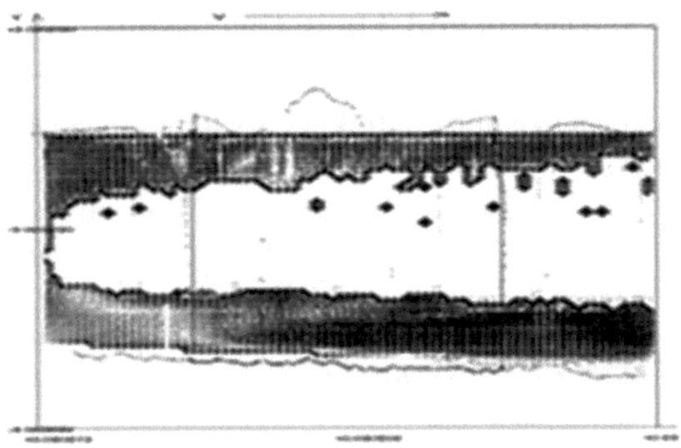

Fig. 11.1 Graphical visualization of the ink splitting process

On fixed net it is used finite-difference methods with putting into operation mobile boundary components for second cylinder boundary and fluid free boundary. Initially, fluid is on the first cylinder without relative movement.

We have investigated the behavior picture of ink final volume between rotary cylinders at Reynolds numbers from 1 up to 100 on two-dimensional net with different steps and with the help of finite-difference methods. We used nets with N_x (N_y) equal 80. We have obtained graphic models which allow to visualize ink transfer process onto paper (Fig. 11.1).

We consider a special case, when a rotating cylinder surface of a printing unit contact zone is experiencing elastic deformations. We define the dynamic equation of the deformation under pressure on a moving boundary for some elastic layer (paper, rubber) with density ρ, Yung' modulus E and thickness L. We make the sumption: low speed of moving boundary is examined in comparison with speeds of elastic perturbation spreading which quickly is stabilized and is not form in a solving of the wave haracteristics. So, the equation is

$$\rho LS\ddot{x}_c/2 = -\frac{E}{L}\left(\frac{\Delta r}{2} + x_c\right) \cdot S + (P_1 - P_0) \cdot S, \tag{11.2}$$

where Δr—value of a boundary deformation at a time t_1.

Integrating (11.2) we define a new boundary shift $\Delta r + 2x_c$ at a time $t_2 = t_1 + \Delta t$ with a certain amount $\dot{x}_{c,1}$ from the previous calculation, taking into account the initial value $\dot{x}_{c,1} = 0$ at $t_1 = 0$ and $\Delta r = 0$. After conversion the equation is:

$$\frac{d\dot{x}_c^2}{dx_c} = -\frac{4E}{\rho L^2}\left(\frac{\Delta r}{2} + x_c\right) + 4\,(P_1 - P_0)\,\frac{1}{\rho L}. \tag{11.3}$$

The sign "+" or "−" we select according to the sign $\left[(P_1 - P_0) - \frac{E}{L}\frac{\Delta r}{2}\right]$, which indicates a direction of a boundary deformation.

Table 11.1 Results of tests

Sample's number	Roughness (Ra), μm	Average pore diameter, nm	Ink quantity inside substrate pores %	Ink quantity on to substrate, %	Total ink quantity on print after engagement zone, %
1	0.231	55	3.05	52.03	55.08
2	0.427	110	5.23	47.08	52.31
3	0.495	968	37.33	35.21	72.54

We have obtained numerical solutions of flow characteristics showing the formation possibility of local zones of ink breaking off from substrate surface when the latter leaves the contact zone. It allows receiving a more perfect concept concerning processes taking place in ink at the moment of its transfer onto paper and at the moment of leaving contact zone.

With the help of developed software we have calculated the transfer coefficient of dispatched ink onto paper passing between impression and offset cylinders. We have established that the coefficient of ink splitting depends on printing conditions and microgeometry of substrate (Table 11.1).

11.4 Conclusions

Practical realization of the model is applicable to analyze of the influence of the main printing factors during design stage of printing machines based on the selected criterion set for the print system components.

Reference

1. L.G. Varepo, A.V. Panichkin, V.I. Bobrov, Numerical simulation of the transfer of ink in the printing areas in sheetfed offset printing. Problems printing and publishing No. 5, pp. 30–36 (2012)

Chapter 12
Microstructural Investigation of MIG-Brazed 304L Stainless Steel Joints

Mehmet Ekici, Faruk Varol, Ugur Ozsarac and Salim Aslanlar

Abstract In this paper, austenitic 304L stainless steel plates having 1.0 mm thickness were joined by copper-based CuAl8 wire in gas metal arc brazing technique with 100 % Ar gas as a protective atmosphere. Specimens were prepared in butt and overlap joint positions. The metal inert gas (MIG) brazing operations were done with seven different arc voltages and weld currents as 40, 45, 50, 55, 60, 65 and 70 A. CuAl8 wire composed largely of copper serves as the filler metal was used. Owing to the low fusion temperature of the filler metal (900–1,100 °C), no fusion of the base metal takes place in MIG-brazing. The 1,500 °C fusion temperature of the base metal was not reached. A solid joint between the workpieces was brought about by diffusion. Having accomplished the brazing operations, micro and macrostructures of joints were investigated by optical microscope and scanning electron microscope (SEM) in order to see the joinability of stainless steels by gas metal arc brazing technique.

M. Ekici
Vocational School of Yalova, University of Yalova, Yalova, Turkey
e-mail: mekici@yalova.edu.tr

F. Varol
Vocational School of Karasu, University of Sakarya, Sakarya, Turkey
e-mail: fvarol@sakarya.edu.tr

U. Ozsarac (✉) · S. Aslanlar
Department of Metallurgical and Materials Engineering, University of Sakarya, Sakarya, Turkey
e-mail: ozsarac@sakarya.edu.tr

S. Aslanlar
e-mail: aslanlar@sakarya.edu.tr

E. K. Polychroniadis et al. (eds.), *International Multidisciplinary Microscopy Congress*,
Springer Proceedings in Physics 154, DOI: 10.1007/978-3-319-04639-6_12,

12.1 Introduction

Stainless steel sheets are increasingly used for several applications because of their high corrosion resistance, beautiful appearance and good weldability [1]. Austenitic stainless steel constitutes the largest stainless family in terms of alloy type and usage. The standard austenitic stainless steel weld metals contain two phases (austenite + ferrite) similar to an "as cast" microstructure. The austenitic stainless steels are used for a very broad range of applications when an excellent combination of strength and corrosion resistance in aqueous solutions at ambient temperature is required. In general, austenitic stainless steels are easily weldable [2, 3]. However, high amount of heat is given into the base metal during welding operation. Therefore the grain size is affected due to high heat input, so just melting the filler metal during the joining process seems to be a better remedy, and the brazing method has been developed. The brazing is a process by which metallic materials are joined with the aid of melted filler (solder), and the melting temperature of which lies below that of the base metal which is wetted, without being smelted. In Gas-shielded Metal Arc (GMA) Brazing, copper-based alloys are often used as the filler, the melting temperature of which is below that of the steel to be joined. In an ideal situation, the parent metal is not melted at the edges. Although the arc, which burns under a shielding gas atmosphere, is used to warm the parent metal and melt the filler, the process is much more like soldering. Suitable shielding gases for GMA-Brazing are inert gases like Argon but gas mixtures consisting of Argon and small amounts of active gases like Oxygen are more common. Because of that the process is called GMA Brazing including MIG- and MAG-Brazing [4]. The main advantages of gas metal arc brazing over existing commercially available joining processes are its low cost and potential for being easily incorporated as a workstation in the production chain. This will significantly reduce the inventory cost for parts that otherwise would have to be prepared separately and stored for assembly line needs. Other advantages include no vacuum requirement even when joining super-alloys or titanium-alloys, controlled joint interfaces with limited reactions, few microstructure or property changes in the base materials after joining due to rapid processing, and potential for localized heating for joining large machine parts [5, 6].

Stainless steels are covered by tenacious oxide layers, making them highly resistant to oxidation. These passive films are also a barrier for wetting and brazing by liquid metals and alloys. However, above a certain temperature, heating in high vacuum leads to good wetting (contact angle $\theta \leq 90°$) thus enabling these materials to be joined by brazing. The wetting temperature is an important technical parameter since an excessively high brazing temperature can affect the microstructure and properties of the steel. Moreover, in heterogeneous brazing, a too high brazing temperature can also affect the microstructure of the metallic partner of the steel [7]. There are many soldering and gas metal arc brazing studies in literature. For example, Lopez and Vianco tried to select the correct elements in order to improve the solderability affected by a surface finish of layered nickel, palladium and gold [8]. Markovich et al. joined titanium and chromium-based materials by means of

Table 12.1 Spectrometric analysis of 304L steel sheets

Chemical composition	C	Si	Mn	Cr	Ni	N	P	S
wt. %	0.021	0.38	1.8	18.2	8.1	0.054	0.33	0.002

Table 12.2 The chemical composition of filler material used during brazing operations

Chemical composition	Al	Mn	Fe	Ni	Sn	Cu
wt. %	8.05	0.4	0.18	0.08	0.45	Balance

brazing technique heating by using ion bombardment. They used both silver-based BAg-7 and aluminum-based BAlSi-4 filler metals in manufacturing small-size parts of electronic devices [9]. In addition, Vianco explained the basics of hand soldering as an engineering technology [10]. Quintino et al. [11], Joseph et al. [12] and also Guimaraes et al. [13] have been studied MIG-brazing of galvanized thin sheet joints for automotive industry. Shahi and Pandey also researched claddings of stainless steels by gas metal arc brazing method [14]. There are less study dealing with the MIG-brazing of stainless steels in literature. Kozlova et al. [7] investigated the wetting and brazing performances of stainless steel samples by copper–silver based filler materials. Özsaraç et al. [4] researched the effect of joining parameters on beam geometry and mechanical properties in MIG-brazing of galvanized steels. This present work reports the results of gas metal arc brazing process made by SG-CuSi3 wire used as filler metals in joining of 304L austenitic stainless steel sheets and focuses on the microstructure characterization of brazed joints. Microstructural characteristics of the joint region were examined in detail.

12.2 Experimental Work

12.2.1 Materials

The 304L austenitic stainless steel plates having 1.0 mm thickness were joined by copper-based CuAl8 wire in gas metal arc brazing technique. Specimens were prepared in two different joining forms as butt joint and overlap joint. Both types of brazing operations were done with seven different weld currents as 40, 45, 50, 55, 60, 65, and 70 A. The chemical composition of steel sheets used in welding operations was given in Table 12.1 and the chemical composition of filler material used during brazing operations was given in Table 12.2.

Table 12.3 MIG-brazing parameters for butt joints

Specimen	Arc current (A)	Wire speed (m/min)	Arc voltage (V)	Gas pressure (l/min)	Shielding gas	Heat input (Q)
1	40	2.0	11.6	12	Argon	0.23
2	45	2.2	11.7	12	Argon	0.233
3	50	2.5	11.8	12	Argon	0.21
4	55	2.8	12	12	Argon	0.24
5	60	3.1	12.1	12	Argon	0.20
6	65	3.4	12.3	12	Argon	0.19
7	70	3.7	12.4	12	Argon	0.16

Table 12.4 MIG-brazing parameters for overlap joints

Specimen	Arc current (A)	Wire speed (m/min)	Arc voltage (V)	Gas pressure (l/min)	Shielding gas	Heat input (Q)
1	40	2.0	11.6	12	Argon	0.14
2	45	2.2	11.7	12	Argon	0.15
3	50	2.5	11.8	12	Argon	0.14
4	55	2.8	12	12	Argon	0.15
5	60	3.1	12.1	12	Argon	0.21
6	65	3.4	12.3	12	Argon	0.16
7	70	3.7	12.4	12	Argon	0.13

12.2.2 MIG-Brazing Operations

The gas metal arc brazing (MIG-brazing) operations were performed in a welding machine which was specially designed for brazing purposes in D.C. (+) position. The shielding gas was chosen as Argon inert gas and 100 % Argon tubes were used during metal inert gas (MIG-brazing) operations. In literature, it was recommended that gas mixtures of CO_2, NO and O_2 with varying amounts can be added into Argon in order to obtain arc stability, but we preferred pure Argon to check the joinability of 304L thin sheets for automotive applications. Our specially designed and manufactured MIG-brazing machine ensures controlled transfer of the molten metal through the electric arc directed to high quality requests and joints of stainless steel sheets with filler electrodes. All MIG-brazing operations were prepared in an inverter-based FLEX 3000 C welding machine which was programmed using the integrated MigaCARD which holds both program data and machine software. All specimens and filler metal were cleaned with alcohol and acetone sequentially prior to brazing. The gas metal arc brazing operations were performed in 40, 45, 50, 55, 60, 65 and 70 A welding currents and other parameters chosen were given in Table 12.3 for butt joints and in Table 12.4 for overlap joints.

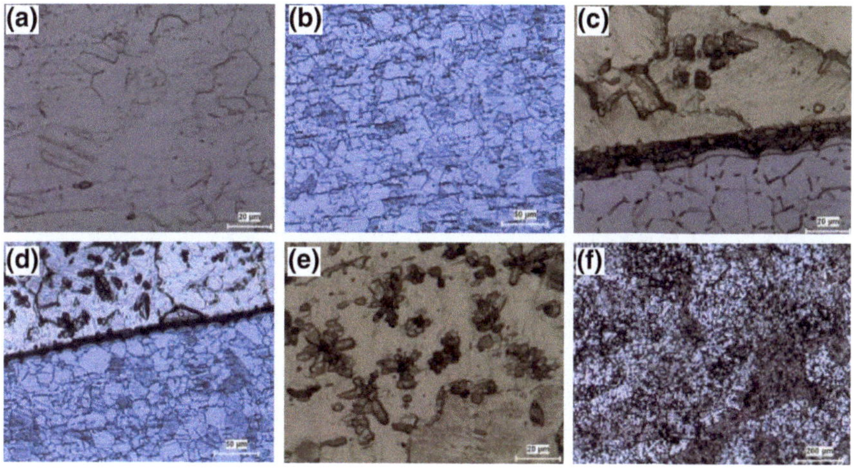

Fig. 12.1 Microstructures of MIG-brazed 304L samples joined at 60 A in butt form, **a** and **b** base metal, **c** and **d** HAZ/brazing interface, and **e** and **f** brazing zones

12.3 Results and Discussion

Optical microstructures of MIG-brazed 304L steel samples were given in Fig. 12.1. The sectioned MIG-brazed surfaces of samples were ground on SiC paper to a 2000-grit finish and polished by diamond suspension of 1 μm particle size, and then ultrasonically cleaned in a acetone bath. The metallographic characterization of base metal is performed by using the following etchings: (i) Electrolytic etching, with oxalic acid according to ASTM A 262–91 Practice A that reveals the grain boundaries of the austenite and the δ-ferrite partially decomposed and with an orientation in the rolling direction, (ii) Murakami's reagent does not attack the austenite and reveals the partial decomposition of the δ-ferrite into carbides.

SEM micrograph of 304L sample joined at 50 A in butt form, and EDS analysis of points from 1 to 4 were given in Fig. 12.2. This figure shows the heat affected zone (HAZ) and brazing region of joined samples. In Point-1 the wt. % Ni was found as 4.39 which are lower than that of base metal. Fortunately, the wt. % Cr amount is same with the initial value (18.2 wt. %). An oxide particle was detected in Point-2 having 7.975 wt. % O elements. It was thought that the oxidation was occurred during the brazing process. The Point-3 region is very close to brazing zone. Here Cr-ingredient decreases to 16.901 wt. %, however, wt. % Ni rises to 8.597 wt. % which is higher than that of initial value of base metal. The rise in Ni peaks proves this situation. The Point-4 was taken from inside the brazing region. Again according to EDS analysis 84.146 wt. % Cu was measured in this point. Similarly, Al is the second abundant element with the amount of 9.812 wt. % in point-4. This analysis is appropriate with the ingredient of filler wire. The braze metal has a cellular-dendritic structure [11]. The structure is finer for lower heat inputs. The microstructure of the HAZ varies

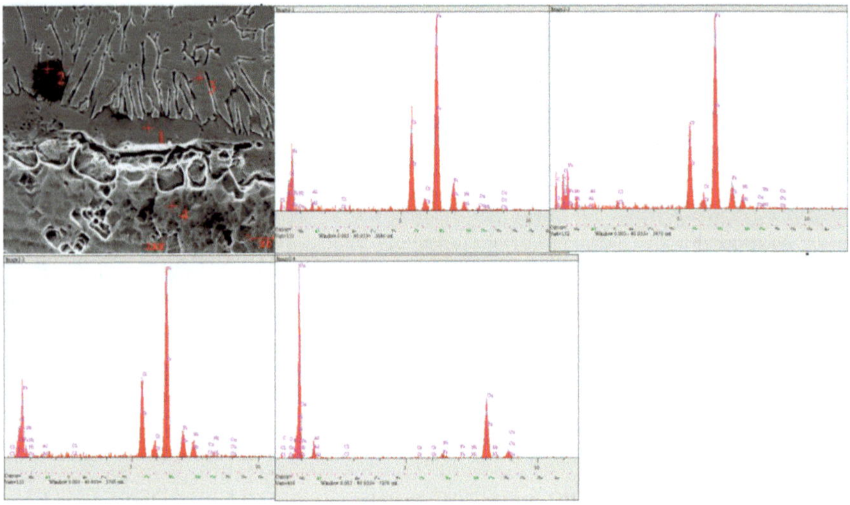

Fig. 12.2 SEM micrograph of 304L sample joined at 50 A in butt form, and EDS analysis of points from 1 to 4

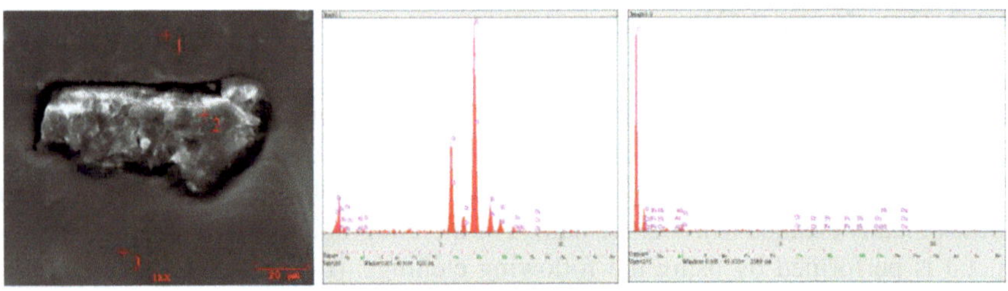

Fig. 12.3 SEM micrograph of 304L sample joined at 55 A in overlap form, and EDS analysis of points from 1 and 2

with heat input. For high heat inputs the microstructure consists of ferrite and perlite with pro-eutectoid ferrite in a platelike morphology usually called Widmanstätten. The structure is finer when the heat input decreases. Coarse-grained bainite was observed in the joints produced with low heat input, due to higher cooling rates, in a non-equilibrium solid state transformation [11, 15]. Our findings are parallel with the studies of Quintino et al. [11], Rangel et al. [15] and Özsaraç et al. [4].

In Fig. 12.3, SEM micrograph of 304L sample joined at 55 A in overlap form, and the EDS analysis of points from 1 to 3 were given. Here is the center of the overlap region. Point-1 is in the base metal of stainless steel sheets. There is an oxide particle in the center of brazing region and marked as Point-2. Its EDS analysis was obtained as 53.535 wt. % C, 39.061 wt. % O, 1.895 wt. % Si, 1.131 wt. % Ni and 2.258 wt. % Cu. Ni ingredient was taken from base metal and Cu from filler wire. It was thought that the tensile strength of the joints was negatively affected by such an oxide inside the brazing zone. Point-3 is the brazing zone of the joint with the 87.424 wt. % Cu and 9.976 wt. % Al composition.

12.4 Conclusion

The results of gas metal arc brazing process made by CuAl8 wire used as filler metals in joining of 304L austenitic stainless steel sheets showed that this process can be applied successfully with the mentioned parameters. The diffusion-brazing is a double-stage joining process, which combines the beneficial features of diffusion bonding and transient liquid-phase bonding techniques. This process eliminates the adverse influence of single joining technique.

References

1. Ó. Martín, P. Tiedra, M. López, M. Juan, C. García, F. Martín, Y. Blanco, Quality prediction of resistance spot welding joints of 304 austenitic stainless steel. Mater Design. **30**, 68–77 (2009)
2. D. Özyürek, An effect of weld current and weld atmosphere on the resistance spot weldability of 304l austenitic stainless steel. Mater. Design. **29**, 597–603 (2008)
3. U. Özsaraç, M. Baytemir, S. Aslanlar, A. Oğur, Joining of AISI 310 and AISI 430 stainless steels by resistance spot welding and the mechanical properties of joint, in *Welding Technology VIII. National Congress, Proceedings CD*, 1–11, 18–19, Ankara, Turkey (Turkish) Nov 2011
4. U. Özsaraç, A. Karadag, F. Varol, E. Ferik, M. Ekici, S. Aslanlar, Effect of joining parameters on beam geometry and mechanical properties in mig-brazing of galvanized steels, in *IMMC16, 16th International Metallurgy & Materials Congress, Proceedings CD*, 1283–1294, 13–15, İstanbul, Turkey Sept 2012
5. Y.Z. Shui, L.R. Feng, Q.I. Kai, Growth behaviour of interfacial compounds in galvanized steel joints with cusi3 filler under arc brazing. Trans. Nonferrous Met. Soc. China. **16**, 1391–1396 (2006)
6. L.R. Feng, Y.Z. Shui, Q.I. Kai, Interfacial structure and joint strengthening in arc brazed galvanized steels with copper based filler. Trans. Nonferrous Met. Soc. China. **16**, 397–401 (2006)
7. O. Kozlova, R. Voytovych, M.F. Devismes, N. Eustathopoulos, Wetting and brazing of stainless steels by copper-silver eutectic. Mater. Sci. Eng. A. **495**, 96–101 (2008)
8. E.P. Lopez, P.T. Vianco, Select the right surface finish to improve solderability. Weld. J. **90**(9), 44–46 (2011)
9. M. Markovich, G. Fischer, A.E. Shapiro, Brazing titanium and chromium using ion bombardment heating. Weld. J. **90**(9), 41–43 (2011)
10. P.T. Vianco Hand soldering basics. Weld. J. **90**(9), 47–5 (2011)
11. L. Quintino, G. Pimenta, D. Iordachescu, R.M. Miranda, N.V. Pépe, MIG Brazing of galvanized thin sheet joints for automotive industry. Materi. Manufact. Process. **21**, 63–73 (2006)
12. A. Joseph, C. Webb, M. Haramia, D. Yapp, Variable polarity (AC) arc weld brazing of galvanized sheet, in *Proceedings of the 56th IIW International Conference, Bucharest, Romania*, 6–11, IIW Doc.XII-1779-03, July 2003
13. A.S. Guimaraes, M.T. Mendes, H.R.M. Costa, J.S. Mchado, N.K. Kuromoto, An evaluation of the behaviour of a zinc layer on a galvanised sheet joined by MIG brazing. Weld. Int. **21**(4), 271–278 (2007)
14. A.S. Shahi, S. Pandey, Effect of auxiliary preheating of the filler wire on quality of gas metal arc stainless steel claddings. J. Mater. Eng. Perform. **17**, 30–36 (2008)
15. A.F. Rangel, L.A. Matlakhova, R.P. Darochaparanhos, A.N. Matlakhov, Evaluation of MIG-brazing welded joint by means of metallographic techniques. Weld. Int. **20**(11), 889–893 (2006)

Chapter 13
Microstructural Properties of Ceramics Produced from Granite and Zeolite

Nuray Karakuş and A. Şükran Demirkıran

Abstract Granite is a common type of intrusive, felsic, igneous rock which is granular and phaneritic in texture. This rock consists mainly of quartz, mica, and feldspar. Zeolites are microporous, aluminosilicate minerals commonly used as commercial adsorbents. Zeolites are widely used in industry for water purification, as catalysts, for the preparation of advanced materials and in nuclear reprocessing. The aim of this study was to investigate the possibility to use natural granite and zeolite in the manufacturing of ceramic products. Different compositions using natural zeolite and granite from Turkey were prepared to produce a traditional ceramic body. Zeolite and granite raw materials for the 4 different compositions were mixed; %wt. 25 Ze-75 Gr, 50 Ze-50 Gr, 75 Ze-25 Gr, 100 Ze. Mixing was performed by ball milling for 10 h with alumina balls. Samples of 20 mm in diameter were shaped by dry pressing at the pressure of 400 MPa. Prepared samples were sintered for 1 h at 1150 and 1200 °C in atmospheric conditions. Afterwards, products were characterized by SEM, EDS. Also, bulk densities, micro hardness and % water absorptions values of the samples were measured.

13.1 Introduction

Zeolites are a group of over 40 crystalline, hydrate aluminosilicate minerals with a structure based on a three-dimensional network of an aluminum and silicon tetrahedral, linked by sharing oxygen atoms [1]. These units in which either Si or Al are sitting in their centers form cages and channels of molecular dimensions.

N. Karakuş (✉) · A. Ş. Demirkıran
Engineering Faculty, Department of Metallurgy and Materials Engineering, Sakarya University, 54187 Sakarya, Turkey
e-mail: nurayc@sakarya.edu.tr

A. S. Demirkıran
e-mail: dkiran@sakarya.edu.tr

E. K. Polychroniadis et al. (eds.), *International Multidisciplinary Microscopy Congress*, Springer Proceedings in Physics 154, DOI: 10.1007/978-3-319-04639-6_13,

The presence of $(AlO_2)^-$ units cause residual negative charge on the oxygen framework. These negative charges are compensated by the cations located in the pores bonded to aluminum by electrostatic forces assuring their easy exchange for other ions [2]. They contain exchangeable alkaline and alkaline earth metal cations such as Na^+, K^+, Ca^{2+}, and Mg^{2+} that maintain charge neutrality [3]. Due to specific pore sizes and large surface areas, zeolites have the potential in a wide range of applications such as molecular sieves, adsorbents, and catalysts [4]. The price of conventional raw materials used in porcelain production is higher than that of zeolite. Also, feldspar and silica are both harder raw materials than zeolite and their crushing and millings are costly compare to zeolite [5].

Granite is an intrusive holocrystalline rock characterized by medium and fine grain structure which usually consists of orthoclase, acidic plagioclase and quartz. Magnesia and ferruginous compounds (hornblends, biotite, muscovite, and others) can also be present in granites as accessory minerals [6]. Granite grains and dust are considered waste materials during production of granite products. These wastes are occurred during preparation and polishing processes. These processes produce a large amount of waste materials [7]. Different industrial wastes, some volcanic ash, zeolite rich rocks, nepheline synite, gabbro, granite, basalt and basaltic tuffs were successfully used to produce vitrified ceramic tiles on firing at temperature ranging between 900 and 1300 °C [6].

The main objective of this paper is to produce ceramic bodies with granite and zeolite or to examine the effect of granite on the properties of ceramic bodies produced from zeolite.

13.2 Experimental Procedure

Natural granite and zeolite as raw materials were used to prepare ceramic bodies in this work. The chemical compositions of used powders were given in Table 13.1. The granite and zeolite (clinoptilolite) were grinded by dry ring milling and sieved to pass through −75 μm. The prepared powders were mixed as %wt. 25 Ze-75 Gr, 50 Ze-50 Gr, 75 Ze-25 Gr, 100 Ze. The codes of samples are given in Table 13.2. Mixing was performed by ball milling for 10 h with alumina balls. The samples were shaped by uniaxial dry pressing under the pressure of 400 MPa. After shaping, samples were dried at 80 °C for 24 h in an oven. Dried samples were sintered in an electrical furnace with a heating rate of 10°C/min at 1150 and 1200 °C for 1 h. Then, the sintered samples were cooled down to room temperature in the furnace. Bulk densities, micro hardness (Vickers) and % water absorptions values of the samples were measured. Also, the microstructural characterizations were realized using a JEOL JSM 6060 LV scanning electron microscopy (SEM) with EDS attachment.

Table 13.1 The chemical compositions of used powders

Component (%)	SiO_2	Al_2O_3	TiO_2	Fe_2O_3	CaO	MgO	Na_2O	Others
Granite	69.07	17.10	0.17	2.12	2.35	0.47	4.15	0.6
Zeolite	61.89	24.80	0.60	1.15	0.32	0.24	3.59	7.53

Table 13.2 The codes of samples

Compositions	%25 Zeolite-%75 Granite	%50 Zeolite-%50 Granite	%75 Zeolite-%25 Granite	%100 Zeolite
Codes	Z25	Z50	Z75	Z100

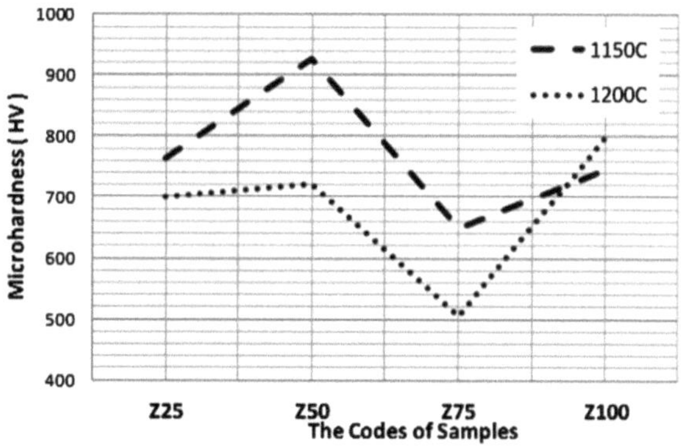

Fig. 13.1 Microhardness changes of sintered samples at 1150 °C and 1200 °C for 1 h

13.3 Results and Discussion

Zeolite-granite bodies were characterized by hardness measurement for mechanical properties and these measurements were carried out with the Vickers hardness method. The microhardness results were given in Fig. 13.1 An overall evaluation, the hardness was obtained to decrease with increased temperature. The maximum hardness was found in Z50 coded sample at 1150 °C as 927 HV. In this temperature, the microhardness of Z100 coded sample was lower than that. So, the presence of granite (wt.% 50) was increased the hardness.

The water absorption (%) values of sintered samples at 1150 and 1200° for 1 h were given in Fig. 13.2a. Here, the amounts of water absorption are more in Z25 coded sample at 1200 °C than 1150 °C but are the same values in others samples approximately. Also, the water absorption values <0.5 % are reached, so complying the standard limit for ceramics [8].

The bulk density curves of all compositions were shown in Fig. 13.2b. The bulk densities of sintered samples were measured using Archimedes method. As seen from the figure, the density was decreased with increased temperature from 1150 to 1200 °C. The presence of granite in the samples led to the presence of additional

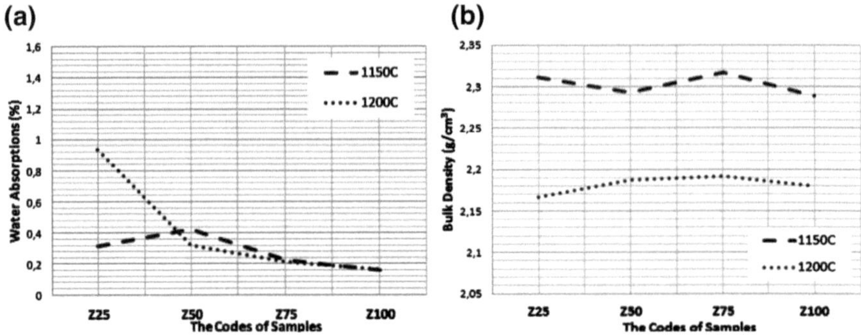

Fig. 13.2 **a** The water absorptions (%) values **b** the bulk density values of sintered samples

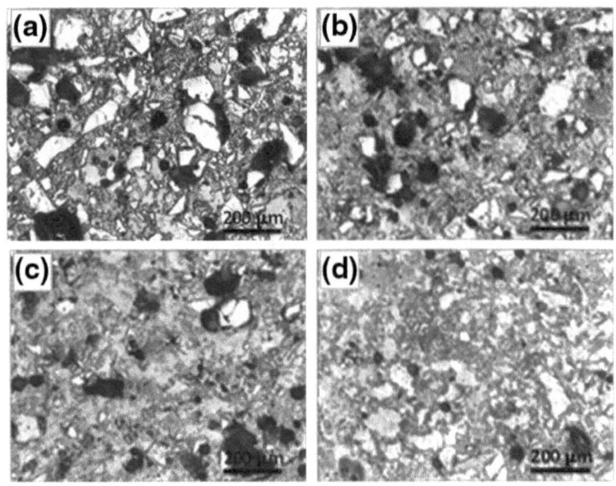

Fig. 13.3 The optic micrographs of samples sintered at 1150 °C for 1 h **a** Z25, **b** Z50, **c** Z75 and **d** Z100

fluxing ions, which activates produces a decreasing of the sintering temperature. It is possible to conclude that the granite content increases the liquid phase. The formed liquid phase fills the open pores, improves the bulk density and decreases the apparent porosity of the fired samples [6, 9].

The optic micrographs for examination of the microstructure are shown in Figs. 13.3 and 13.4. Compared to figures, the amount of porosity is more in the sintered samples at 1200 °C. This result was confirmed with the bulk density measure given in Fig. 13.2b. Investigating the results of all the analysis, the sample Z50 was determined as the optimum composition.

In the SEM analysis of sintered sample Z50 at 1150 °C given in Fig. 13.5 were seen very small cracks, quartz grains and porosity. It is thought that, these cracks are due to thermal expansion differences of different phases occurred in during sintering. These results are similar to the literature [6, 10]. The EDS analysis was applied to the same sample for different regions (Fig. 13.6) and Si, Al, K, Fe, Mg elements were detected. These elements belong to chemical compositions given in Table 13.1.

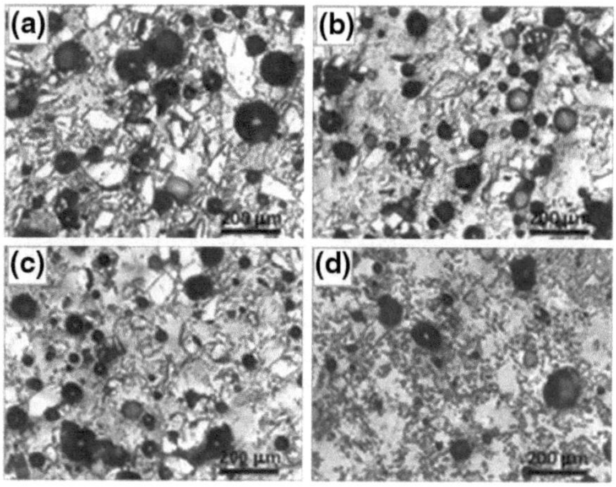

Fig. 13.4 The optic micrographs of samples sintered at 1200 °C for 1 h **a** Z25, **b** Z50, **c** Z75 and **d** Z100

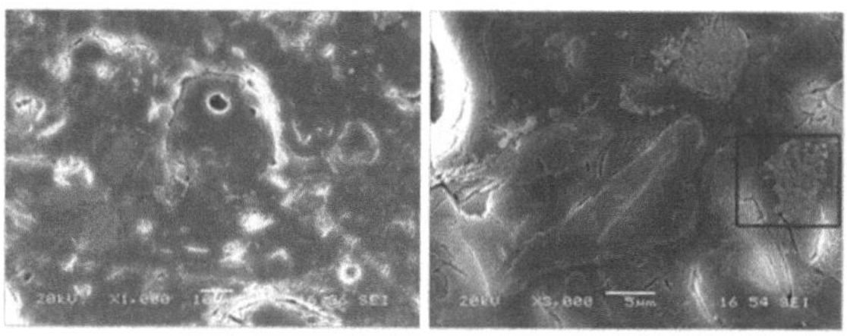

Fig. 13.5 The SEM analysis of the sample Z50 after 1150 °C-1 h

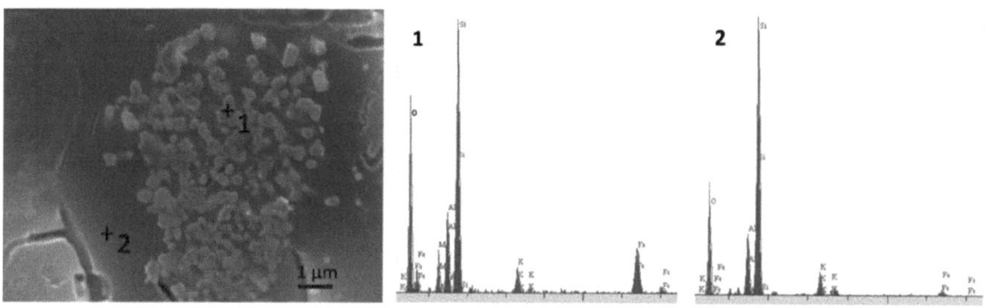

Fig. 13.6 The EDS analysis of the sample Z50 after 1150 °C-1 h

In this work, the use of granite could help to achieve the complete bodies densification at low sintering temperature. In fact during sintering both the granite and zeolite could favor the formation of a liquid phase that could contribute to obtain the final ceramic products at lower temperature.

13.4 Conclusions

The economic advantages obtained from this study confirm the feasibility of use of natural granite and zeolite in the ceramic bodies. The maximum hardness was found in Z50 coded sample at 1150 °C as 927 HV. The amounts of % water absorption were more in Z25 coded sample at 1200 °C than 1150 °C. The bulk densities were decreased with increased temperature from 1150 to 1200 °C. The amount of porosity as the optic micrographs was less in the sintered samples at 1150 °C. In the SEM analysis of sintered sample Z50 at 1150 °C were seen very small cracks, quartz grains and porosity. In the EDS analysis of sintered sample Z50 at 1150 °C were detected Si, Al, K, Fe, Mg elements. Also, this product was achieved in a low temperature. As the best result, sintered at 1150 °C for 1 h sample Z50 was determined.

References

1. T. Wajima, Y. Ikegami, Synthesis of zeolitic materials from waste porcelain at low temperature via a two-step alkali conversion. Ceram. Int. **33**, 1269–1274 (2007)
2. L. Guczi, I. Kiricsi, Zeolite supported mono-and bimetallic systems: structure and performance as CO hydrogenation catalysts. Appl. Catal. A. **186**, 375–394 (1999)
3. E.A. Ortega, C. Cheeseman, J. Knight, M. Loizidou, Properties of alkali-activated clinoptilolite. Cem. Concr. Res. **30**, 1641–1646 (2000)
4. T. Wajima, Y. Ikegami, Synthesis of crystalline zeolite-13X from waste porcelain using alkali fusion. Ceram. Int. **35**, 2983–2986 (2009)
5. A.Ş. Demirkiran, R. Artir, E. Avci, Electrical resistivity of porcelain bodies with natural zeolite addition. Ceram. Int. **36**, 917–921 (2010)
6. H.F. El-Maghraby, M.M. El-Omla, F. Bondioli, S.M. Naga, Granite as flux in stoneware tile manufacturing. J. Eur. Ceram. Soc. **31**, 2057–2063 (2011)
7. A.E.M.Abd Elmoaty Mechanical properties and corrosion resistance of concrete modified with granite dust. Constr. Build. Mater. **47**, 743–752 (2013)
8. R. Gennaro, M. Dondi, P. Cappelletti, G. Cerri, M. Gennaro, G. Guarini, A. Langella, L. Parlato, C. Zanelli, Zeolite-feldspar epiclastic rocks as flux in ceramic tile manufacturing. Microporous Mesoporous Mater. **105**, 273–278 (2007)
9. S.M. Naga, F. Bondioli, M.M.S. Wahsh, M. El-Omla, Utilization of granodiorite in the production of porcelain stoneware tiles. Ceram. Int. **38**, 6267–6272 (2012)
10. A.S. Demirkiran, R. Artir, E. Avci, Effect of natural zeolite addition on sintering kinetics of porcelain bodies. J. Mater. Process. Technol. **203**, 465–470 (2008)

Chapter 14
Characterization of CaO-ZrO_2 and CaO-ZrO_2-Al_2O_3

M. Ipek

Abstract This study is focused on production of partially stabilized zirconia containing 8 mol % CaO and examination of effect of Al_2O_3 adding on it. Zirconia nanopowders were synthesized by the precipitation method from zirconium sulfate precursors. Precipitates were dried at 80 °C for 72 h, and $CaCO_3$ powders were added to dried powders. Then this powder mixture was calcinated at 1300 °C for 1 h. And the second powder mixture was prepared by adding wt.%10 Al_2O_3 to calcinated powders. Each powder mixture were pressed under uniaxial compression of 300 MPa and sintered at 1600 °C for 1 and 2 h in an open atmospheric electrical resistance furnace. Relative densities of sintered samples measured in terms of Archimedes' principle were about 99 % for CaO-ZrO_2 and 97.5 % for CaO-ZrO_2-Al_2O_3. XRD analysis revealed that sintered CaO-ZrO_2 has monoclinic, tetragonal or/and cubic zirconia and sintered CaO-ZrO_2-Al_2O_3 has similar phases to CaO-ZrO_2 has, differently include $CaAl_{12}O_{19}$ phase. SEM-EDS analyses showed that the sintered samples have inhomogeneous grain structure and calcia was only dissolved in large zirconia grains. This means that calcia promoted grain growth of zirconia particles.

14.1 Introduction

In the last three decades, among various structural ceramics, zirconia-based ceramics have been widely researched, because of their better toughness and strength properties. The applications of ZrO_2-based ceramics include mechanical seals, engine components, cutting tools, sensors, and thermal barrier coatings and lately, as biomedical implants [1].

M. Ipek (✉)
Engineering Faculty, Department of Metallurgy and Materials Engineering, Esentepe Campus, Sakarya University, 54187 Sakarya, Turkey
e-mail: mipek@sakarya.edu.tr

E. K. Polychroniadis et al. (eds.), *International Multidisciplinary Microscopy Congress*, Springer Proceedings in Physics 154, DOI: 10.1007/978-3-319-04639-6_14,

Pure zirconia is monoclinic (*m*) at room temperature and atmospheric pressure. With increasing temperature the material transforms to tetragonal (*t*), approximately at 1170 °C and then to a cubic (*c*) fluorite structure starting about 2370 °C with melting at 2716 °C. These lattice transformations are martensitic, characterized by (1) being diffusionless (i.e. involving only coordinated shifts in lattice positions versus transportof atoms), (2) occurring athermally implying the need for a temperature change over a range rather than at a specific temperature and, (3) involving a shape deformation. This transformation range is bounded by the martensitic start and martensitic finish temperatures. Volume changes on cooling associated with these transformations are substantial enough to make the pure material unsuitable for applications requiring an intact solid structure: $c \rightarrow t$ approximately 2.31 %; $t \rightarrow m$ approximately 4.5 %. Sintered structures transforming from *t* to *m* on cooling from sintering temperatures (approximately 1300–1500 °C) undergo spallation with portions crumbling into multi-grained powders [2, 3].The phase changes may be suppressed by stabilization of the high temperature forms by alloying with other oxides. Certain cubic oxides such as CaO, MgO, Y_2O_3, CeO_2 and other rare-earth oxides stabilize the high temperature cubic phase to room temperature and at the same time promote a decrease in the transformation temperatures. However, if insufficient stabilizing oxide is added than the amount required for full stabilization, a partially stabilized zirconia (PSZ) is formed. It is generally accepted that the most useful mechanical properties are obtained by partial stabilization so that a two or three phases (cubic, tetragonal and monoclinic) microstructure results. The good mechanical properties of PSZ ceramics are primarily as a result of stress induced martensitic phase transformation of metastable tetragonal phase to the stable monoclinic phase, hence the name transformation toughening [4–6].

The goal of this study is to fabricate a ceramic based on CaO partially stabilized zirconia (Ca-PSZ) from precipitated zirconia and $CaCO_3$ powders and to investigate effect of wt. %10 Al_2O_3 adding to Ca-PSZ.

14.2 Experimental Details

Zirconia powders were synthesized by precipitation method from $Zr(SO_4)_2$ salt. Zirconium sulfate salt was dissolved in water and NH_4OH was added into solution for precipitation. Precipitate was dried at 80 °C and a required amount of $CaCO_3$ was added to dried powders and mixed by conventionally ball milling for 2 h. Then, the powder mixture was calcinated at 1300 °C for 1 h for transformation of $CaCO_3$ to CaO, thus the first powder mixture was prepared (Ca-PSZ). In order to obtain the second powder mixture, wt. %10Al_2O_3 powders were added to calcinated powders (Ca-PSZ+A). Each powder mixture was consolidated by uniaxial pressing at 300 MPa without any binder. The compacts were finally sintered in an open atmospheric electric resistance furnace at 1600 °C for 1 and 2 h with heating and cooling rates of 5 °C/min.

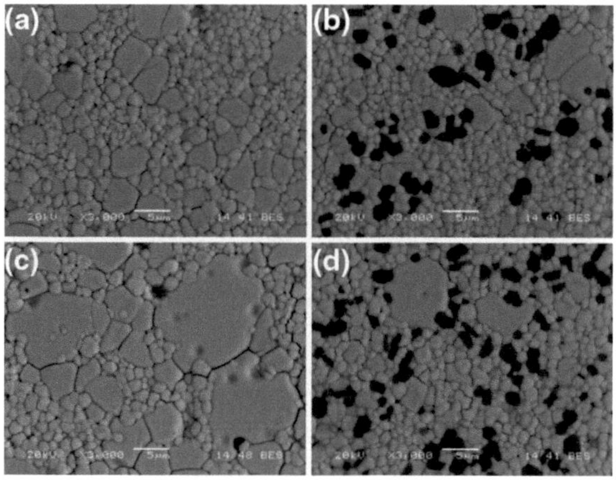

Fig. 14.1 SEM micrographs of samples sintered at 1600 °C **a** and **b** Ca-PSZ, **c** and **d** Ca-PSZ+A

Phase analysis of calcinated powders and sintered samples were performed via Rigaku XRD instrument by using Cu Kα radiation with a wavelength of 1.5418° A over a 2θ range of $10° \leq 2\theta° \leq 90°$. The morphology and the microstructure of the products were examined by means of scanning electron microscopy, energy dispersive spectroscopy (SEM-EDS) (JOEL JSM- 6060 LV Model). Relative densities of sintered samples were determined by Archimedes' method.

14.3 Results and Discussion

SEM micrographs of sintered samples at 1600 °C for 1 and 2 h (which were polished and thermally etched at 1500 °C for 1 h) are illustrated in Fig. 14.1. Ca-PSZ and also Ca-PSZ+A samples have heterogeneous grain size distribution in such a way that average grain size of 1 and 5 μ has two different grain size in Ca-PSZ sintered 1 h. The amount and size of coarse grains increased with increasing sintering time. Similar abnormal grain growth is present in Ca-PSZ+A samples but amount of coarse grains areas are much smaller than that of Ca-PSZ sample. In Ca-PSZ+A samples, light gray areas are zirconium and dark gray areas are aluminum rich grains. Abnormal grains in all sintered samples include calcium with together zirconium, and dissolution of calcium in zirconia promotes grain growth (Fig. 14.2a, Marks 1, 3, 5 and Fig. 14.2b, Marks 1, 2, 3, 4) [6, 7]. Furthermore, it is interesting to note that calcium was dissolved primarily in alumina grains as it can be seen in Fig. 14.2b, Marks 7, 8. Abnormal growing grains in Ca-PSZ+A samples are less than that of Ca-PSZ due to dissolving of Ca in alumina.

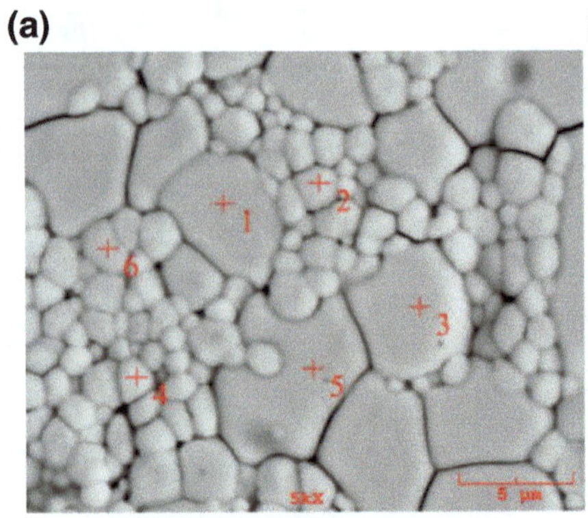

Mark	wt.%		
	Zr	Ca	O
1	74.97	3.71	21.32
2	81.78	-	18.72
3	76.58	4.14	19.28
4	77.81	-	22.19
5	73.63	3.56	22.82
6	78.79	-	21.21

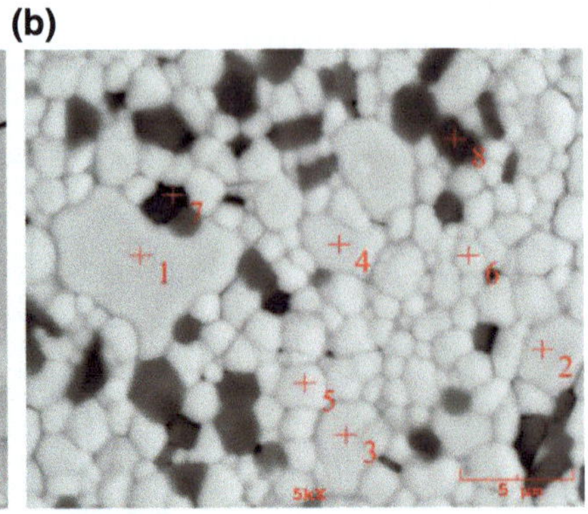

Mark	wt.%			
	Zr	Ca	Al	O
1	74.29	3.93	-	21.78
2	76.38	2.99	-	20.63
3	75.59	3.35	-	21.06
4	76.18	3.81	-	20.01
5	81.97	-	-	18.03
6	77.52	-	-	22.48
7	21.42	6.05	45,73	26.80
8	19.05	5.54	43,02	32.39

Fig. 14.2 SEM micrographs and EDS point analysis of sintered specimens **a** Ca-PSZ, **b** Ca-PSZ+A

While relative density of Ca-PSZ sintered for 1 h is 99.5 % that of Ca-PSZ sintered for 2 h is 98.8 %. The reason of decrease in relative density may be increasing abnormal grain growth with increasing sintering time. Relative density of Ca-PSZ+A is 97.5 % and this value is slightly lower than relative density of Ca-PSZ.

XRD analysis shows that Ca-PSZ has monoclinic, cubic and/or tetragonal phases (Fig. 14.3). According to open literature [2, 8–10], Ca-PSZ has monoclinic and cubic zirconia but XRD instrument for same peaks belong cubic phase described tetragonal phase. For that reason, in XRD analysis, it was indicated the tetragonal phase with together cubic phase. Alumina added samples have monoclinic, cubic and/or tetragonal zirconia and $CaAl_{12}O_{19}$ phase. According to CaO-Al_2O_3 binary phase diagram, $CaAl_{12}O_{19}$ occur with peritectic reaction at 1833 °C in alumina rich area [11]. Also monoclinic zirconia ratio is too much comparing to Ca-PSZ samples. As a mentioned above, CaO dissolves in alumina and dissolving CaO amount in zirconia matrix is not enough for the stabilization of cubic and/or tetragonal phase.

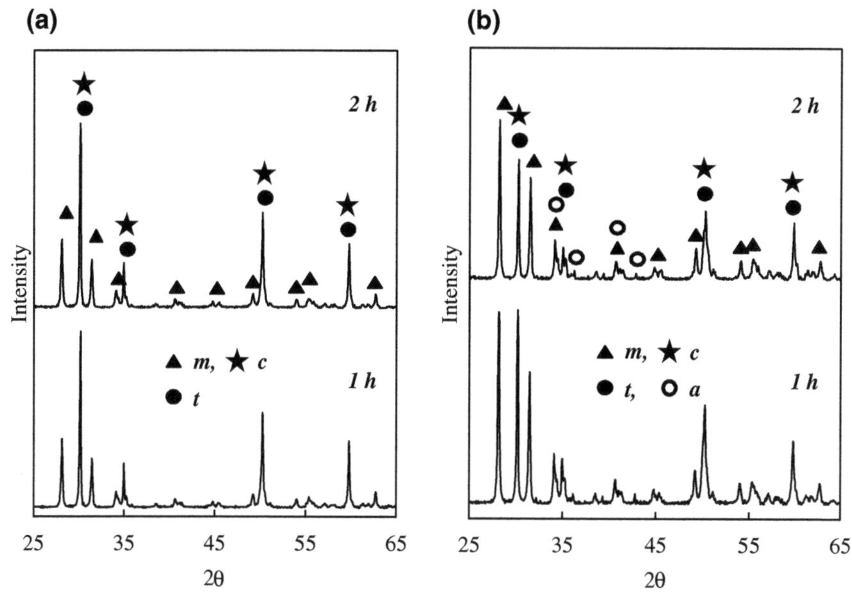

Fig. 14.3 XRD patterns of sintered samples, **a** Ca-PSZ and **b** Ca-PSZ+A (*m*: monoclinic, *c*: cubic and *t*: tetragonal zirconia, *a*: $CaAl_{12}O_{19}$)

14.4 Conclusion

The results obtained from this study are listed below.

1. Relative densities of sintered samples were about 99 % for $CaO\text{-}ZrO_2$ and 97.5 % for $CaO\text{-}ZrO_2\text{-}Al_2O_3$.
2. Sintered samples have inhomogeneous grain structure and calcium was only dissolved in large zirconia grains. This means that calcia promoted abnormal grain growth of zirconia particles. Abnormal growing grains in $CaO\text{-}ZrO_2\text{-}Al_2O_3$ samples are less than that of Ca-PSZ due to dissolving of Ca in alumina.
3. Sintered $CaO\text{-}ZrO_2$ has monoclinic, tetragonal and cubic zirconia and sintered $CaO\text{-}ZrO_2\text{-}Al_2O_3$ has similar phases to $CaO\text{-}ZrO_2$ has, differently include $CaAl_{12}O_{19}$ phase. Monoclinic zirconia ratio in $CaO\text{-}ZrO_2\text{-}Al_2O_3$ sample is too much comparing to $CaO\text{-}ZrO_2$. Calcium primarily dissolves in alumina and therefore dissolved calcium in zirconia matrix is not enough for the stabilization of cubic and/or tetragonal phase.

Acknowledgments The author thanks expert Fuat Kayis for performing XRD studies and technician Ersan Demir for experimental assistance at Sakarya University. Also, special appreciations are extended to Prof. Dr. Cuma Bindal and Prof. Dr. Sakin Zeytin for their notable support.

References

1. S. Nath, N. Sinha, B. Basu, Microstructure, mechanical and tribological properties of microwave sintered calcia-doped zirconia for biomedical applications. Ceram. Int. **34**, 1509–1520 (2008)
2. R.H.J. Hannink, Growth morphology of the tetragonal phase in partially stabilized zirconia, J. Mater. Sci. **13**(8) 2487–2496 (1978)
3. J.R. Kelly, I. Denry, Stabilizedzirconia as a structuralceramic: an overview. Dent. Mater. **24**, 289–298 (2008)
4. E.A. Korableva, V.S. Yakushkina, O.S. Grishin, V.V. Vikulin, O. P. D'yachenko1, A structural study of Y_2O_3-partially stabilized zirconia ceramics. Refrac. Ind. Ceram. **46**, 21–23 (2005)
5. I. Denry, J.R. Kelly, State of the art of zirconia for dental applications. Dent. Mater. **24**, 299–307 (2008)
6. R.K. Govila, Strength characterization of MgO-partially stabilized zirconia. J. Mater. Sci. **26**, 1545–1555 (1991)
7. A.K. Shukla, V. Sharma, N.A. Dhas, K.C. Patil, Oxide-ion conductivity of calcia- and yttria-stabilized zirconias prepared by a rapid-combustion route. Mater. Sci. Eng. **B40**, 153–157 (1996)
8. E.A.A. Mustafa, Ca-PSZ prepared via polymeric sol-gel route. Ceram. Int. **26**, 215–220 (2000)
9. V.V. Silva, F.S. Lameiras, Synthesis and characterization of composite powders of partially stabilized zirconia and hydroxyapatite. Mater. Charact. **45**, 51–59 (2000)
10. Y.L. Bruni, L.B. Garrido, E.F. Aglietti, Reaction and phases from monoclinic zirconia and calcium aluminate cement at high temperatures. Ceram. Int. **38**, 4237–4244 (2012)
11. http://www.crct.polymtl.ca/fact/phase_diagram.php?file=Al-Ca-O_CaO-Al2O3.jpg&dir=FToxid

Chapter 15
Study of Microstructure of the Al-Fe Alloys After Hot Rolling Deformation

Magdalena Barbara Jabłońska, Kinga Rodak and Iwona Bednarczyk

Abstract The aim of the paper is a microstructure analysis of alloys from the Al-Fe system after hot rolling tests, conducted by using a scanning transmission electron microscopy STEM and scanning electron microscope equipped with EBSD detector. Hot rolling was carried out at Technical University of Ostrava, Faculty of Metallurgy and Material Engineering, Institute of Modelling and Control of Forming Processes. The samples were heated to a temperature of 1200 °C. The EBSD and STEM techniques have been applied in order to determine the influence of chemical composition and deformation parameters on structural changes. The microstructure analysis has included parameters such us: grain/sub-grain size, area fraction of grains/subgrains, misorientation angles, grains/subgrains shape aspect ratio and dislocations structure. The research structure techniques in scanning-transmission electron microscopy revealed numerous FeAl28 alloy phase separations of secondary nucleating sites favoured energetically, which are the boundary of grains/subgrains and dislocations. These changes in the structure of the test results have been confirmed by EBSD, which revealed the presence of grains/subgrains misorientation angle boundaries above 15°.

15.1 Introduction

Over the last 10 years Department of Materials Science at Silesian University of Technology as well as leading international centres has conducted research in order to know the structural phenomena which occur during hot plastic treatment aimed at elaboration of heat and plastic treatment technology of selected alloys based on intermetallic phases from the Al-Fe system. Those alloys are classified as a group

M. B. Jabłońska (✉) · K. Rodak · I. Bednarczyk
Institute of Materials Science, Silesian University of Technology, Krasińskiego 8, 40-019 Katowice, Poland
e-mail: magdalena.jablonska@polsl.pl

E. K. Polychroniadis et al. (eds.), *International Multidisciplinary Microscopy Congress*, Springer Proceedings in Physics 154, DOI: 10.1007/978-3-319-04639-6_15,

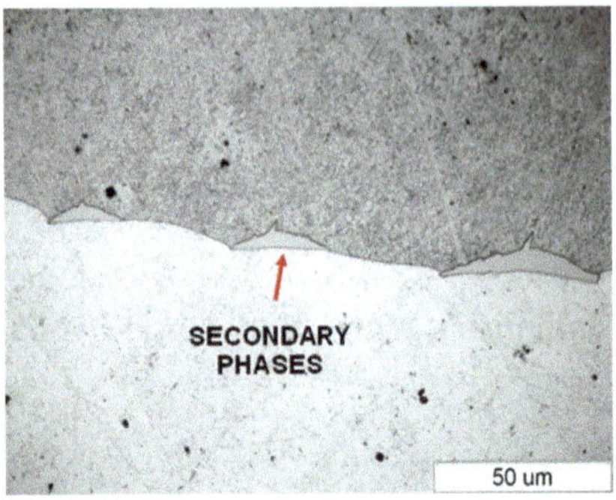

Fig. 15.1 Microstructure of the FeAl28 after annealing at 900 °C for 24 h (LM)

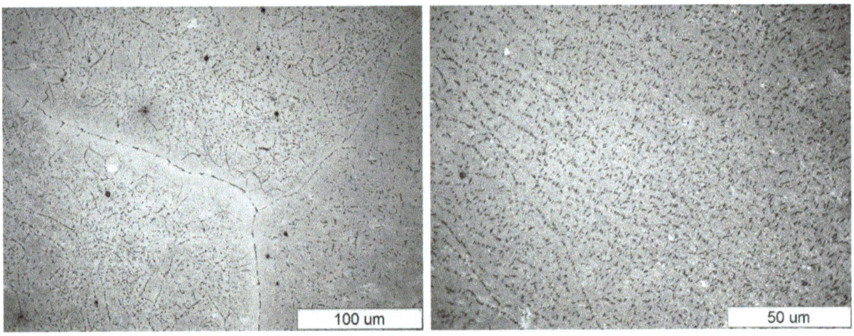

Fig. 15.2 Microstructure FeAl28 alloy after hot-rolling (LM)

of high-temperature creep-resisting materials of advantageous physicochemical and mechanical properties at an elevated and high temperatures [1–6]. Their properties, such as: low density, high melting temperature, high strength and good oxidizing resistance together with good crack resistance, create wide prospects for their industrial applications, for components of machines working at a high temperature and corrosive environment. The major problem restricting their universal application is their low plasticity and their brittle cracking susceptibility. Consequently, the research of intermetallic phase based alloys focuses on their plasticity is required [6–8]. One of the factors responsible for the unfavourable technological properties of Al-Fe alloys is a tendency to coarsely crystalline primary structure. The results indicate that grain refinement by thermo-mechanical processing has positive effect on the improvement in both strength and durability of these alloys [2, 8–12]. Therefore, the important issue is the evaluation of the behaviour of the material during plastic working. It should be noted that a significant influence on the properties of Al-Fe alloys by thermo-mechanical processing is the presence long-range ordering and the strengthening by thermal vacancy [4–6, 13, 14].

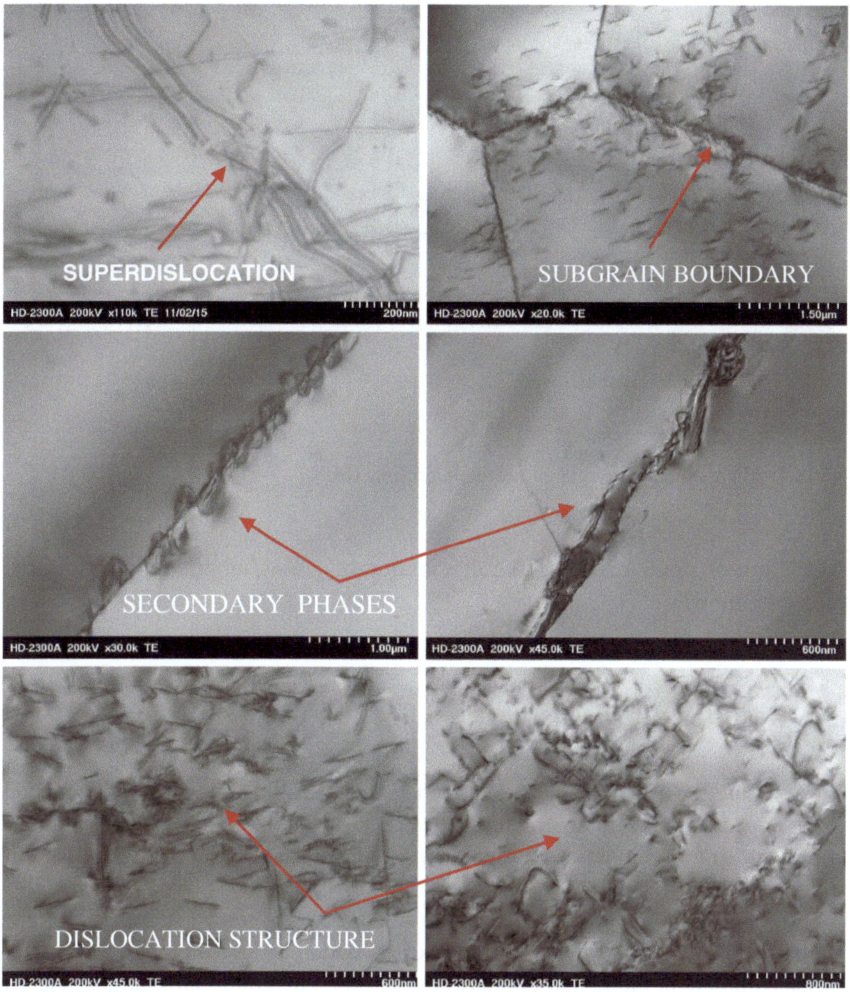

Fig. 15.3 Substructure FeAl28Al alloy after hot—rolling (STEM)

15.2 Materials and Methodology Research

The examined material consisted of alloy from Al-Fe system with a 28 % at. Al. Achieved ingots of alloy FeAl28 after casting process underwent heat treatment which was based on homogenising in temperature of 900 °C in 24 h and cooling in furnace. Hot rolling was carried out at Technical University of Ostrava, Faculty of Metallurgy and Material Engineering, Institute of Modelling and Control of Forming Processes. The samples were deformed at a temperature of 1200 °C. Structural tests were conducted with the use of light microscopy techniques and scanning transmission electron microscopy STEM. With the use of EBSD technology the

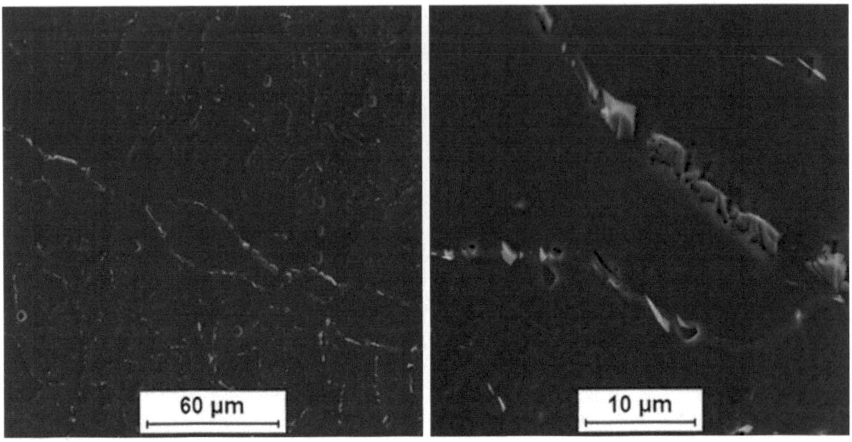

Fig. 15.4 Microstructure FeAl28 alloy after hot-rolling (SEM)

distribution of grain sizes and the distribution of grains/subgrains misorientation were determined.

15.3 Results and Discusion

After conduction of initial heat treatment and hot-rolling process the secondary phases appeared in the microstructure of tested alloy FeA128. Those phases cumulate on primary grain boundaries and inside of them surrounding newly formed grains (subgrains) (Fig. 15.1).

The FeA128 microstructure after the rolling process is shown in Fig. 15.2. In the microstructure disclosed the occurrence of secondary phase precipitates. Precipitates accumulate in the grain boundaries, and inside the grains surrounding the new emerging grains (subgrains).

Conducted substructure analysis with the use of STEM scanning transmission microscope has confirmed the presence of secondary phases located favored areas such as subgrains boundaries. In substructure of tested alloy there is also the presence of dislocation structure observed (Fig. 15.3). The conducted microanalysis of secondary phase division has confirmed that analysed phases contain more Al than matrix (Figs. 15.4 and 15.5). Based on the results of microanalysis it is not possible to identify the phase type, because of too small phase size which is smaller than electron beam impact. Literature implies that the phase occurring for that alloy group (23–36 % at), may be identified as Fe_3Al, but it would need to be confirmed in further experiments. Figure 15.6 presents microstructure map for alloy FeA128 achieved in EBSD analysis. Presented histogram of surface participation of grains shows the dominance of grains from intervals: from 1100 to 1200 μm which are more than

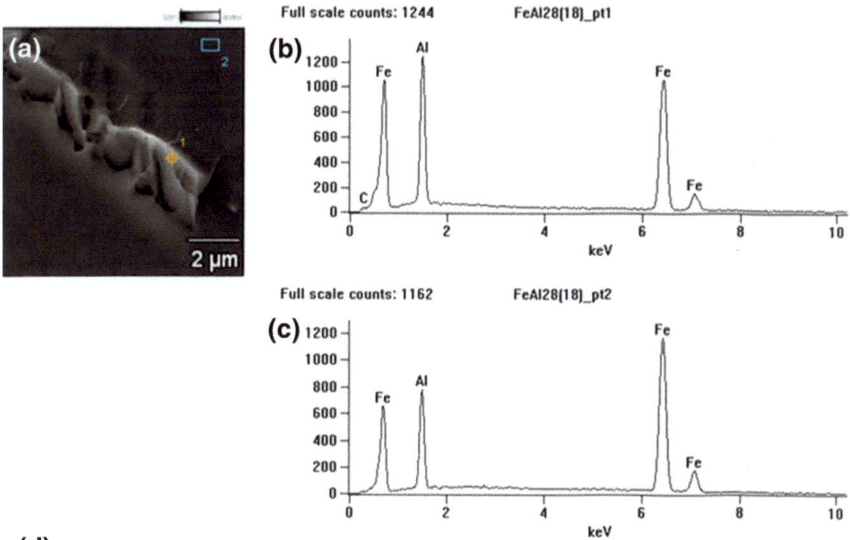

(d)

Atom %	*Al-K*	*Fe-K*
FeAl28(18)_pt1	34.9	65.1
FeAl28(18)_pt2	24.7	75.3
Weight %	***Al-K***	***Fe-K***
FeAl28(18)_pt1	20.6	79.4
FeAl28(18)_pt2	13.7	86.3

Fig. 15.5 **a** The microstructure of FeA128 alloy with marked places to study the chemical composition, **b** and **c** energy dispersive X-ray spectra (EDS) with marked locations in figure (**a**), **d** the results of chemical microanalysis of selected points of Fig. 15.4a

20 % of the analyzed surface (Fig. 15.5a). The remaining participations are grains of sizes of 1600 μm (Fig. 15.5a). Average diameter of grains/subgrains equals D = 697 μm. In the assessment of the disorientation angles and histogram of frequency of grain presence it was concluded that there were only boundaries present which had high angle above 15° (Fig. 15.5b).

A process of formation of new recrystallized grains on the contact line of three grains or on the wide-angle boundary can also be observed. Results achieved in conducted tests significantly enrich the amount of knowledge connected with the behavior of alloy on intermetallic phase matrix under the influence of hot plastic deformation.

(a)

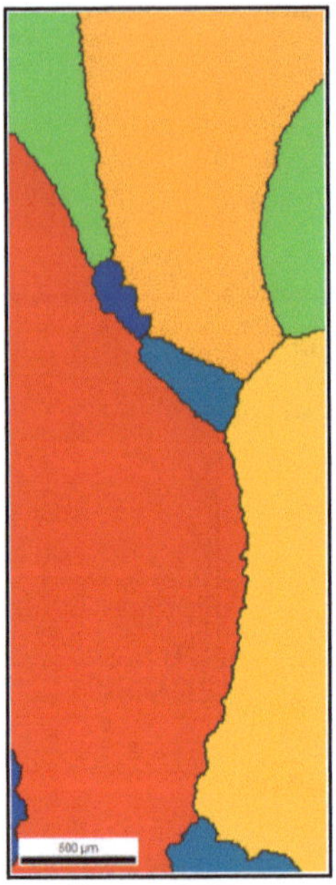

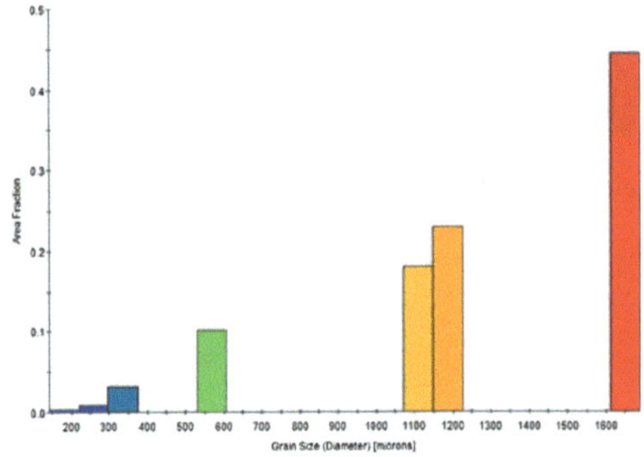

$D_{śr(xy)}$=697µm

(b)

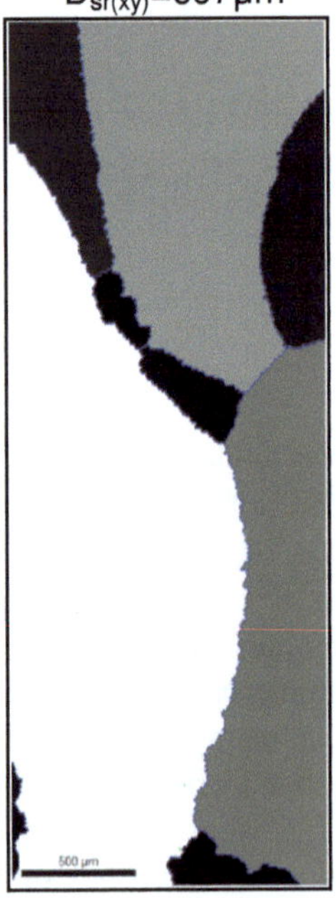

Grain boundaries			Fraction
	Angle		
	Min	Max	
(green)	5°	15°	0,15
(blue)	15°	180°	0,85

Fig. 15.6 **a** EBSD microstructure map for FeAl28, **b** microstructure map of misorientation angles between subgrains/grains

15.4 Conclusion

In the paper were carried out the microstructure analysis of FeAl28 alloy after hot rolling process. Conducted substructure analysis with the use of STEM technique has confirmed the presence of secondary phases located in favored places in the structure such as subgrains boundaries. In substructure of these alloy there is also the presence of dislocation structure observed. During the deformation a process of formation of new recrystallized grains on the contact line of three grains or on the wide-angle boundary can be observed.

References

1. N.S. Stoloff, Iron aluminides : present status and future. Mater. Sci. Eng. A **258**, 1–14 (1998)
2. P. Kratochvıl, I. Schindler, Hot rolling of iron aluminide Fe28.4Al4.1Cr0.02Ce (at%). Intermetallics **15**, 436–438 (2007)
3. I. Schindler, P. Kratochvıl, P. Prokopcakova, P. Kozelsky, Forming of cast Fe—45 at.% Al alloy with high content of carbon. Intermetallics **18**, 745–747 (2010)
4. J. Kansy, A. Hanc, D. Giebel et al., Vacancy formation in Fe-Al of B2 and DO3 alloys. Acta Phys Pol. A **113**(5), 1409–1413 (2008)
5. A. Śmiglewicz, M. Jabłońska, Thermal expansion of the alloys from Al—Fe system. DSL 2011-defect and diffusion. Forum **326–328**, 587–592 (2012)
6. L.M.C. Pike, T. Liu, The effect of vacancies on the environmental yield strength dependence of boron-free and boron-doped Fe-40Al. Intermetallics **8**, 1413–1416 (2000)
7. R.G. Baligidad, A. Radhakrishna, Effect of hot rolling and heat treatment on structure and properties of high carbon Fe-Al alloys. Mater. Sci. Eng. **A308**, 136–142 (2001)
8. M. Jabłońska, E. Bernstock, A. Jasik, Microstructure and mechanical properties of intermetallics on the base of Fe-Al alloy obtained by casting. Arch. Mater. Sci. Eng. **28**, 625–628 (2007)
9. H. Yamaguchi, K. Ito, High-temperature structural intermetallics. Acta Mater. **48**, 307–322 (2000)
10. I. Bednarczyk, D. Kuc, G. Niewielski, The structure of FeAl and Fe_3Al-5%Cr intermetallic phase—based alloys after hot deformation processes. Arch. Mater. Sci. Eng. **30**, 5–8 (2008)
11. E. Bernstock-Kopaczynska, I. Bednarczyk, M. Jabłońska, The influence of thermo-mechanical treatment on the structure and plasticity of FeAl intermetallic phase-base alloys. Arch. Civil Mech. Eng. **8**(3), 15–22 (2008)
12. I. Bednarczyk, M. Jabłońska, Plasticity of low aluminium alloys from Fe-Al. Arch. Metall. Mater. **1**(57), 271–276 (2012)
13. D. Li, D. Lin, Y. Liu, Effect of temperature on the tensile properties and dislocation of FeAl alloys. Mater. Sci. Eng. **A249**, 206–216 (1998)
14. J. Konrad, S. Zaefferer, A. Schneider, D. Raabe, G. Frommeyer, Hot deformation behavior of a Fe_3Al-binary alloy in the A2 and B2-order regimes. Intermetallics **13**, 1304–1312 (2005)

Chapter 16
Microstructure Characterization of High Mn-Al Steel After Hot Compression Tests

Magdalena Barbara Jabłońska

Abstract In last year's numerous research centres have been focused to the development of Mn-Al belong to the novel AHSS steels for the automotive and railway industry. The X55MnAl25-5 high manganese steel belong to these group demonstrate a dominant stress mechanism known as Twining Induced Plasticity (TWIP), which occurred as a main mechanism of plastic deformation, strongly dependent on Stacking Fault Energy (SFE). Moreover, these steels reveal a high yield stress and very high total elongation from. Properties of these steels predispose it to be used as constructional elements in automobile industry, which are characterized by a great energy reserve, absorbed during plastic deformation (vehicle collision) and high rigidity. This paper presents the results of applied methods of SEM/EBSD techniques for characterize the structure of high manganese steel after deformation by compression tests, with various deformation parameters like temperature and strain rate. Applied methods revealed the influence of deformation parameters on grain size, fraction of twinned grains and misorientation angles. The obtained results can be used for the development of this steels and their plastic deformation models.

16.1 Introduction

The high manganese steels belong to the novel AHSS group for the automotive and railway industry, exhibit attractive strength–ductility combinations in association with high strain hardenability [1–3]. Plastic deformation of fcc austenite depends greatly on the stacking fault energy (SFE). It is generally accepted that, as the SFE increases, the dominant deformation mechanism of austenite is progressively changed from transformation-induced plasticity (TRIP) to TWIP and to

M. B. Jabłońska (✉)
Institute of Materials Science, Silesian University of Technology, Krasińskiego 8, 40-019 Katowice, Poland
e-mail: magdalena.jablonska@polsl.pl

E. K. Polychroniadis et al. (eds.), *International Multidisciplinary Microscopy Congress*, Springer Proceedings in Physics 154, DOI: 10.1007/978-3-319-04639-6_16,

dislocation gliding and during the dynamic deformation to shear-band-induced plasticity (SIP) and microband-induced plasticity (MBIP) [3–8]. TWIP steels are currently being studied both extensively and intensively in various aspects i.e. production methods, thermomechanical processing, thermal and mechanical properties, welding and forming by various methods. Research of this group of steels particularly have grown importance in the last decade, with the advent of automotive applications but they were also studied, the primary purpose of replacing austenitic stainless steels for cryogenic applications [1–4, 7–12]. The TWIP steels reveal a high yield point approx. 600–1100 MPa and high elongation from 60 to 95 %. These attractive properties can be achieved by controlled thermomechanical treatment. The special feature of these steels for automotive applications is a great energy reserve, absorbed during plastic deformation thus improving the safety of the vehicle [1, 3, 6]. Also because of the aluminium content can be reached at reducing vehicle weight, which increases the attractiveness of these steels to apply as constructional elements of a vehicle, such as: strain hardening of a side pillar, bumper beam, strain hardening of a roof and pier [3, 8, 11].

16.2 Materials and Methodology Research

The research was carried out on high-manganese X55MnAl25-5 steel with chemical compositions gathered in Table 16.1. The steel was smelted in a vacuum induction furnace and cast using gravity casting technique and ingots with a cross-section of 20 × 40 mm and with a length of 120 mm. The ingots were homogenized in an air furnace at 1200 °C, hot rolled at 1100 °C to obtain a bars with 15 mm diameter. The rolling direction was the same as for the hot rolling. Obtained material was subjected to annealing at the temperature of 1100 °C for 1 h with following water cooling.

Following the annealing, the material was used for the preparation of the samples for the compression test, which consisted of axisymmetrical strain on Gleeble 3800 simulator, with simultaneous structure-freeze by rapid quenching. The samples were cylindrical, measuring $\phi = 10$ mm and h = 12 mm.

The compression samples were conducted at temperatures ranging from 800 to 1100 °C at a strain rate $1\,s^{-1}$, until the true strain values reached $\varepsilon = 1.0$. The compression test results, such as the sample temperatures T [°C], stresses σ [MPa], forces [N] and strains ε, processed with the calculation sheet, provided the means for determining the flow curves in the stress σ-strain ε system.

After hot compression tests the material was analyzed by SEM using EBSD sensor and STEM. The surface of metallographic microsections for observation using EBSD method was prepared with ion polishing technique. The results obtained from individual observation spots were presented as maps of crystallographic orientation, maps with angles of grains misorientation, and histograms of grains size and misorientation. The specimens for STEM were cut out of the cold rolled samples along the longitudinal direction. The specimens were mechanically fine polished to a thickness of about 0.03–0.08 mm. Thin films were prepared using

Table 16.1 Chemical composition of the investigated steels (% at.)

[% at.]	C	Mn	Si	Al	Fe
X55MnAl25-5	0.53	23.7	0.25	5.38	Bal.

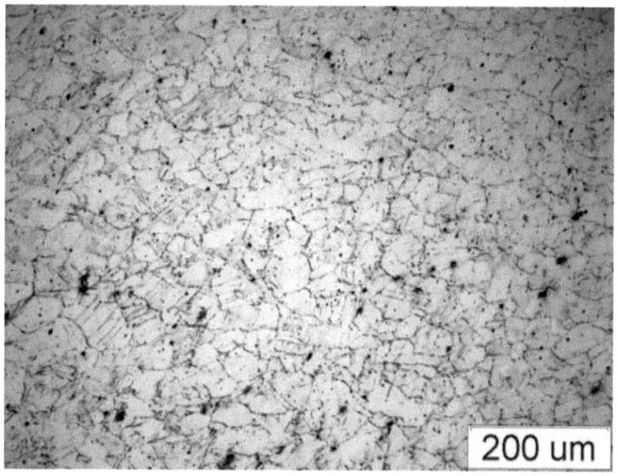

Fig. 16.1 Microstructure of the examined steel after forging—initial state

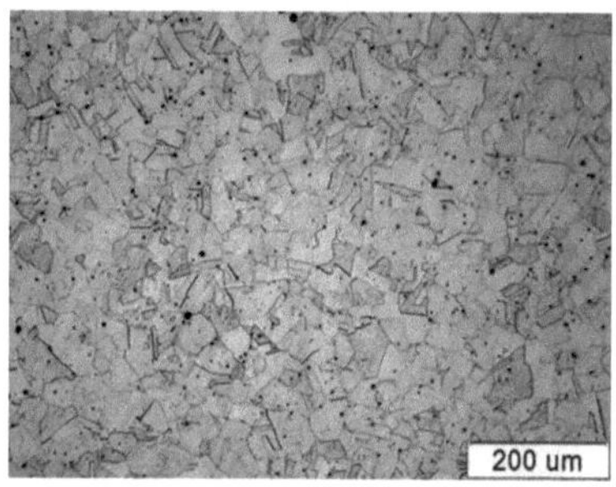

Fig. 16.2 Microstructure of the examined steel after forging and saturation—1100/1 h

the twin-jet electro-polishing equipment in a Struers TenuPol Controller Unit at a voltage of 50 V and temperature of 15 °C. The electrolyte contained 5 vol% of chloric (VII) acid and 95 vol% of acetic acid.

16.3 Results and Discusion

Microstructure of the examined material i.e. a bar after forging and after saturation process shown in Figs. 16.1 and 16.2. The X55MnAl25-5 steel, a monophase austenitic structure was obtained as expected, with characteristic annealing twins present after the saturation process.

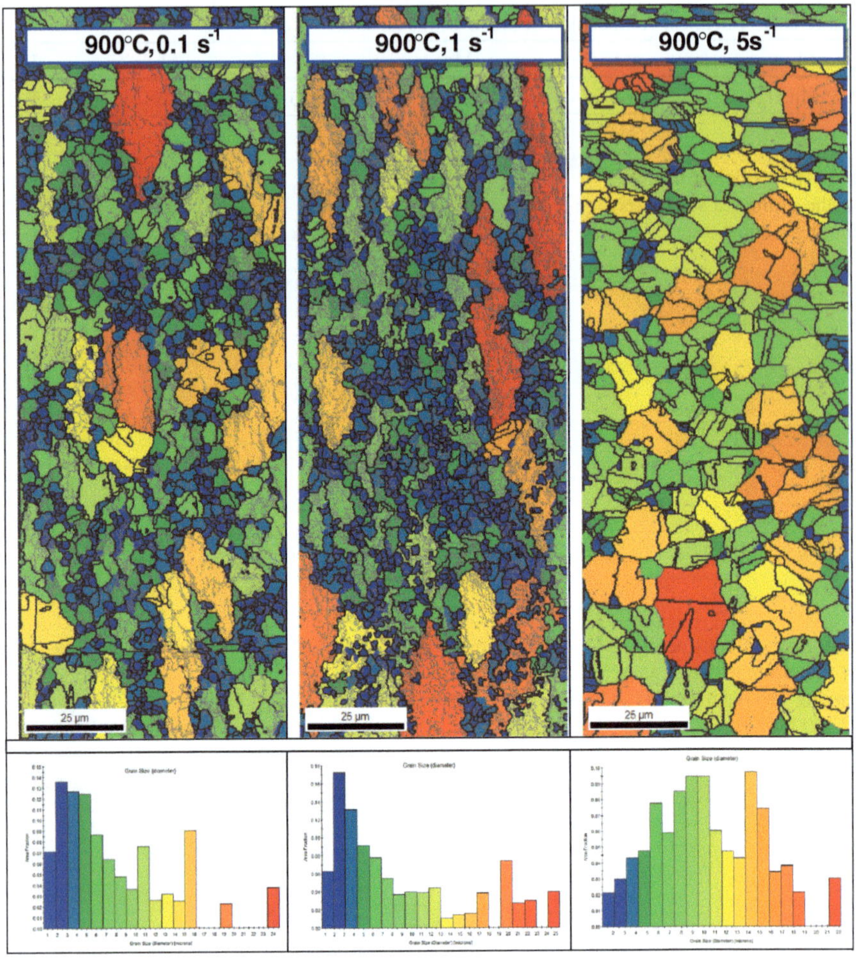

Fig. 16.3 EBSD microstructure maps and grains distribution for X55MnAl25-5 steel deformed at 900 °C

Results of the microstructure investigation of the X55MnAl25-5 steel using EBSD technique are shown in Figs. 16.3 and 16.4. For presentation was selected results of EBSD analysis after deformation at temperature 900 and 1000 °C because in this two temperatures recrystallization process occurs to effectively. Elongated primary grains were found after the deformation at 900 °C at strain rate 0.1 and 1 s^{-1}. After deformation at 900 °C at strain rate 5 s^{-1} grains growth was observed. On the grains distribution diagrams for X55MnAl25-5 steel deformed at 900 °C at strain rate 0.1 s^{-1} grain size up to 5 μm are about 41 % of the area. Increasing the strain rate to 1 s^{-1} increases area fraction of grains with size up to 4 μm to 50 %. Deformation at strain rate 5 s^{-1} causes grain growth and reducing the share of grain size

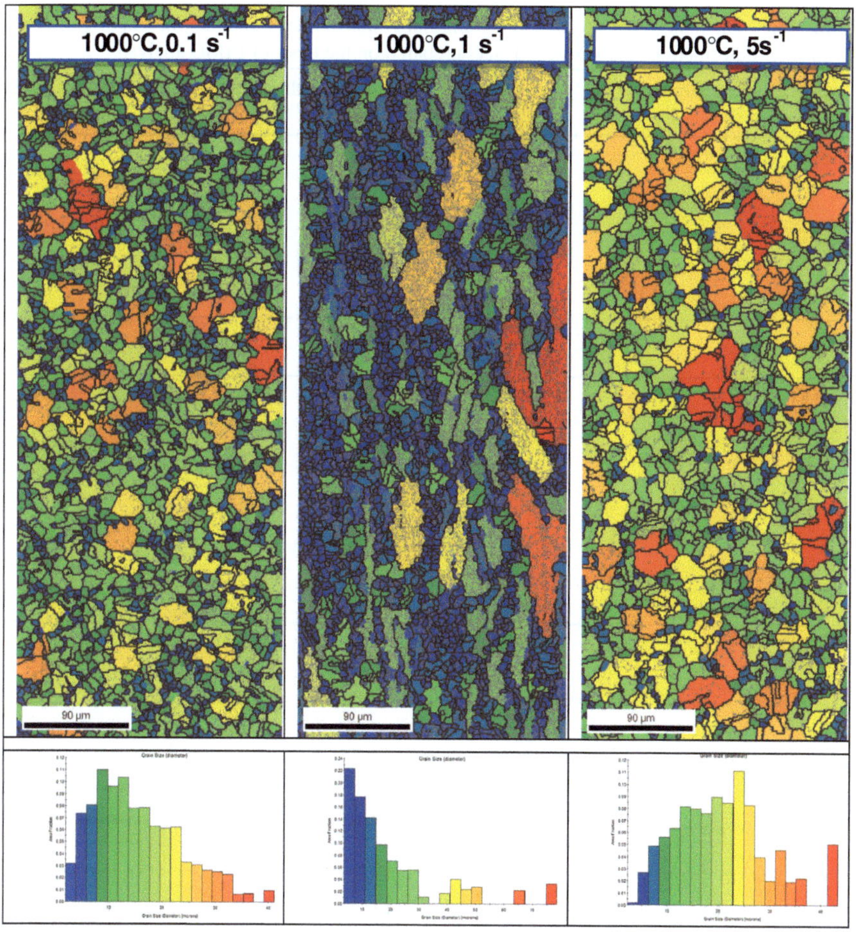

Fig. 16.4 EBSD microstructure maps and grains distribution for X55MnAl25-5 steel deformed at 1000 °C

of 4 μm up to 15 %. The recrystallization process of is therefore a most effective at temperature 900 °C and strain rate 1 s^{-1}. Recrystallized structure without the primary grains we receive after deformation at strain rate 5 s^{-1}. The situation is similar for the deformation at temperature 1000 °C, wherein the increasing of deformation temperature influences the growth of grain size in comparison to the deformation at 900 °C. On the grains distribution diagrams for X55MnAl25-5 steel deformed at 1000 °C at strain rate 0.1 s^{-1} grain size up to 10 μm are about 30 % of the area. Increasing the strain rate to 1 s^{-1} increases area fraction of grains with size up to 10 μm to 4 %. Deformation at strain rate 5 s^{-1} causes grain growth and reducing the share of grain size of 4 μm up to 14 %. Analysis of the average grain size after deformation tests (Table 16.2) has shown that the most effective grain refinement process takes place

Table 16.2 The data containing average grains/subgrains diameter and the fraction of grains with twins after deformation X55MnAl25-5 steel prepared on the basis of EBSD analysis

Temperature	Strain rate (s^{-1})	Average grains/subgrains diameter (μm)	The fraction of grains with twins
900 °C	0.1	2.41 (with primary grains)	0.22
	1	2.81 (with primary grains)	0.10
	5	4.61	0.39
1000 °C	0.1	7.70 (with primary grains)	0.17
	1	7.27 (with primary grains)	0.20
	5	12.00	0.36

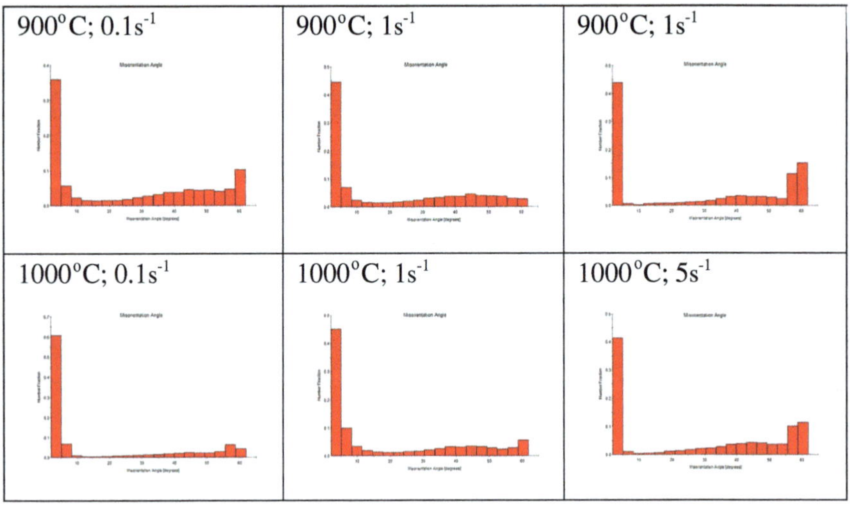

Fig. 16.5 EBSD misorientation angles distribution's for X55MnAl25-5 steel deformed at 900 °C and 1000 °C

during the deformation at 900 °C and strain rate 5 s^{-1}. Not observe already there the primary grains and average grains/subgrains diameter amounts to 4.61 μm.

Below on Fig. 16.5 the results of the EBSD analysis as misorientation angles distribution's for X55MnAl25-5 steel deformed at 900 and 1000 °C was presented.

The Fig. 16.3 clearly shows, that in samples after deformation at 900 °C, 0.1 s^{-1}, and 1 s^{-1} especially, the elongated grains with some fine recrystallized grains coexist. From misorientation angle maps (Fig. 16.5) is known, that the large misorientation angles among boundaries is characteristic for recrystallization process. The misorientation angle map shows an obvious increase of misorientation between 30 and 55°. Twin boundaries with misorientation angle 60° existed in the samples deformed at 900 °C, 0.1 s^{-1}, and 900 °C, 5 s^{-1} especially. This suggest that at high strain rate the recrystallization process is restrained.

Many large grains observed for samples deformed at 900 °C, 0.1 s^{-1}, 900 °C, 1 s^{-1} are divided into several parts by low-angle boundaries, indicating the existence of many dislocation cells and subgrains. From Fig. 16.5 is know that a large fraction of low-angle misorientation is about 5°. Large fraction of low-angle disorientation observed in sample deformed at 900 °C, 5 s^{-1} is a results of dynamic recovery. Figures 16.4 and 16.5 shows that many fine grains with high-angle boundaries are form in samples deformed at 1000 °C, 1 s^{-1}. Only few twin boundaries existed in this sample, suggesting the minimal contribution of twinning under this condition. The misorientation angle maps shows that the fraction of low-angle misorientation is high (Fig. 16.5), implying more deformation structures form.

Due to the fact that the X55MnAl25-5 steel is characterized by a formation mechanical twins during the deformation process results also include the quantitative calculation of the fraction of grains with twins (including the mechanical twins). As shown in Table 16.2 the highest number of twins calculated as fraction of grains with twins 39 % was recorded during the deformation at 900 °C and strain rate 5 s^{-1}.

16.4 Conclusion

The quantitative microstructural studies revealed microstructure rebuilding process, as a results of dynamic recovery and recrystallization. Fine grains with high-angle boundaries are well developed, indicating the occurrence of recrystallization. New fine grains with misorientation between 30° and 55° confirm, that the new grains are results from recrystallization. Most of grains are elongated and indicating relatively high defect density and/or strain in grains. These findings confirm that the fine grains are attributed to dynamic recrystallization during deformation because the grains are not free of strain/defect.

References

1. O. Grässel, L. Krüger, G. Frommeyer, L.W. Meyer, High strength Fe-Mn-(Al, Si) TRIP/TWIP steels developments properties-application. Int. J. Plast. **16**, 1391–1409 (2000)
2. S. Allain, J.-P. Chateau, O. Bouaziz, S. Migot, N. Guelt, Correlations between the calculated stacking fault energy and the plasticity mechanisms in Fe-Mn-C alloys. Mater. Sci. Eng. A **387–389**, 158–162 (2004)
3. S. Allain, J.-P. Chateau, O. Bouaziz, S. Migot, N. Guelton, G. Frommeyerr, U. Brux, Microstructures and mechanical properties of high-strenght FeMn-Al-C light TRIPLEX steels. Steel Res. Int. **77**, 627–633 (2006)
4. C. Scott, S. Allain, M. Faral, N. Guelton, The development of a new Fe-Mn-C austenitic steel for automotive applications. Rev. Metall. **6**, 293–302 (2006)
5. S.A. Hamada, Manufacturing, mechanical properties and corrosion behaviour of high Mn Twip steels, Acta Universitatis Ouluensis **C281**, 1–51 (2007)

6. H. Hofmann, M. Menne, S. Göklü, H. Richter, Properties of austenitic high manganese steels with induced plasticity (LIP steels), in *Proceedings of International Conference on Steel Future for the Automotive Industry*, Wiesbaden, Germany, pp. 73–80 (2005)
7. P. Yang, Q. Xie, L. Meng, H. Ding, Z. Tang, Dependence of deformation twinning on grain orientation in a high manganese steel. Scripta Materialia **55**, 629–631 (2006)
8. E. Mazancova, I. Ruziak, I. Schindler, A Influence of rolling conditions and aging process on mechanical properties of high manganese steels, Arch. Civil Mech. Eng. **12** 142–147 (2012)
9. A. Grajcar, W. Borek, Thermo-mechanical processing of high-manganese austenitic TWIP-type steels. Arch. Civil Mech. Eng. VIII **4**, 29–38 (2008)
10. S. Wiewiórowska, Determination of content of retained austenite in steels with TRIP effect deformed at different strain rates. Steel Res. Int. **81**(9), 262–265 (2010)
11. M. Jabłońska, A. Śmiglewicz, Analysis of substructure of high-Mn steels in the context of dominant stress mechanism. Defect and Diffusion Forum. Diffus. Solids Liquids VIII **334–335**, 177–181 (2013)
12. M. Jabłońska, A. Śmiglewicz, G. Niewielski, M. Hetmańczyk, Heat treatment of high manganese type X57MnAl27-5 austenitic steel. Mater. Sci. Eng. Conf. Ser. **22**, 012–015 (2011)

Chapter 17
Microstructural Examination of Oxidized $Fe_{(14-x)}Nb_5B_x$ Alloy Produced from Powders Prepared by Mechanical Alloying

A. Şükran Demirkıran, Saduman Sen, Ozkan Ozdemir and Ugur Sen

Abstract In the present study, ferrous niobium, ferrous boron and iron were used as starting powders. The mixture of the powders which were calculated to give the designed compositions was prepared by using planetary high energy ball mill. Mechanically alloyed powders were pressed and sintered at 1350 °C for 120 min in Ar atmosphere. The cyclic oxidation experiments were carried out in an electrical furnace at 650, 750 and 850 °C in open atmosphere for 96 h. The specimens were periodically weighed for the determination of weight change. Before and after oxidation, the present phases of the samples were determined by X-ray diffraction analysis (XRD). The microstructural characterizations were realized using scanning electron microscopy (SEM) with EDS attachment.

17.1 Introduction

Recently, Fe–Nb–B alloys are one of the important alloys investigated due to their magnetic properties such as effective permeability and saturation magnetic flux density. These materials may have application in magnetic parts and devices such as inductors, low- and high-energy frequency transformers, alternating current machines, motors, generators and sensors. For reason, in the last decade, many studies have been done on Fe–Nb–B ternary alloys leading to the development of good soft magnetic materials, nowadays used commercially [1, 2]. Fe–Nb–B alloys include $Nb_3Fe_3B_4$ and NbFeB ternary intermetallic compounds. Ordered intermetallics with high melting points are often proposed as candidate materials for high temperature structural applications. The interest in intermetallics stems from their exceptional hardness and retention of strength at elevated temperatures as a consequence of their

A. Ş. Demirkıran (✉) · S. Sen · O.Ozdemir · U. Sen
Department of Metallurgy and Materials Engineering, Engineering Faculty,
Sakarya University, Esentepe Campus, 54187 Sakarya, Turkey
e-mail: dkiran@sakarya.edu.tr

E. K. Polychroniadis et al. (eds.), *International Multidisciplinary Microscopy Congress*,
Springer Proceedings in Physics 154, DOI: 10.1007/978-3-319-04639-6_17,

Table 17.1 The chemical compositions of starting powders

	Components (wt. %)						
	Nb	B	Fe	Al	C	Si	Others
Fe–B	–	18	81	0.2	0.39	0.099	0.311
Fe–Nb	66	–	29	2.67	0.17	1.65	0.51
Pure Fe	–	–	99.99	–	0.01	–	–

ordered structures [3]. Fe–Nb–B alloys including Fe, much more than 50 % by at., Nb, much more than 25 % by at. and balance boron, include NbFeB ternary compound and this phase content increases with increase in Nb and B content in the alloy [4]. Increases of this phase make the composite hardness of these alloys harder. The main aim of the present study is to investigate synthesis and oxidation of Fe–Nb–B in-situ composite produced by mechanical alloying and sintering process.

17.2 Experimental Procedures

In this study, ferrous boron, ferrous niobium and pure iron powders used as starting powders. The chemical compositions of the starting powders were given in Table 17.1. The powders were grinded by dry ring milling and then sieved to pass through $-38\,\mu$m. Then, the prepared powders were mixed as atomic % Fe $=$ 60, Nb $=$ 25 and B $=$ 15. This mixture is alloyed by using mechanical alloying. Mechanical alloying was carried out in a planetary high-energy ball mill using stainless steel balls. The milling was performed at velocity of 355 rpm and the ball-to-powder mass ratio was 5:1 for 20 h.

The samples were produced at the dimensions of 20 mm in diameter and 3 mm in thickness by classic powder metallurgy methods. The samples were shaped by uniaxial dry pressing under the pressure of 60 MPa and polyvinyl alcohol was used as a binder material for shaping. After shaping, samples were dried at 80 °C for 24 h in an oven. Dried samples were sintered in an electrical furnace with a heating rate of 6 °C/min at 1350 °C for 120 min in the Ar atmosphere. Then, the sintered samples were cooled down to room temperature in the furnace. The produced samples were ground on emery papers and then washed in acetone. The cyclic oxidation experiments were carried out in an electrical furnace with a heating rate of 10 °C/min at 650, 750 and 850 °C in open atmosphere for 96 h. The specimens were periodically weighed. The weight of samples was measured by an electronic balance with accuracy at 0.1 mg. Before and after oxidation, the present phases of the samples were determined by a RIGAKU DMAX 220 PC diffractometer. The microstructural characterizations were realized using a JEOL JSM 6060 LV scanning electron microscopy (SEM) with EDS attachment.

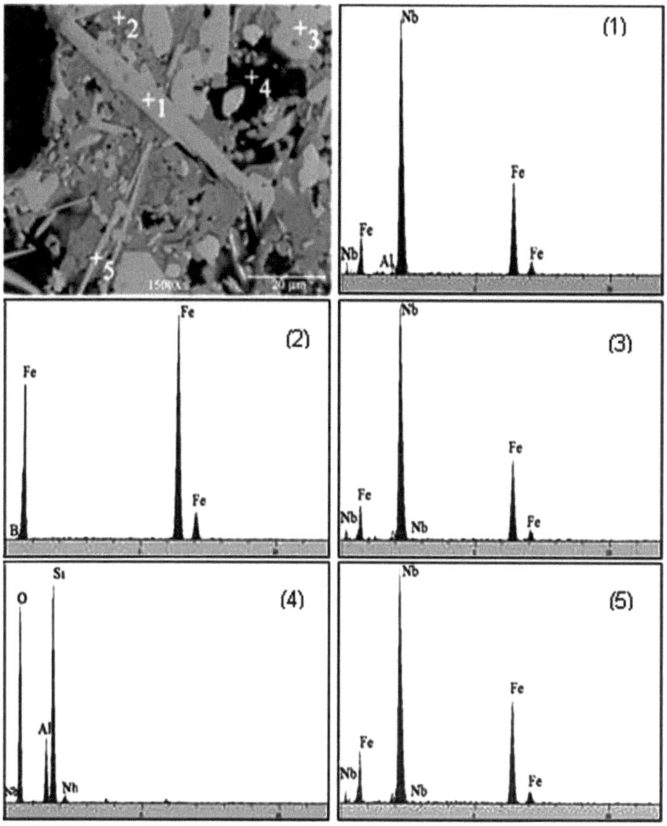

Fig. 17.1 SEM micrograph and EDS analysis of produced samples at 1350 °C temperature with a sintering time of 120 min

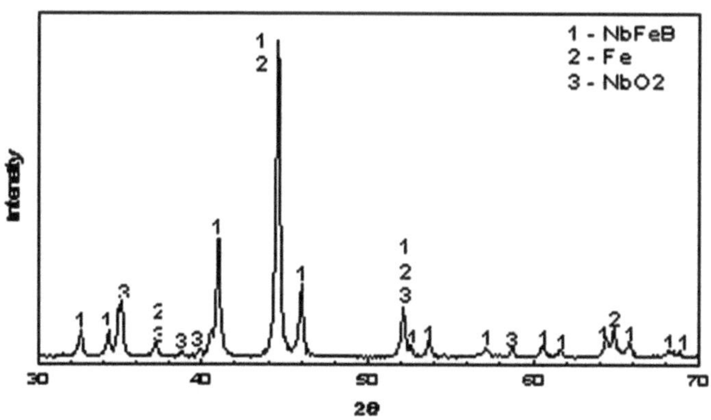

Fig. 17.2 XRD pattern of sintered samples at 1350 °C with a period of 120 min

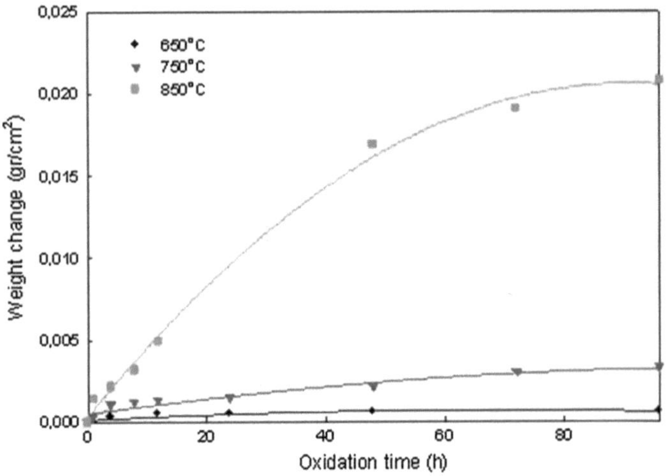

Fig. 17.3 The variation in weight gain as a function of duration time during oxidation test

17.3 Results and Discussion

Before oxidation tests, SEM-EDS observations of the produced samples showed that there are three different forms on the microstructure which were grey forms, dark forms and white blocky and needle forms (Fig. 17.1). EDS analysis showed that the dark forms consist of Si, Al and O elements (marked as 4 on the microstructure). X-ray diffraction analysis could not have any phase including these elements. So, these forms might be as glassy phase. It is probable that boron has a trigger effect on the formation of glassy phase.

Gray color form taken place between white and dark forms might be iron matrix which includes dominantly Fe element (marked as 2 on the microstructure). White form consists of Nb and Fe elements (marked as 1, 3 and 5 on the microstructure). Therefore, white blocky and needle forms might be NbFeB intermetallic which as confirmed by X-ray diffraction analysis (Fig. 17.2). In order to understand microstructural changes and to correlate with the other results, XRD analysis was performed. The phases formed in the produced samples are NbFeB (major phase), NbO_2 (trace phase) and Fe as shown in Fig. 17.2.

Figure 17.3 shows the weight gain as a function of oxidation time and temperature. As it can be seen in Fig. 17.3, oxidized samples exhibited a relatively low oxidation rate at 650 and 750 °C, but the rate increased rapidly at 850 °C. As seen in this figure, oxidation layer increased with increasing oxidation temperature. This trend verifies that the oxidation is a diffusion-controlled process. There is a nearly parabolic relationship (parabolic rate law) between weight gain and process time especially at high temperatures. High temperature oxidation of alloys is a complex phenomenon. The reaction rate is governed by several factors including temperature, time, gas

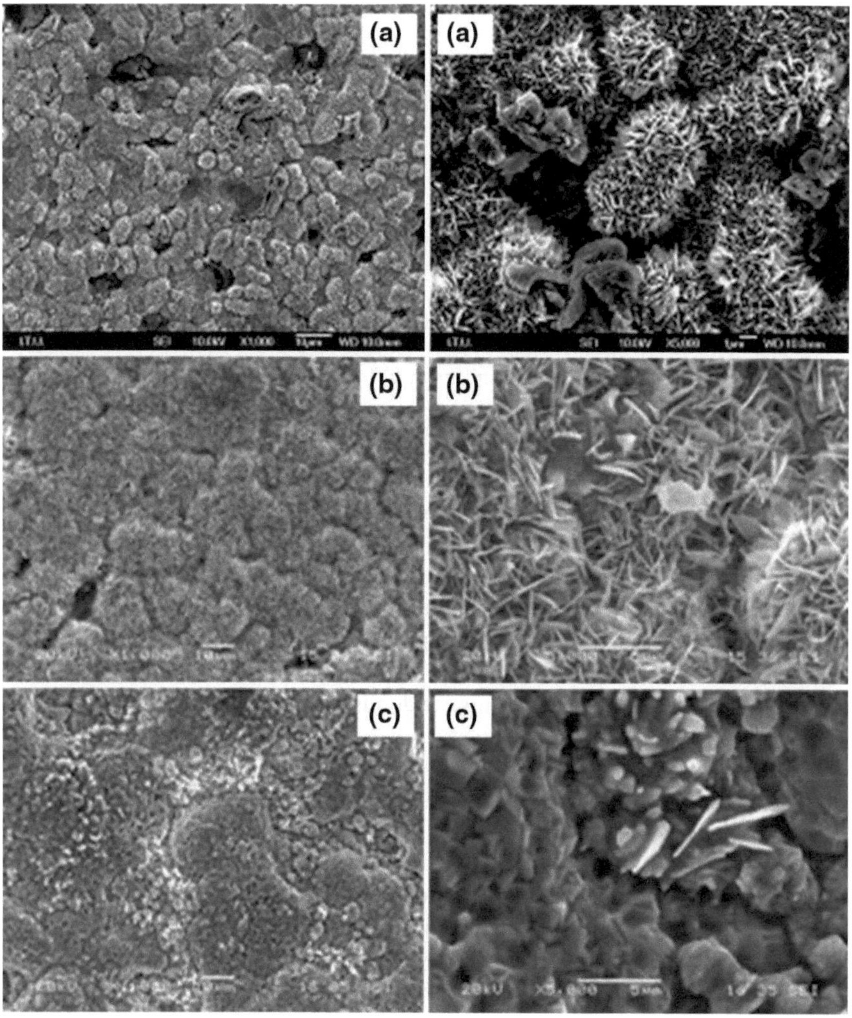

Fig. 17.4 Microstructures of oxidized sample for 12 h **a** 650 °C, **b** 750 °C and **c** 850 °C

composition, and the structure and composition of the reaction products (i.e. scales). Therefore, it is unlikely that any single rate law, such as parabolic law, will be obeyed exactly [5].

Microstructures of oxidized materials at 650, 750 and 850 °C for 12 h are shown in Fig. 17.4a–c. There are two types of oxides that can be seen: whisker oxides and nodule oxides. With increasing temperature, oxide morphology was coarsened and observed a morphological change from whisker oxides to nodule oxides of oxidized structure. In addition, no spallation was observed on these types of oxide scales.

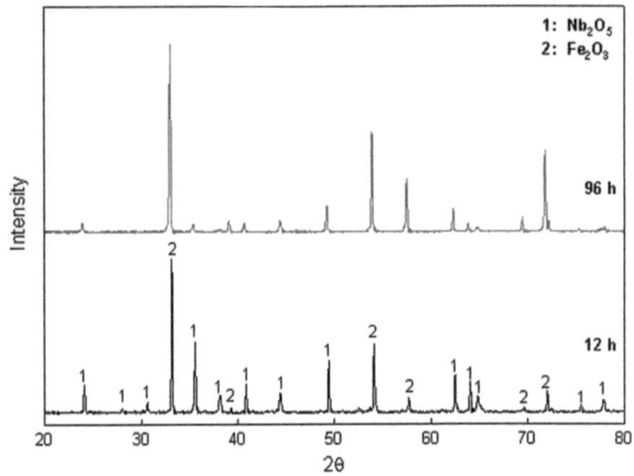

Fig. 17.5 XRD patterns of test materials oxidized at 850 °C for 12 and 96 h

It was also observed that the higher the oxidation temperature, the more the weight gains, as expected. The weight gain of the sample during oxidation is proportional to the growth in thickness of the oxide layer. Figure 17.5 shows XRD pattern of oxide layer formed on the surface of test material oxidized at 850 °C for 12 and 96 h. The oxides formed on the surface of test sample at 850 °C for 12 h have been identified as Fe_2O_3 and Nb_2O_5 phases. As seen in Fig. 17.5, while very little Nb_2O_5 phase is present in the period of 96 h, Fe_2O_3 is the dominant phase.

17.4 Conclusions

Microstructural examinations of the produced $Fe_{(14-x)}Nb_5B_x$ alloys showed that there were three different forms on the microstructures which were gray forms (iron matrix), dark forms (glassy phase) and white blocky and needle forms (borides). X ray diffraction analysis showed that the $Fe_{(14-x)}Nb_5B_x$ alloys include NbFeB, NbO_2 and Fe. In addition, there is a glassy structure including Si, Al and O elements in the samples. Oxidation layer increased with increasing oxidation temperature. Two types of oxides were seen such as whisker oxides and nodule oxides. With increasing temperature, oxide morphology was coarsened and observed a morphological change from whisker oxides to nodule oxides of oxidized structure. Fe_2O_3 as dominant phase at oxidized samples has been determined.

Acknowledgments This work was supported by Turkey National Boron Research Institute (BOREN-2008-Ç0145). We would like to thank this organization for the financial support. Authors thank to Eng. Pelin AKSOY due to experimental assistance at Sakarya University.

References

1. J. Torrens-Serra, J. Rodriguez-Viejo, M.T. Clavaguera-Mora, Influence of composition in the crystallization process of $Fe_{75-x}Nb_{10}B_{15+x}$ metallic glasses. J. Non-Cryst. Solids **353**, 842–844 (2007)
2. J.J. Sunol, A. Gonzalez, J. Saurina, L. Escoda, P. Bruna, Thermal and structural characterization of Fe-Nb-B alloys prepared by mechanical alloying. Mater. Sci. Eng. A **375–377**, 874–880 (2004)
3. D.R. Johnson, Intermetallic-based composites. Curr. Opin. Solid State Mater. Sci. **4**, 249–253 (1999)
4. V. Raghavan, *Phase Diagrams of Ternary Iron Alloys* (Indian Institute of Metals, Calcutta, 1992), pp. 372
5. C.H. Xu, W. Gao, H. Gong, Oxidation behaviour of FeAl intermetallics. The effects of y and/or Zr on İsothermal oxidation kinetics. Intermetallics **8**, 769–779 (2000)

Chapter 18
Characteric Proporties of NiTi Shape Memory Alloy Powders with Powder Injection Molding

Sinan Aksöz and Bülent Bostan

Abstract In this paper, NiTi Shape Memory Powders were investigated on producticability with powder injection moulding. Powders which are used in this study were sized respectively; 10 and 35.6 μm. Characteristics proporties of Powders (Powders characteristic features) were researched by SEM (Scanning Electron Microscopy), EDS (Element Diffraction Spectrometer), DSC (Differential Scanning Calorimetry) and powder size analizing method. For injection moulding, two types of binder were determined which wasn't used for this type of powders. TG (Gravimetrik Analysis), DTA (Differantial Thermal Analysis) and DTG (Differential Transformation Gravimeter) analize system were used for analizing the binders. Analizing results showed that spherical and transformation temperetures of powders were were good enough.

18.1 Introduction

Shape Memory Alloys (SMA) such as NiTi have enabled technology development in various areas, such as microrobotics and manipulation for instance. These alloys undergo a reversible solid-state displacive crystalline and phase transformation dominated by shear between a high symmetry parent phase (austenite in form of ordered BCC superlattice β phase in the case of Ni-50.0 % Ti) and a low symmetry product phase (martensite in the form of monoclinic distortion of a B19 lattice) [1].

S. Aksöz (✉)
Department of Mechanics and Materials Technology, Atatürk High Vocational School, University of Gazi, Ankara, Turkey
e-mail: sinanaksoz@gazi.edu.tr

B. Bostan
Department of Materials and Metallurgy Engineering, University of Gazi, Ankara, Turkey
e-mail: bostan@gazi.edu.tr

E. K. Polychroniadis et al. (eds.), *International Multidisciplinary Microscopy Congress*, Springer Proceedings in Physics 154, DOI: 10.1007/978-3-319-04639-6_18,

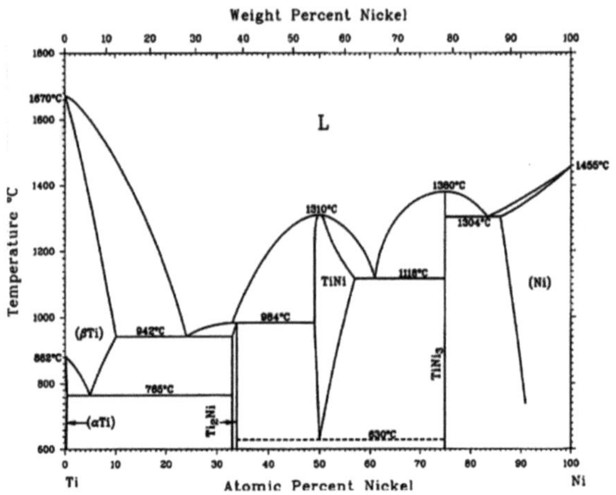

Fig. 18.1 Ni–Ti phase diagram [5]

Using titanium and alloys in biomedical productions continues to gain note, because of Titanium and alloys unique properties, including high specific strenght, low density and lightweight feel, excellenet corrosion resistance, and biocompatility [2]. Nickel Titanium (also known as Nitinol or NiTi) is in a unique class of shape memory alloys. A thermoelastic martensitic phase transformation which the material contents is responsible for its extraordinary properties. Nitinol properties include the shape memory effect, superelasticity, and high damping capability. These technical properties of NiTi shape memory alloys can be modified to a great extent by changes in composition, mechanical working, and heat treatment [3].

Ni–Ti binary equilibrium was seen in Fig. 18.1. In phase diagram thermodynamically stable phases exist in the proximity of equiatomic percentages of Ni and Ti. Applying heat treatments can have significant effects on types of microstructural phases in the final products and also on thermomechanical properties of NiTi devices, such as annealing, solution treatment, and aging [4].

Many applications require material properties which polymers cannot provide e.g. if high strenght, high corrosion or thermal resistance are required. This can be solved by powder injection molding which is an economically viable process for complex shaped metal or ceramic parts in large scale series [6].

For this purpose, metal or ceramic powders are mixed with a binder system and injected into the mold which contains the microstructured mold inserts. In order to achive a good filling of the mold inserts, feedstocks of low viscosity and evacuated injection molding are required. Normally, a relatively high mechanical stability of the binder system is necessary for a safe demolding. The molded part is processed further in a furnace to remove the binder system. Subsequently, sintered under defined athmosphere to achive a dense micro component [6].

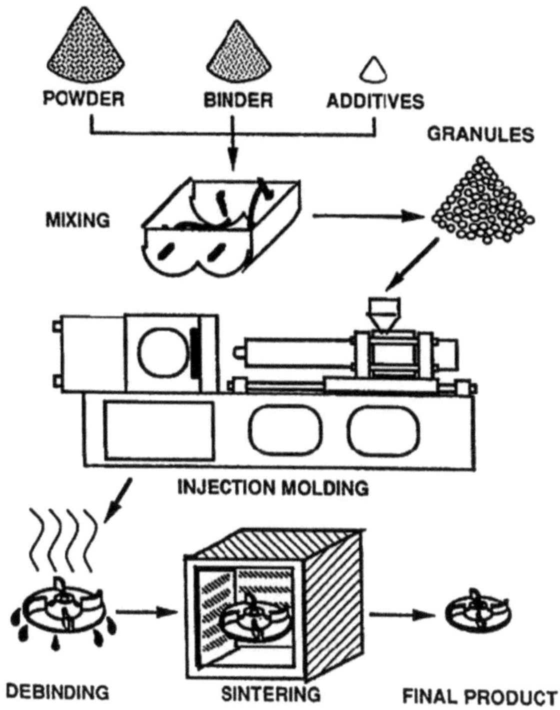

Fig. 18.2 Schematic diagram of MIM process [7]

In this study, NiTi shape memory alloy powders were investigated to productubility of Powder Injection Moulding (Fig. 18.2). For this porpose, powders were analized for size, DSC and EDS analizer.

18.2 Experimental Study

18.2.1 Powder Characterization

In this section, prealloyed Ni-rich TiNi powder (Ti-55.52 % at Ni 99.9 % purity, supplied by Nanoval GmbH and Co.KG) were used as the raw material. Two sized powders were obtained to fabric. These powders general structure were examined by using SEM images (Figs. 18.3 and 18.4). The examinationed of the overall structure of powders founded spherical in SEM images. For having a good productubility, powders were spherical and sized 35 and 10 μm with a low sattallites.

Powders were analyzed with the size analizer and could be seen there that aproximately the size was given by manufacturer. While this fabricate gave in receipt 10,

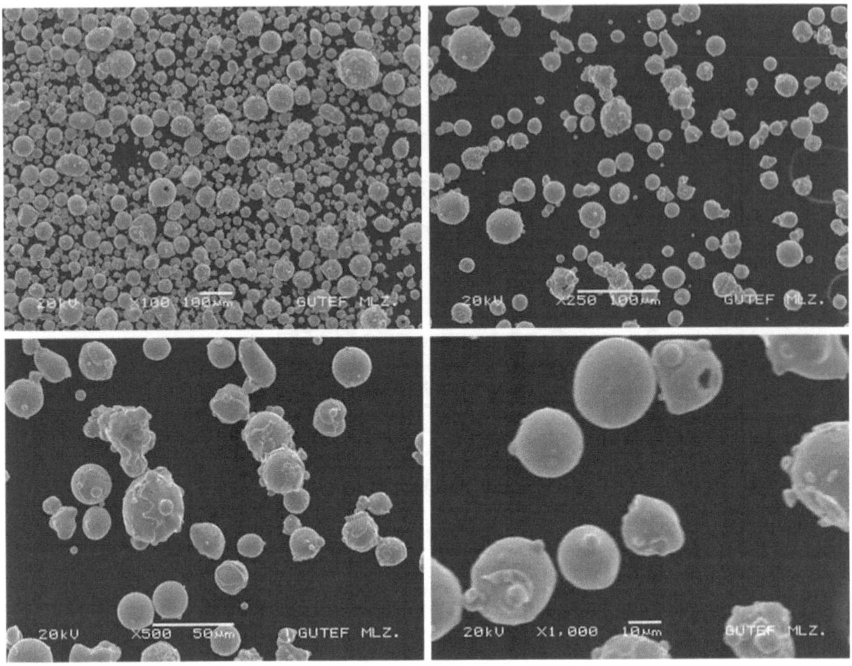

Fig. 18.3 SEM images of 10 µm sized powders

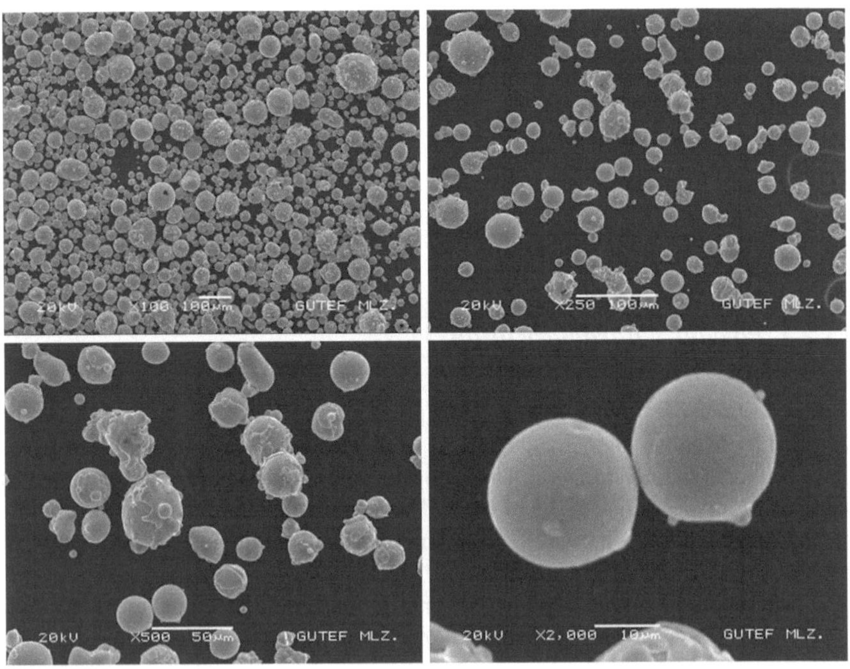

Fig. 18.4 SEM images of 35 µm sized powders

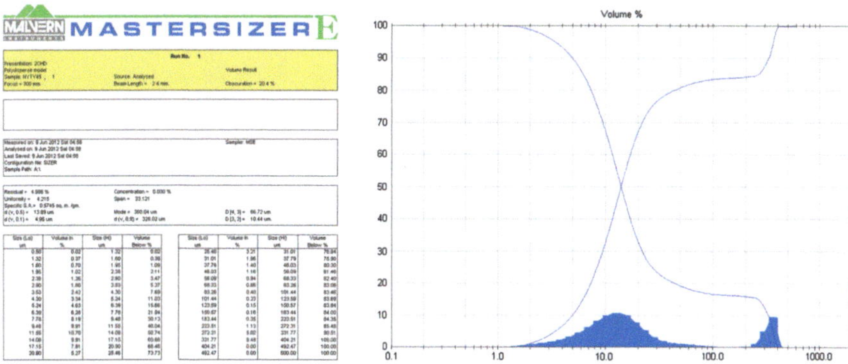

Fig. 18.5 NiTi powders average particle size (10 μm)

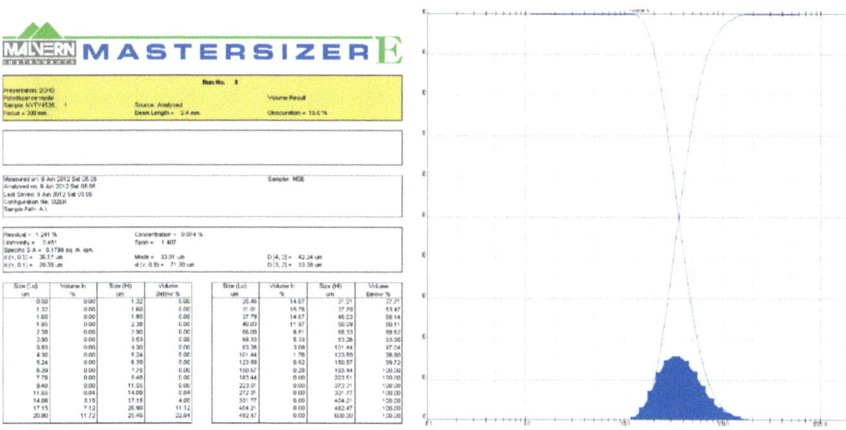

Fig. 18.6 NiTi powders average particle size (35 μm)

13 μm was founded by powder size analizer (Fig. 18.5) and 36.17 μm was founded for 35 μm powders by powder size analizer (Fig. 18.6).

In order to determine the transformation temperature of the SMA, DSC analysis were applied on 10 and 35 μm sized-powders (Figs. 18.7 and 18.8). The onset temperature of the alloy was at −24.35 °C and the end temperature was at −1.97 °C. For 35 μm sized powders onset and final temperature was between −15.17 and 4.17 °C which could be seen in Fig. 18.4.

To obtain the chemical composition of NiTi powders, microanalsis was carried out by using EDS (Figs. 18.10 and 18.12) on the SEM images (Figs. 18.9 and 18.11). The spot analyses (Figs. 18.10 and 18.12) revealed twostructure, Ni and Ti being

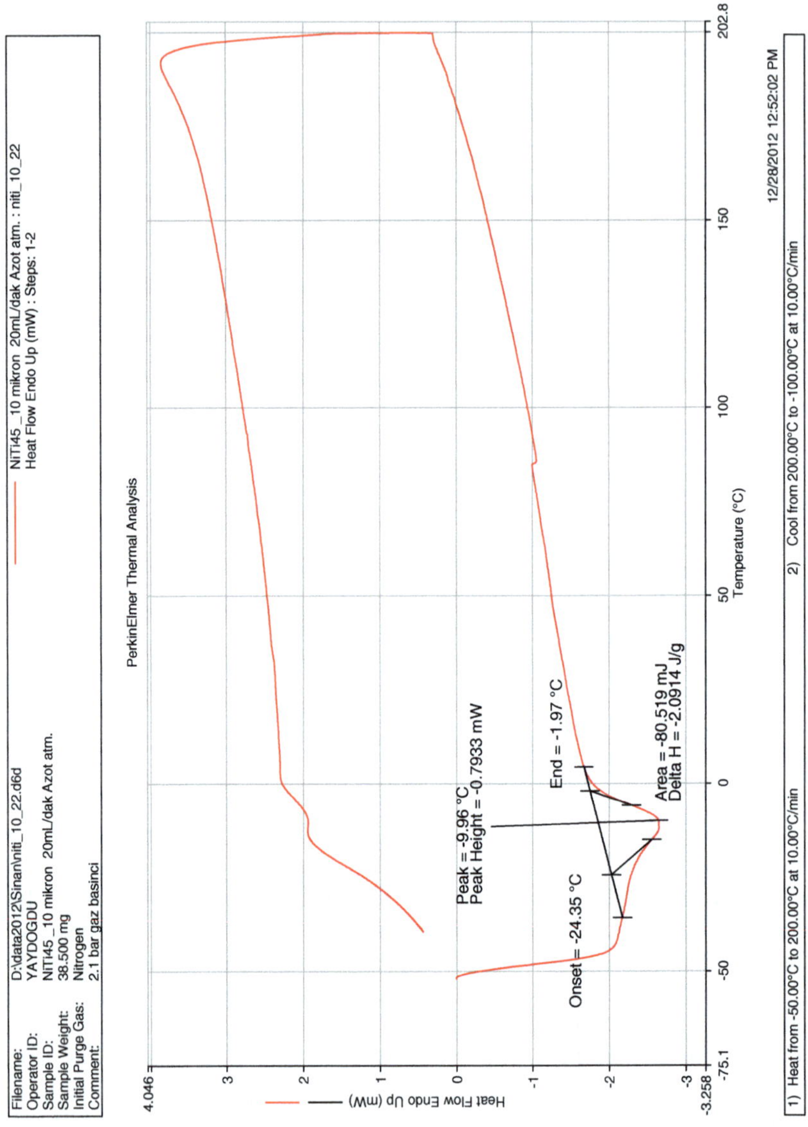

Fig. 18.7 DSC analysis of the 10 μm size powders

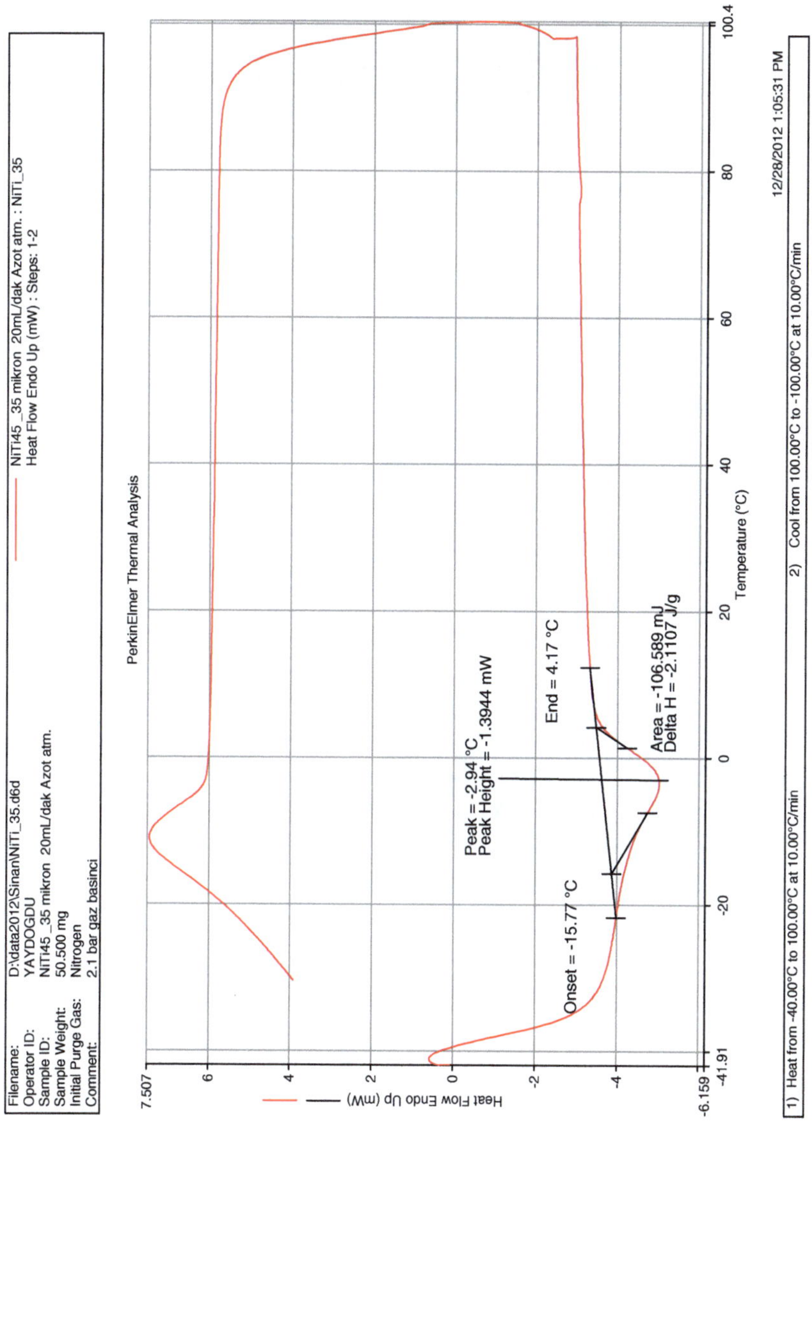

Fig. 18.8 DSC analysis of the 35 μm size powders

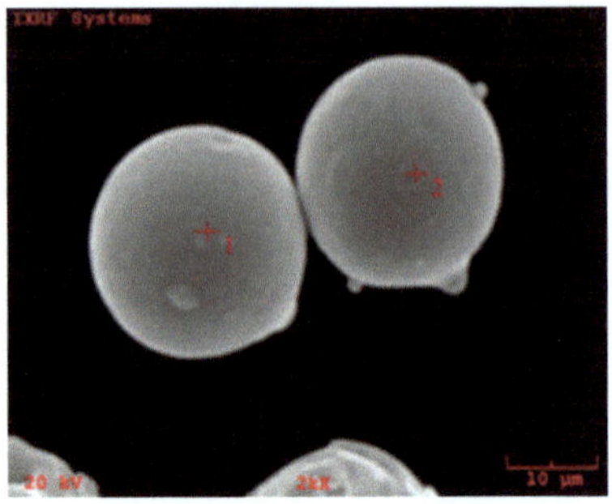

Fig. 18.9 SEM micrograph of the NiTi powders sized 10 µm

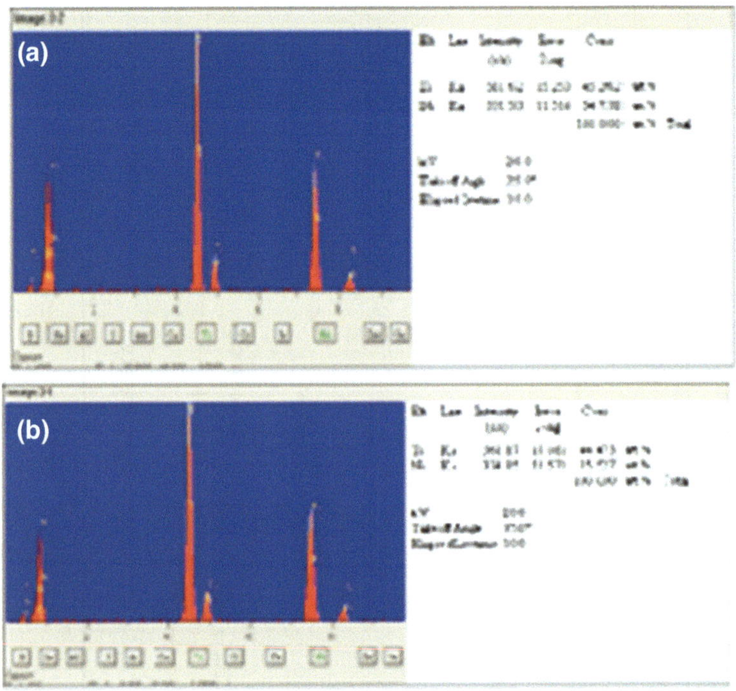

Fig. 18.10 10 µm-sized powders **a** and **b** EDS results

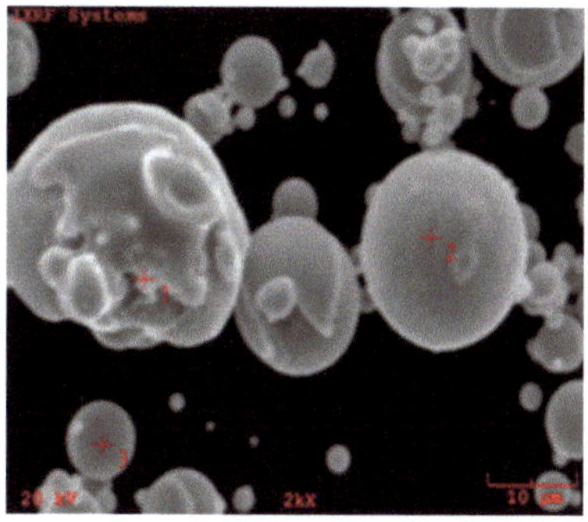

Fig. 18.11 SEM micrograph of the NiTi powders sized 35 μm

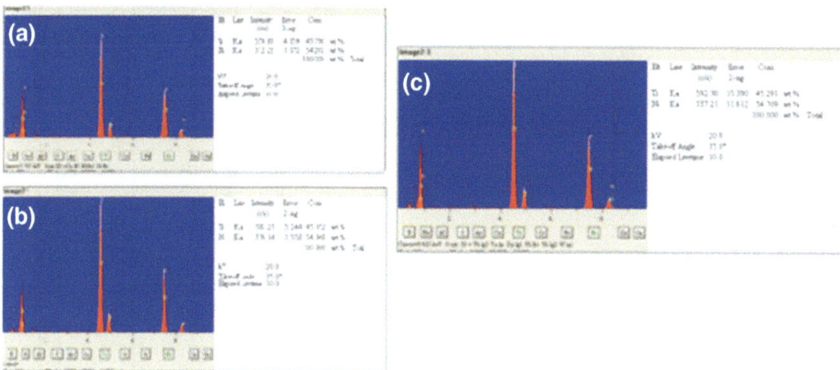

Fig. 18.12 35 μm-sized powders **a**, **b** and **c** EDS results

present in both. According to the EDS peaks, only Ni and Ti peaks were determined. Ni ratio was more than Ti ratio for 10 μm sized powders. When compared Figs. 18.10 and 18.12, Ti ratio was more than in Fig. 18.12.

The binder which was used in this work is composite of 80 % PEG 8000 (Sigma Aldrich), 15 % Polypropylen (Petkim Co.Inc.) and 5 % Stearic Asit (SA). Powder loading for this work is 50, 60 and 70 % and binder used. For first binder system, it was composed of PEG 8000, Polipropylen and Stearik Asit. The first peak which was given by PEG 8000 and Stearik Asit (peak temperature was 69.85 °C). Polipropylen's peak was the second and approximately 161.90 °C (Fig. 18.13). Second binder's tranformation temperature was aproximately the same as the first temperature (Fig. 18.14).

It can be seen that in TG result (Fig. 18.15), material loss was beginnig approximately 205 °C. DTA results showed that aproximately at 70 and 170 °C two transformation obtained. First one is PEG 800 and SA, second one is Polipropylen.

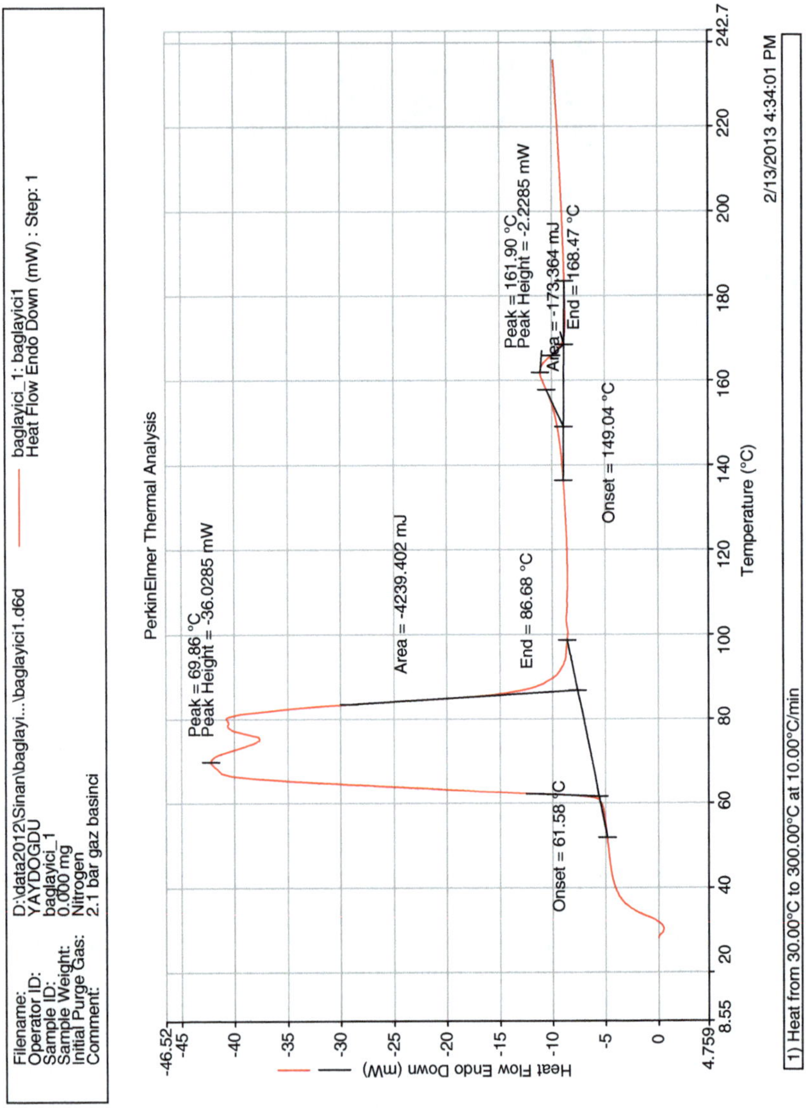

Fig. 18.13 DSC analysis of Binder 1

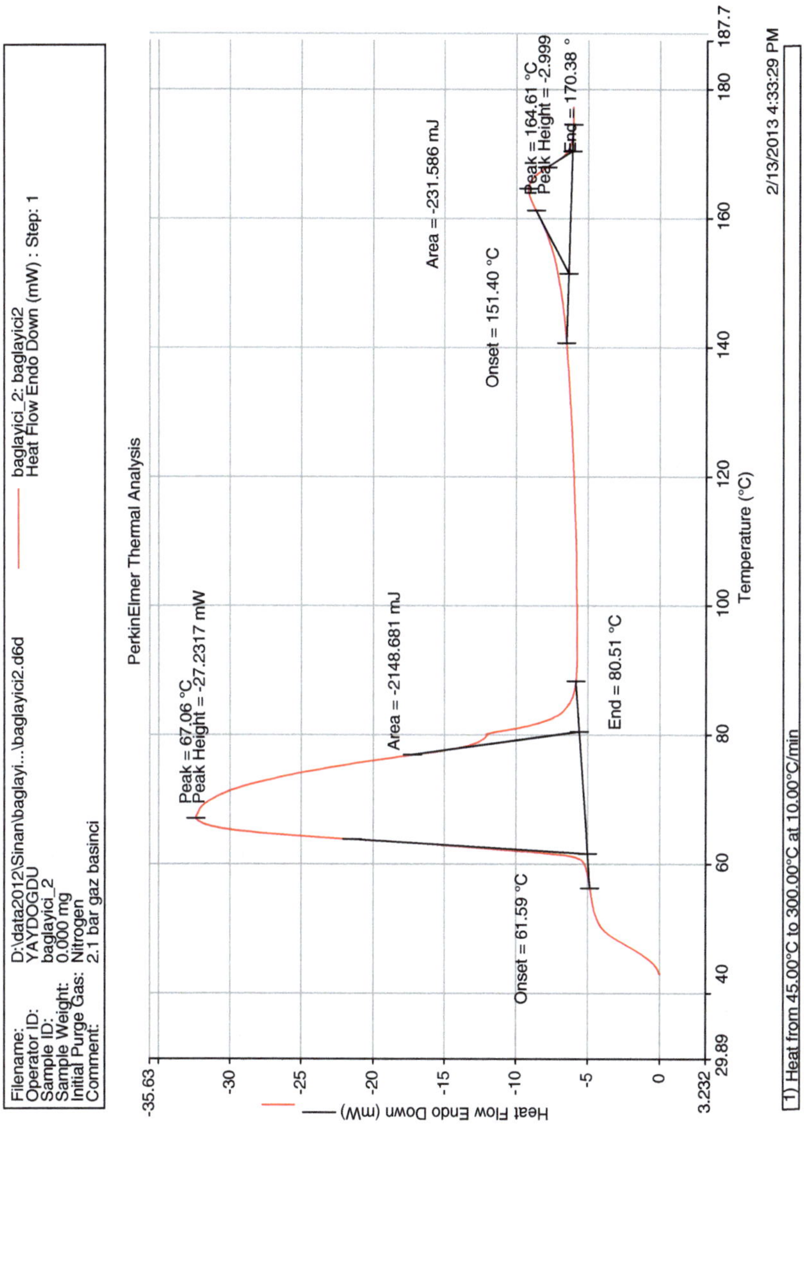

Fig. 18.14 DSC analysis of Binder 2

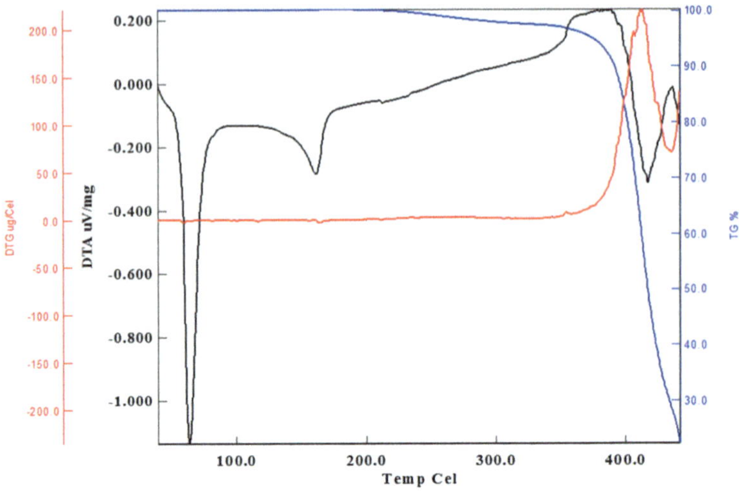

Fig. 18.15 DTG, DTA and TG analysis of Binder 1

18.3 Conclusion

Our proposal, avoiding the exothermic reaction problems with elemental Ni and Ti powders, prealloyed NiTi powders (diameter <20 μm) were used [8]. Prealloyed NiTi powders were used and investigated for mouldability in powder injection moulding method. For productubility, two types of powders were used. Differential Scanning Calorimetry (DSC) results of the prealloyed NiTi powders (approximately; Ni55–Ti) indicate that martensitic transition occurs between −15 and 4 °C for 35 μm sized powders and between −24 and −2 °C for 10 μm sized powders. NiTi alloys generally used for manufacturing implants and powder metallurgy process is the most versatile and practical ones to make NiTi implants were also known [8]. Two type of binders were investigated and used in rheolojical application, including PEG 8000, Polipropplen, Stearik Acid. High quality parts could be produced for medical application with powder injection moulding.

(With powder injection moulding, could be produced high quality parts for medical application)

Acknowledgments The researchers would also like to thank to Scientific Research Project group for their helps (BAP Project No: 41/2012-11).

References

1. M.H. Elahinia, M. Hashemi, M. Tabesh, S.T. Bhaduri, Manufacturing and processing of NiTi implants: a review. Prog. Mater. Sci. **57**, 911–946 (2012)
2. E. Nyberg, M. Miller, K. Simmons, K. Scott Weil, Microstructure and mechanical properties of titanium components fabricated by a new powder injection molding technique. Mater. Sci. Eng. C **25**, 336–342 (2005)
3. http://jmmedical.com/nitinol.html (30.08.2013)
4. M. Bram, A. Ahmad-Khanloua, A. Heckmannb, B. Fuchsa, H.P. Buchkremera, D. Stovera, et al., Powder metallurgical fabrication processes for NiTi shape memory alloy parts. Mater. Sci. Eng. A **3**37, 254–63 (2002)
5. M. Bram, A. Ahmad-Khanloua, A. Heckmannb, B. Fuchsa, H.P. Buchkremera, D. Stovera, et al., Powder metallurgical fabrication processes for NiTi shape memory alloy parts. Mater. Sci. Eng. A **337**, 254–63 (2002)
6. V. Piotter, W. Bauer, T. Benzler, A. Emde, Injection molding of components for microsystems. Microsyst. Technol. **7**, 99–102 (2001)
7. A. Safarian, K. Abrinia, A.H. Behravesh, The practical analysis of fabricationof a copper part by metal injection molding. in *6th International Powder Metallurgy Conference and Exhibition*, Middle East Technical University Ankara, Turkey, 2011
8. M.H. Elahinia, M. Hashemi, M. Tabesha, S.B. Bhaduri, Manufacturing and processing of NiTi implants: a review. Prog. Mater. Sci. **57**, 911–946 (2012)

Chapter 19
Microstructural Examinations of Fe-W-B Base Hard-Faced Steel

Eray Abakay, Bülent Kilinc, Saduman Sen and Ugur Sen

Abstract It is now well established that considerable improvement in the mechanical/chemical properties of near surface regions of materials can be achieved by the process of surface alloying. In the present study, surface alloying treatment with Tungsten and Boron on the surface of AISI 1020 steel was realized by the technique of TIG welding. Ferrous boron alloy and ferrous-tungsten were used for surface alloying treatment. Before the treatment, ferrous alloy was ground and sieved to be smaller than 45 μm. The powders were mixed to be composed of Fe_5WB_4. Prepared powder was pressed on the steel substrate and melted by TIG welding for surface alloying. Coated layers formed on the steel substrate were investigated using by optical and scanning electron microscopy, X-ray diffraction analysis and Vickers micro-hardness tester. It was shown that surface alloyed layer has composite structure including eutectic matrix and blocky boride phase which is white color and well distributed. Borides formed in the coated layers have a sharp corner structure and distributed in the matrix. X-ray diffraction analyses showed that coated layers include Fe_2B, FeB and FeW_2B_2 phases. The hardness of blocky boride phases is 2095 $HV_{0.01}$.

E. Abakay (✉) · S. Sen · U. Sen
Department of Metallurgy and Materials Engineering, Sakarya University Engineering Faculty, 54187 Serdivan, Sakarya, Turkey
e-mail: eabakay@sakarya.edu.tr

S. Sen
e-mail: sdmnsen@sakarya.edu.tr

U. Sen
e-mail: ugursen@sakarya.edu.tr

B. Kilinc
Arifiye Vocational High School, Sakarya University, 54580 Arifiye, Sakarya, Turkey
e-mail: bklilinc@sakarya.edu.tr

E. K. Polychroniadis et al. (eds.), *International Multidisciplinary Microscopy Congress*, Springer Proceedings in Physics 154, DOI: 10.1007/978-3-319-04639-6_19,

19.1 Introduction

Hard facing is a commonly employed method to improve surface properties of agricultural tools, components for mining operation, soil preparation equipments and others [1]. This technique primarily done to enhance the surface properties of the base metal (substrate) and hard-faced materials generally exhibit better wear, corrosion and oxidation resistance than the base metal [2].

High-energy density sources have widely applied hard facing alloys such as electron beam, plasma arc and laser [2–5]. Among the welding deposition techniques, tungsten inert gas arc (TIG or GTAW) welding is a very effective and techno-economical solution for wear applications. This process has following advantages: high deposition rate, high maneuverability, large-scale availability, low cost and compatibility with a wide range of materials [6]. TIG surface alloying associated with rapid heating and cooling rate provided a unique opportunity for the non-equilibrium synthesis of materials and produced rapidly solidified fine microstructures with extended solid solution of alloying elements [7].

Tungsten (W) is an attractive material for high-temperature applications because of its high melting point (3410 °C) and high affinity to produce carbide, nitride and boride phases. Hence, W is an obvious choice for structural applications exposed to very high temperatures. Tungsten borides are known to have high hardness values chemical inertness and electronic conductivity. They have possible industrial applications as abrasive, corrosion-resistant and electrode materials, which are exposed to exceptional environments [8].

In the present study, surface alloying treatment with Tungsten and Boron on the surface of AISI 1020 steel was realized by the technique of TIG welding. Coated layers formed on the steel substrate were investigated using by optical and scanning electron microscopy, X-ray diffraction analysis and Vickers micro-hardness tester.

19.2 Experimental Procedure

The substrate material used for surface alloying was prepared from AISI 1020 steel plates with the dimensions of 20 mm × 60 mm × 5 mm. The nominal chemical composition of AISI 1020 steel (in wt. pct.) was as follows: C, 0.23; Si, 0.18; Mn, 0.52; and Fe (balance). Before the surface alloying treatment, these specimens were ground and cleaned with acetone to remove any oxide and grease and then dried with compressed air. The nominal composition of ferrous tungsten alloys used in the study (%wt.) was follows: W, 75; Si, 0.50; C, 0.20; Mn, 0.25; Cu, 0.15; and Fe, the balance. The nominal composition of ferrous boron alloys used in the study (wt. %) was as follows: 19.63 % B, 0.44 %C, 0.05 %Al, 0.98 %Si and Fe (balance). Ferrous boron and ferrous tungsten were grounded by ring grinder and sieved to be −45 μm particle size. Figure 19.1 shows the scanning electron microscopy (SEM) image and Energy dispersive spectroscopy (EDS) analysis of the ferrous boron and ferrous tungsten powders used in the surface alloying treatment.

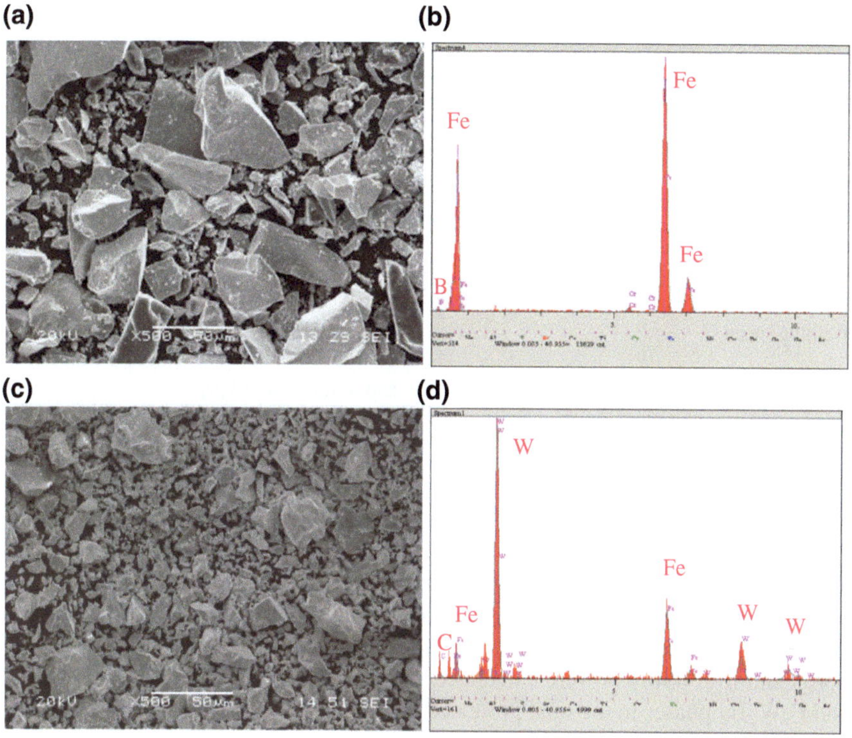

Fig. 19.1 SEM images and EDS analysis of **a**, **b** ferrous boron and **c**, **d** ferrous tungsten powders, respectively

TIG welding was realized by an electric power supply in which the welding torch was moved back and forth at a constant speed. TIG welding was applied to melt the substrate and alloy filler. The welding parameters are listed in Table 19.1.

An X-ray diffractometer (Rigaku XRD/D/MAX/2200/PC) with Cu K_α radiation was used to analyze the constituent phases in the microstructure. Metallographic analysis was performed by optical microscopy (OM) to the samples which were replaced in resin, ground on silicon carbide papers to 1200 grit and then progressively polished with 0.3 μm Al_2O_3 paste.

The etchant was selected as 3 % Nital for determination of produced phases in the surface alloyed layer for metallographic examinations. The microstructures of the cross-section of the alloyed layers were observed by using OM (NICKON EPIPHOT), the hardness of the phases formed in the alloyed layer and transition zone and matrix were measured by using Future-Tech FM 700 μ hardness tester.

Table 19.1 Experimental parameters of TIG surfacing

Parameter	Value
Electrode	Type W-2 pct ThO
Diameter	2.4 mm
Angle	70°
Voltage	20 V
Current	110 A
Heat input	2.2 MJ/m
Protective gas	Type Ar (%99.9 Ar)
Flow	12 L/min
Welding speed	Travel speed 60 mm/ min

Heat input $Q = 60 \times I \times V/S$, I: current, V: voltage, and S: travel speed [9]

19.3 Results and Discussion

TIG welding surface modification was applied on AISI 1020 steel. Alloying treatment was realized using by ferrous tungsten and ferrous boron as filler alloys as to be Fe_5WB_4. The composition of the filler alloy mixtures of ferrous tungsten, ferrous boron and iron powders were calculated and used for the production of Fe_2B, FeB and FeW_2B_2 phases in the final produced alloy that as known the borides of the iron and FeW_2B_2 phase have very high hardness, wear resistance and corrosion resistance [10–12]. In the process, prepared powder mixture was melted on the steel samples and simultaneously rapidly solidified to form a dense coating bonded to the base metal. Produced surface alloy layer compact, dense and well bonded on the steel samples as seen in Fig. 19.3. The figure shows that surface alloyed steel samples include three distinct regions which are: (i) surface alloyed layer which includes Fe, B and W, (ii) transition zone and (iii) steel matrix. X-ray diffraction analysis of the surface alloyed layers showed that the layer includes Fe_2B, FeB and FeW_2B_2 phases as major and, W_2B and W_2B_5 phases as trace. In the TIG surface alloy treatments, solidification of the melted zones is too speed and it is possible that the formation of some deal phases of the used elements can be realized because of sufficient time for the production of stable phases took place in the phase diagram of the Fe-W-B [13]. As seen in Fig. 19.2, W_2B and W_2B_5 trace phases formed in the surface alloyed layer besides Fe_2B, FeB and FeW_2B_2 phases. Islak et al. explained that surface alloyed plain carbon steels with W, B and C elements produced the FeW_2B_2 phase beside iron borides and iron boron carbides [14]. It is agree with the present study. Klueh [15] explained that 3 %W wt. in the steel composition can produce FeW_2B_2 phase with boron of which composition gets over 0.0092 % wt. The thickness of the surface alloyed layer ranged from 2 to 3 mm.

As seen from Figs. 19.3 and 19.4, optical and SEM images of the surface alloyed layer showed that the layer includes eutectic microstructure of the Fe_2B and FeB phases and some boride islands of the FeW_2B_2 phases, separately. EDS analysis of

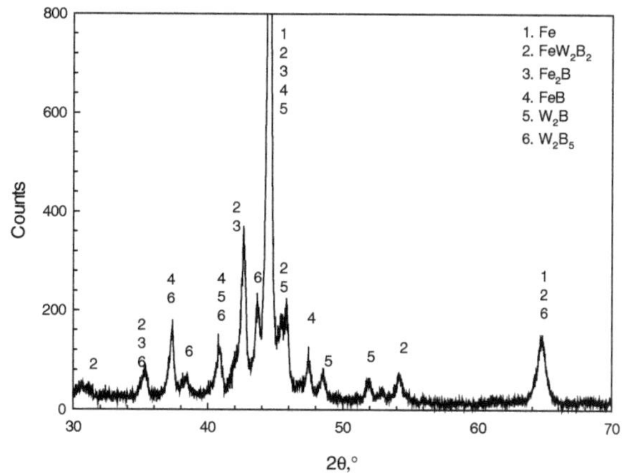

Fig. 19.2 X-ray diffraction analysis of hard faced layer on AISI 1020 steel

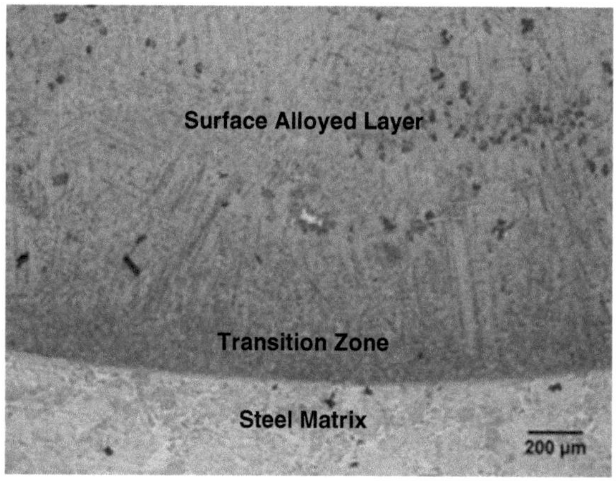

Fig. 19.3 Cross-sectional micrographs of hardfacing layer

the alloyed layer showed that both of the phases took place in the eutectic structure includes iron and boron elements.

The Vickers hardness of the surface alloyed layer is 2095 ± 215 $HV_{0.01}$. The hardness value of the alloyed layer is much higher than that of steel matrix (195 $HV_{0.5}$). As explained that the hardness of the iron borides and FeW_2B_2 phase. Islak et al. showed that increase in boron content in the alloyed layer caused to increase in the hardness [15]. The hardness of the surface alloyed layer in the present study is much higher than that of the Islak et al. study that the surface layer includes much more carbon according to the surface layer realized in the present study [14].

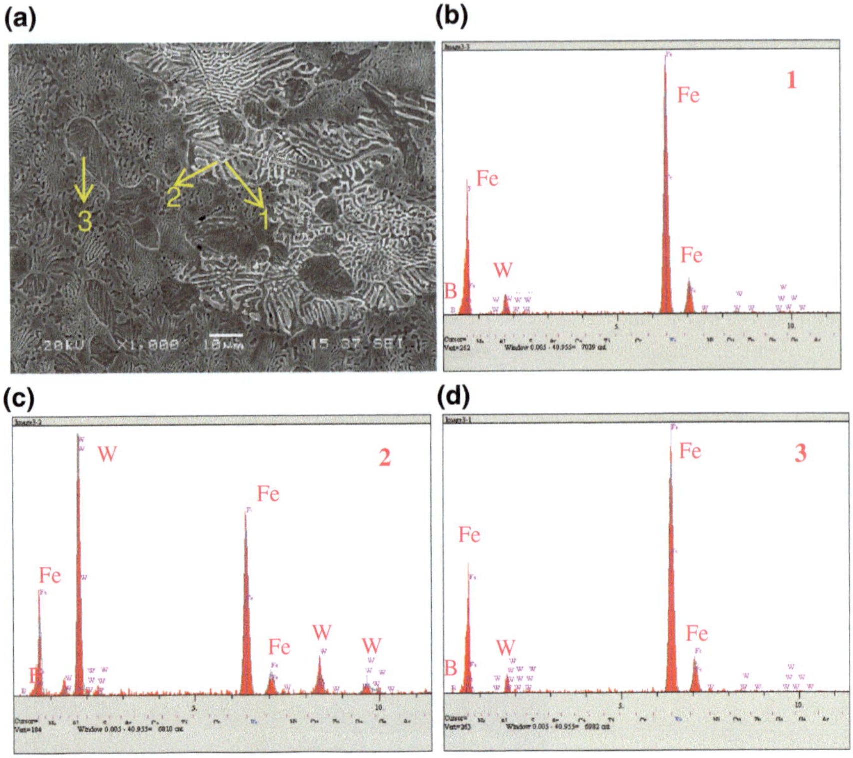

Fig. 19.4 **a** SEM image and **b**, **c** and **d** EDS analysis of the hard facing layer on the 1020 steel, respectively

19.4 Conclusion

The present study has shown that surface alloying with ferrous tungsten and ferrous boron by TIG welding on AISI 1020 steel can be developed effectively and economically. Surface alloyed layer with ferrous tungsten and ferrous boron on the plain carbon steel has eutectic microstructure of iron boride and FeW_2B_2 phases as in-situ composite structure. The phases formed in the alloyed layers are Fe_2B, FeB and FeW_2B_2 phases as major and, W_2B and W_2B_5 phases as trace. The hardness of the boride phases is very high according to the steel matrix.

References

1. C.M. Chang, C.M. Lin, C.C. Hsieh, J.H. Chen, W. Wu, Micro-structural characteristics of Fe-40 wt%Cr-xC hardfacing alloys with [1.0–4.0 wt%] carbon content. J. Alloy. Compd. **487**, 83–89 (2009)
2. V. Balasubramanian, R. Varahamoorthy, C.S. Ramachandran, C. Muralidharan, Selection of welding process for hardfacing on carbon steels based on quantitative and qualitative factors.

Int. J. Adv. Manuf. Technol. **40**, 887–897 (2009)
3. J.C. Oh, S. Lee, Correlation of microstructure with hardness and fracture properties of (TiC, SiC)/Ti-6Al-4V surface composites fabricated by high-energy electron-beam irradiation. Surf. Coat. Technol. **179**, 340–348 (2004)
4. A. d'Oliveira, R.S.C. Paredes, R.L.C. Santos, Pulsed current plasma transferred arc hardfacing. J. Mater. Process. **171**, 167–174 (2006)
5. P.W. Leech, Laser surface melting of a complex high alloy steel. Mater. Des. **54**, 539–543 (2013)
6. R.A. Jeshvaghani, M. Jaberzadeh, H. Zohdi, M. Shamanian, Microstructural study and wear behavior of ductile iron surface alloyed by inconel 617. Mater. Des. **54**, 491–497 (2013)
7. S. Buytoz, Microstructural properties of SiC based hardfacing on low alloy steel. Surf. Coat. Technol. **200**, 3734–3742 (2006)
8. K.A. Khor, L.G. Yu, G. Sundararajan, Formation of hard tungsten boride layer by spark plasma sintering boriding. Thin Solid Films **478**, 232–237 (2005)
9. H.D Solomon, *ASM Handbook Vol 06: Welding, Brazing, and Soldering*. (ASM International, Novelty, 1993), pp. 122–123
10. M. Kulka, A. Pertek, The importance of carbon content beneath iron borides after boriding of chromium and nickel-based low-carbon steel. Appl. Surf. Sci. **214**, 161–171 (2003)
11. O. Ozdemir, M. Usta, C. Bindal, A.H. Ucisik, Hard iron boride (Fe_2B) on 99.97 wt% pure iron. Vacuum **80**, 1391–1395 (2006)
12. I. Campos, M. Palomar, A. Amador, R. Ganem, J. Martinez, Evaluation of the corrosion resistance of iron boride coatings obtained by paste boriding process. Surf. Coat. **201**, 2438–2442 (2006)
13. V. Raghavan, B-Fe-W (boron-iron-tungsten). Phase diagram evaluations: section II. J. Phase Equilib. **24**, 457–458 (2003)
14. S. Islak, S. Buytoz, M. Karagoz, Microstructural development, on AISI, steel by FeW/B_4C composite coating produced by using tungsten inert gas (TIG) process. Indian J. Eng Mater. Sci. **19**, 253–259 (1060)
15. R.L Klueh, *Elevated-temperature Ferritic and Martensitic Steels and their Application to Future Nuclear Reactors*. (Oak Ridge National Laboratory, Oak Ridge National Laboratory, 2004)

Chapter 20
Novel Scintillating Materials Based on Phenyl-Polysiloxane for Neutron Detection and Monitoring

M. Degerlier, S. Carturan, F. Gramegna, T. Marchi, M. Dalla Palma, M. Cinausero, G. Maggioni, A. Quaranta, G. Collazuol and J. Bermudez

Abstract Neutron detectors are extensively used at many nuclear research facilities across Europe. Their application range covers many topics in basic and applied nuclear research: in nuclear structure and reaction dynamics (reaction reconstruction and decay studies); in nuclear astrophysics (neutron emission probabilities); in nuclear technology (nuclear data measurements and in-core/off-core monitors); in nuclear medicine (radiation monitors, dosimeters); in materials science (neutron imaging techniques); in homeland security applications (fissile materials investigation and cargo inspection). Liquid scintillators, widely used at present, have however some drawbacks given by toxicity, flammability, volatility and sensitivity to oxygen that limit their duration and quality. Even plastic scintillators are not satisfactory because they have low radiation hardness and low thermal stability. Moreover organic solvents may affect their optical properties due to crazing. In order to overcome these problems, phenyl-polysiloxane based scintillators have been recently developed at Legnaro National Laboratory. This new solution showed very good chemical and thermal stability and high radiation hardness. The results on the different samples performance will be presented, paying special attention to a characterization comparison between synthesized phenyl containing polysiloxane resins where a Pt catalyst has been used and a scintillating material obtained by condensation reaction, where tin based compounds are used as catalysts. Different structural arrangements as a result of different substituents on the main chain have been investigated by High Resolution X-Ray Diffraction, while the effect of improved optical transmittance on

M. Degerlier (✉)
Science and Art Faculty Physics Department, Nevsehir Haci Bektas University, Nevsehir, Turkey
e-mail: meltem.degerlier@lnl.infn.it

S. Carturan, T. Marchi, M. D. Palma, M. Cinausero, G. Maggioni, A. Quaranta and J. Bermudez
INFN-Laboratori Nazionali di Legnaro, Legnaro, PD, Italy

M. D. Palma and A. Quaranta
Department of Industrial Engineering, University of Trento, Trento, Italy

S. Carturan, G. Maggioni and G. Collazuol
Department of Physics and Astronomy, University of Padova, Padova, Italy

E. K. Polychroniadis et al. (eds.), *International Multidisciplinary Microscopy Congress*,
Springer Proceedings in Physics 154, DOI: 10.1007/978-3-319-04639-6_20,

the scintillation yield has been elucidated by a combination of excitation/fluorescence measurements and scintillation yield under exposure to alpha and γ-rays.

20.1 Introduction

A renewed interest in neutron detection has grown up not only in nuclear physics: neutron monitors are requested in nuclear power plants, spallation neutron sources, homeland security, nuclear medicine and materials analysis [1–4].

Organic scintillators, both liquid and solid, are an optimal choice as neutron detectors due to their low Z composition and high concentration of hydrogen atoms [5]. The most common liquids are based on xylene and pseudocumene solvents with the addition of different dyes and suitable wave shifters, so that emission is shifted to the blue region where the most standard photomultiplier tubes can work [6]. Liquid scintillators, suitable for large volumes and pulse shape analyses, display high quantum efficiency and fast response, though they are highly toxic and flammable [5]. Very recently, linear alkylbenzene (LAB) has been used as aromatic solvent for new liquid scintillators, based on its lower toxicity and volatility, though the issues of high flammability and waste disposal are still critical [7, 8].

On the other hand, plastic scintillators, based on polyvinyltoluene (PVT), though displaying high quantum efficiency, are characterized by low radiation hardness and unfitness for pulse shape discrimination.

PSS-based scintillators have been object of deep investigation, owing to their intrinsic superior chemical and physical properties such as thermal and radiation resistance [9–11].

The aim of the present study is to review recently obtained results in this field and present newly developed formulation of polysiloxanes, which lead to further enhancement in the overall performances of the scintillators, preserving the demonstrated capability to detect both fast and thermal neutrons.

20.2 Materials and Methods

In general, PSS can be vulcanized by addition polymerization, where a vinyl terminated resin reacts with a Si-H containing resin in presence of a Pt based catalyst. As a result of complete cross-linking, the structure is robust, non-sticky though elastic, able to withstand severe changes of temperature. Previous works on PSS formulations and cross-linked by Pt catalyzed addition report in detail the synthetic procedures [9–11].

Still another pathway to achieve vulcanized siloxanes consists in a condensation process, where a base siloxane resin, with terminal hydroxyl groups-OH reacts with a small quantity of cross-linker polydiethoxysilane. The two components are mixed in the weight ratio 100:10, in presence of a tin or titanium based catalyst (in this

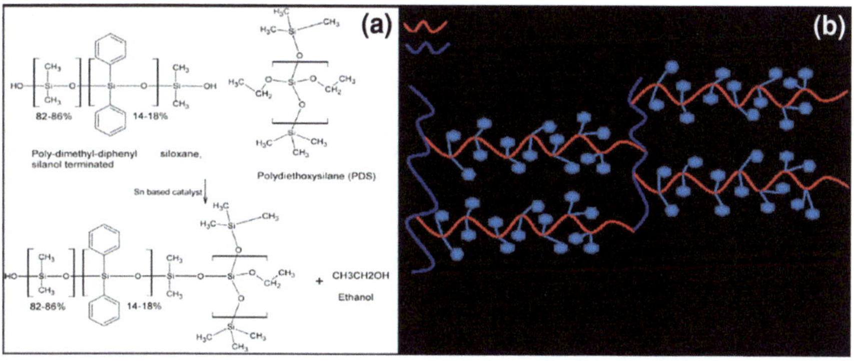

Fig. 20.1 **a** Reaction scheme and components used for the preparation by Sn-catalyzed condensation of the PSS scintillators and **b** schematic drawing of the cross-linked siloxane

case dibutyltindilaurate DBTL was adopted, 1.3 % wt.), as depicted in Fig. 20.1a, whereas in Fig. 20.1b a sketch of the structural arrangement after vulcanization has been figured out.

The use of Sn based catalyst instead of Pt can improve the scintillation efficiency, since Pt can interact with dye molecules, with negative consequences on their optical properties [12] and can easily form Pt nanoclusters, which absorb light in the blue region and, therefore, decrease the light yield of the scintillator.

The condensation reaction was carried out after addition to the silanol terminated resin of the proper amount of either 2,-diphenyloxazole (PPO) or 2-(4-tert-Butylphenyl)-5-(4-phenylphenyl)-1,3,4-oxadiazole (butyl-PBD) as primary dyes, whereas Lumogen Violet (LV, BASF) has been chosen as wave shifting dye. For selected formulations the addition of o-carborane ($C_2B_{10}H_{12}$, Katchem Ltd) was also performed, as described in Table 20.1. The obtained cylindrical samples appeared as perfectly transparent and mechanically stable and firm. According to the details given in Table 20.1, these samples have been labeled as A-cond18, while labels A22 and A100 are referred to samples obtained from the traditional addition Pt-catalyzed reaction, as described in previous papers [9–11, 13].

PPO and butyl-PBD displayed optimal solubility in the aromatic siloxane system, owing to the presence of phenyl substituents along the siloxane chain, which increase chemical affinity with the aromatic structure of both dyes, as can be deduced by their chemical structures (not reported). Concentrations as high as 6 % wt. of PPO have been reached without evident symptoms of precipitation, as reported in Table 20.1. Good dyes dispersion inside the base matrix is a key requirement for the achievement of an organic scintillator since aggregation invariably induces light scattering and yield loss. On the other hand, the good compatibility between the emission range of the primary dye and the excitation maximum of the waveshifter is also crucial in order to shift the final emission in the wavelength range of maximum sensitivity of photo-sensors. In this case, LV absorbs where PPO and butyl-PBD display the maximum emission (320–360 nm) in the UV and re-emits in the violet-blue region

Table 20.1 Composition of the PSS obtained from addition and condensation reaction and tested as scintillators in this work

Resin (A)	Dyes (% wt.)		Resin B	Catalyst
A100	LV 0.02 %	PPO: 1–6 % b-PBD: 0.1–0.5 %, 1 %	Methylhydrosiloxane hydride terminated	Pt 1.2 μL/g
A22	LV 0.02 %	PPO: 1–3 % b-PBD: 0.1–0.5 %, 1 %	Methylhydrosiloxane hydride terminated	Pt 1 μL/g
Acond18	LV 0.02 %	PPO: 1–6 % b-PBD: 0.1–0.5 %	Poly-diethoxysilane(PDS)	DBTL 1.3 % wt.
Acond18	LV 0.02 %	PPO: 1% o-carborane: 4–6 %	Poly-diethoxysilane(PDS)	DBTL 1.3 % wt.

(about 430 nm), where the maximum sensitivity of standard PMTs falls. On the other hand, very recently encouraging results have been presented as related to coupling of red-emitting polysiloxane scintillators with less costly and more durable avalanche photo diodes [14].

Scintillation measurements were performed by exciting the samples with a ^{241}Am alpha source (3 kBq, 5.484 MeV) and a ^{60}Co γ-ray source. Pulse height spectra were obtained by coupling the scintillator samples to an H6524 Hamamatsu PMT. The sample yields were compared with that one obtained from a reference EJ-212 plastic scintillator (Eljen Technology Products) in the same experimental conditions [9], whereas for B doped samples the yield was also compared with EJ-254, which contains 5 % wt. of natural boron.

High Resolution X-Ray Diffraction (HRXRD) patterns were recorded using a Philips X'Pert PRO MRD diffractometer equipped with a Cu Kα1 radiation (~8 keV) as the probe.

20.3 Results and Discussion

In Table 20.2 the light yield obtained as described in the experimental section from the newly developed scintillators, cross-linked by condensation, is reported. The value is expressed as light yield percentage with respect to the yield measured from either the standard EJ212 or EJ254. The comparative yields demonstrate that, in the case of condensation samples, very good results can be obtained, with relative light output up to 90 % of commercial plastic scintillators, in the case of ^{nat}B doped samples, even when the doping percentage is higher in our samples (6 % with respect to 5 %). Some previous results under neutron irradiation have been reported in [13].

The results after irradiation with the ^{241}Am α-source (top) and with the ^{60}Co γ-source (bottom) are shown in Fig. 20.2 (top panel) as a function of growing percentage

Table 20.2 Light output (L.O.) for the newly developed condensation scintillators, compared to standard EJ212 or EJ254 scintillators

Base Resin A:B:Sncat (wt. ratio)	Acond 100:20:1.3	Acond 100:20:1.3	Acond 100:20:1.3	Acond 100:20:1.3
PPO (% wt.)	1	1	1	1
LV (% wt.)	0.02	0.02	0.02	0.02
^{nat}B (% wt.)		4%	5%	6%
^{241}Am				
%L.O vs. EJ212	56 ± 16	49 ± 15	42 ± 13	48 ± 15
%L.O vs. EJ254		73 ± 25	63 ± 22	71 ± 24
^{60}Co				
%L.O vs. EJ212	68 ± 15	70 ± 17	61 ± 15	68 ± 15
%L.O vs. EJ254		90 ± 22	79 ± 20	87 ± 19

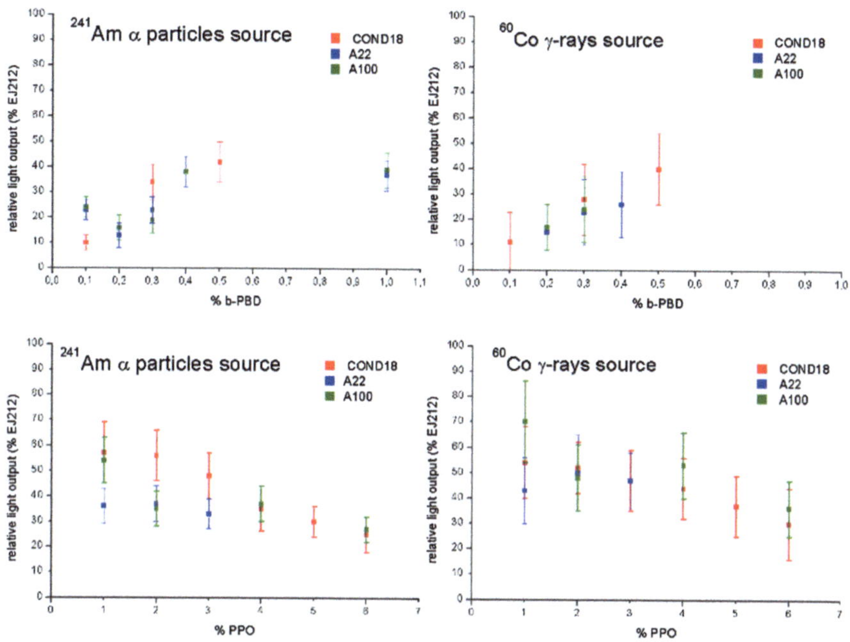

Fig. 20.2 Light yield under α and γ irradiation of samples added with b-PBD (*top panel*) and PPO (*bottom panel*): A18Cond, A100, A22. See Table 20.1 and text for more information

of b-PBD. An increase of the light output up to 0.5 % of b-PBD can be observed, reaching a value similar to that one obtained with the sample at the limit of saturation: over this limit b-PBD evidence slack of solubility. A general better result can be observed for the A18cond samples with respect to the samples made by addition: this is more easily noticed for the 0.3 % percentage case.

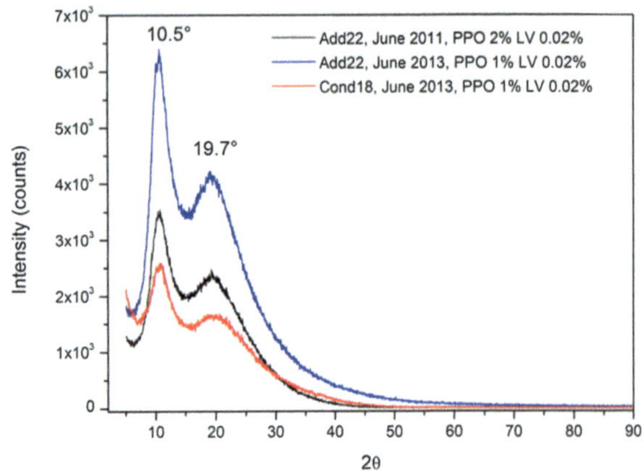

Fig. 20.3 X-ray diffraction spectra of samples, which differ for type and date of synthesis

This may be explained with the higher transparency of the samples prepared by condensation, while a light yellow appearance can be seen in the addition samples, probably due to platinum nanoclusters formation as explained previously.

In Fig. 20.2 (bottom panel) the light output yield is shown for the second series of samples as a function of growing PPO percentage. The percentage of primary dye is in this case much larger than in the case of b-PBD, owing to higher solubility. Samples containing very high PPO concentration (up to 6 %) still display good light yield under exposure to α-particle and γ-rays irradiation (~40 % vs. EJ212). It has to be noticed that only up to 3 % of PPO can be dispersed in the A22 resin, while in A100/A18cond the solubility is still good up to 6 % in weight. PPO showed solubility problems in PSS over 10 % wt. as white residue can be observed mainly on the surface due to the diffusion of saturated PPO.

The results on the Light Output obtained with these samples have similar values of those obtained with samples made two years ago and recently re-measured. On one side reproducible results can be obtained following the synthesis procedure, on the other side negligible ageing effects are observed even at high levels of PPO concentrations, which are useful for PSD [15].

The structure of selected samples has been also investigated by X-ray diffraction (Fig. 20.3): samples A22 were prepared in different periods (June 2011 and June 2013) whereas Cond18 were synthesized by condensation (June 2013). No sharp peaks are evident, indicating the absence of 3D crystalline order in all the samples, as expected. Among the three samples, only negligible changes are evident with the different polymerization routes. Two components are observed: the peak at 19.7° is related to the Si-O amorphous phase of silica-like structure in siloxane, the peak at 10.7° is related to an ordered domain (lamellae with folded chains) without a close crystalline packing.

20.4 Conclusions

In this paper the main results with polysiloxane based scintillators are summarized in order to evidence the good light response under α-particle and γ-rays irradiation as compared to standard plastics. Different primary dyes have been added to both addition and condensation cross-linked siloxanes showing in all cases good solubility, mechanical robustness and overall scintillating performances. In particular, systems based on polycondensation proved to be highly promising as for optical transparency, dye dispersion capability and long-lasting light output. Very high PPO concentrations have been achieved within this systems preserving all the mentioned features: PSD capabilities of these scintillators are actually being tested with both fast and slow neutrons.

References

1. R.T. Kouzes, E.R. Siciliano, J.H. Ely, P.E. Keller, R.J. McConn, Nucl. Sci. Symp. Conf. Rec. **2**, 1115–1119 (2007)
2. L. Ovechkina, K. Riley, S. Miller, Z. Bell, V. Nagarkar et al. Physics Procedia **2**, 161–170 (2009)
3. Z. Bell et al. IEEE Trans. Nucl. Sci. **51**, 1773–1776 (2004)
4. E.V. Van Loef, J. Glodo, U. Shirwadkar, N. Zaitseva, K.S. Shah, IEEE Novel Organic Scintillators for Neutron Detection. Nuclear Science Symposium Conference Record (NSS/MIC). ISSN: 1095–7863 (2010), pp. 1007–1009
5. K.S. Krane, *Introductory Nuclear Physics*. (Wiley, New York, 2008)
6. http://www.hamamatsu.com
7. T. Kogler, R. Beyer, E. Birgersson, R. Hannaske, A.R. Junghans et al., Nucl. Instr. Meth. **A701**, 285–293 (2013)
8. P. Lombardi, F. Ortica, G. Ranucci, A. Romani, Nucl. Instr. Meth. **A701**, 133–144 (2013)
9. A. Quaranta, S. Carturan, T. Marchi, V.L. Kravchuk, F. Gramegna et al., IEEE Trans. Nucl. Sci. **57**, 891–900 (2010)
10. S. Carturan, A. Quaranta, T. Marchi, F. Gramegna, M. Degerlier, M. Cinausero, V.L. Kravtchuk, M. Poggi, Rad. Prot. Dosim. **143**, 471–476 (2011)
11. A. Quaranta, S. Carturan, M. Cinausero, T. Marchi, F. Gramegna, M. Degerlier et al., Mat. Chem. Phys. **137**, 951–958 (2013)
12. M. Buffa, S. Carturan, A. Quaranta, G. Maggioni, G. Della Mea, Opt. Mat. **34**, 1219–1224 (2012)
13. A. Quaranta, S. Carturan, T. Marchi, M. Buffa, M. Degerlier et al., J. Non Cryst. Solids **357**, 1921–1925 (2011)
14. M. Dalla Palma, A. Quaranta, T. Marchi, G. Collazuol, S. Carturan, F. Gramegna, M. Cinausero, *Oral presentation at ANIMMA2013, submitted to Transactions on Nuclear Science*. (2013)
15. T. Marchi et al., in submission

Chapter 21
Structural Transformations in Carbon Black Under the Action of Powerful Energy Beams

M. V. Trenikhin, Yu. G. Kryazhev, O. V. Protasova, I. V. Muromtsev, V. A. Drozdov, V. A. Likholobov, A. D. Teresov, N. N. Koval′ and G. M. Seropyan

Abstract Exposure of carbon black to pulsed electron and laser beams leads to structural transformations of carbon and formation of nanocapsules and rose-like particles with highly ordered graphite like structure.

21.1 Introduction

In recent years, much attention of researchers has been given to structural changes in carbon nanomaterials (CNM) treated with high-power concentrated energy fluxes. Certain of related studies revealed structural transformations of primary carbon particles in CNM on laser or electron beam irradiation.

For example, laser irradiation causes considerable changes in CNM morphology and crystalline structure. It is also observed that primary carbon particles are rapidly heated to temperatures at which carbon atoms and clusters can be sublimated. Fast cooling on pulsed laser irradiation involves crystallization of carbon into diamond- and graphite-like nanostructures [1–3].

M. V. Trenikhin (✉) · O. V. Protasova · I. V. Muromtsev
Institute of Hydrocarbon Processing, Siberian Branch, Russian Academy of Sciences, 54, Neftezavodskaya st., Omsk, Russia 644040
e-mail: tremv@yandex.ru

Y. G. Kryazhev · V. A. Drozdov · V. A. Likholobov
Omsk Scientific Center, Siberian Branch, Russian Academy of Sciences, 54, Neftezavodskaya st., Omsk, Russia 644040

A. D. Teresov and N. N. Koval′
Siberian Branch, Russian Academy of Sciences, Institute of High Current Electronics, Tomsk, Russia 634055

G. M. Seropyan
F. M. Dostoevsky State University, Omsk, Russia 644077

E. K. Polychroniadis et al. (eds.), *International Multidisciplinary Microscopy Congress*, Springer Proceedings in Physics 154, DOI: 10.1007/978-3-319-04639-6_21,

Some researchers studied structural transformations of nanotubes by an electron beam in-situ with an electron microscope [4]. The electron beam, as shown by transmission electron microscopy, causes globular carbon to transform into onion-like structures whose subsequent heating to 700 °C results in nanodiamonds [5].

Recent papers report that pulsed electron beam irradiation of carbon black (CB) causes transformation of nanoglobular carbon into rose-like particles with highly ordered graphene layers [6].

The present paper reports on comparative study of structural transformations of carbon black on irradiation by submillisecond pulsed electron beams with an energy density of 10–50 J/cm^2 and on pulsed laser irradiation with an energy density of ~2 and 6000 J/cm^2.

21.2 Experiments

In experiments, we used carbon black with high dispersion (particle diameter 10–30 nm) and low dispersion (particle diameter 100–300 nm).

The primary CB particles consist of planar or curved structural elements. The latest advances in electron microscopy make it possible to describe carbon black on its lattice scale and to obtain statistical data on the length and curvature of individual graphene layers as well as on the interlayer spacing (d_{002}).

We analyzed the lattice fringe (graphene length) and calculated the degree of its curvature using the Gatan Digital Micrograph software package. The fringe curvature was determined as the ratio of spacing between two fringe endpoints to fringe length [7].

The carbon materials were irradiated with a low-energy (20 keV) pulsed electron beam at an energy density of 10–50 J/cm^2 (pulse duration 100 μs, number of pulses up to 100, electron beam diameter 2 cm) and with a Nd:YAG laser at a wavelength of 1065 nm and energy density of 2 and 6000 J/cm^2 (laser beam diameter 6 and 0.1 mm, respectively). The experimental arrangement and procedure used for laser and electron beam irradiation was described in the previous papers [6, 8]. The internal structure of primary CB particles was examined by transmission electron microscopy on a JEM 2100 JEOL device.

The crystallinity of the irradiated CB materials was determined by analysis of electron diffraction patterns and fast Fourier transform (FFT) of electron microscopic images. Next, FFT data processing was applied to determine the spacing between graphene layers. The average spacings for the initial and irradiated carbon samples were determined through construction of radial intensity profiles by the FFT method [9].

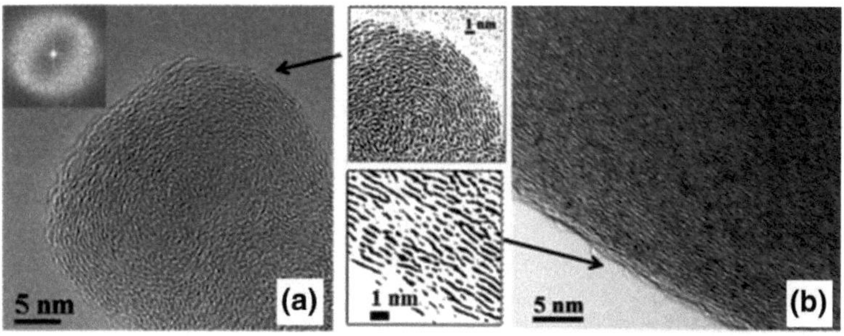

Fig. 21.1 Structure of the primary particles: high dispersive carbon black (**a**), low dispersive carbon black (**b**)

21.3 Results and Discussion

21.3.1 Initial Carbon Black

According to the TEM data obtained, the initial CB powder consists mostly of regular spherical particles or globules. They have an amorphous structure with low ordered graphene layers spaced by 0.37–0.40 nm (Fig. 21.1).

Using the Digital Micrograph software, we determined the length of each fringe and constructed histograms for these data.

Figure 21.2 show the fringe length and curvature distributions. For highly dispersed particles, the length of graphene layers is 0.4–1.0 nm. For low dispersed particles, the fringe length is somewhat longer and is 0.4–2.0 nm. The curvature of graphene layers for highly dispersed particles is higher than that for low dispersed particles.

21.3.2 Electron Beam Irradiation

It is found that on electron beam irradiation, particles of differing morphology are formed depending on the degree of exposure and dispersion of the initial samples. Increasing the energy density decreases the graphene layer spacing to values characteristic of the graphite structure.

On electron beam irradiation with a relatively low energy density (10–20 J/cm^2), such particles escape detection and disordering (amorphization) of the external part of initial globules is observed. At an energy density of 25–30 J/cm^2, hybrid structures with a rose-like central part and amorphous outer shell are formed.

Electron beam irradiation of the low dispersed carbon black causes structural changes in the carbon material. Examinations show that there arise rose-like structures of diameter 100–500 nm which contain a series of parallel graphene layers spaced by 0.36 ± 0.01 nm.

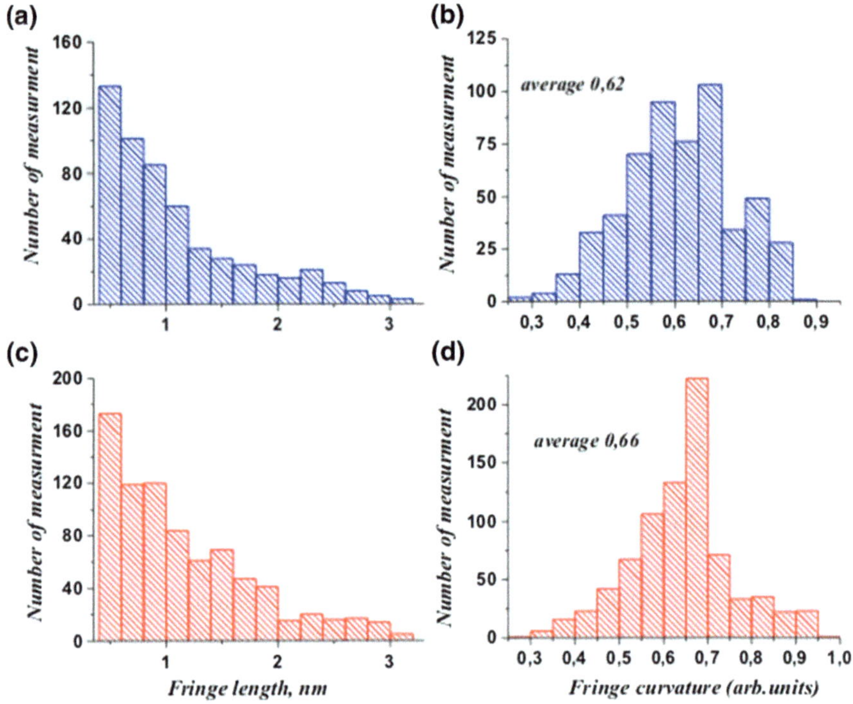

Fig. 21.2 Fringe length distribution of the high dispersive (**a**) and low dispersive (**b**) carbon black, fringe curvature distribution of the high dispersive (**c**) and low dispersive (**d**) carbon black

Electron beam irradiation of the highly dispersed samples at an energy density of 50 J/cm^2 also leads structural changes, but the morphology differs from that described above. In this case, nanocapsules of diameter 2–20 nm are formed and the spacing between graphene layers is 0.37 ± 0.01 nm.

21.3.3 Laser Irradiation

The TEM data obtained for laser irradiated carbon black also reveal deep structural transformation of low dispersed CB samples into rose-like particles (Fig. 21.3a). The TEM images taken after irradiation at an energy density of 6000 J/cm^2 demonstrate the presence of highly ordered graphene layers spaced by 0.35 nm.

Using digital image processing, we calculated the spacing between graphene layers in the initial and irradiated samples. Figure 21.4a shows the radial intensity profile for low dispersed carbon black irradiated at an energy density of 6000 J/cm^2. The spacing between graphene layers is 0.38 nm for the initial samples, 0.36 nm for the samples irradiated at an energy density of 2 J/cm^2, and 0.35 nm for those irradiated

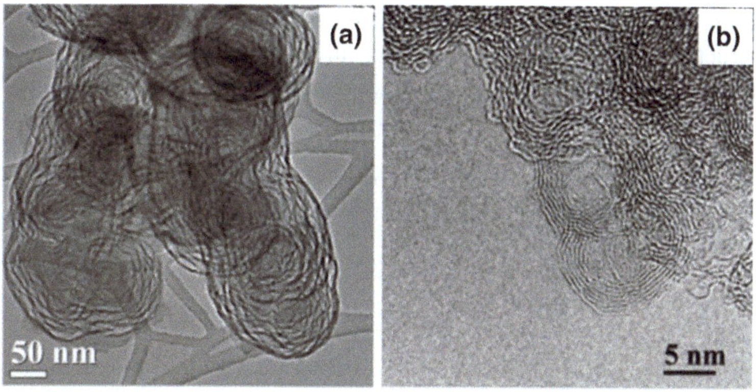

Fig. 21.3 Morphology particles of the carbon black after laser irradiation: low dispersive (**a**) and high dispersive samples (**b**)

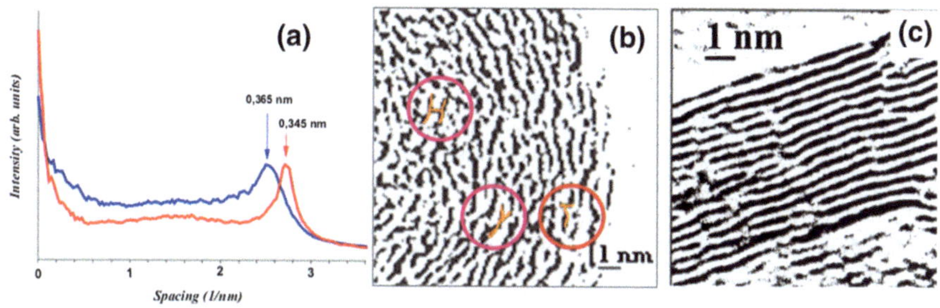

Fig. 21.4 Radial intensity profile samples after irradiation with energy density 6000 J/cm^2 (the spacing between graphene layers 0.345 nm) and 2 J/cm^2 (the spacing between graphene layers 0.365 nm) (**a**); localization places sp^3 bonded carbon atoms in initial CB (**b**); graphene layers structure after irradiation (**c**)

at 6000 J/cm^2. It is seen that increasing the radiant energy results in particles with a graphite-like structure.

TEM examination of the highly dispersed CB samples after laser irradiation reveals the formation of nanocapsules 10–30 nm in diameter with a shell consisting of 5–10 graphene layers (Fig. 21.3b). Electron diffraction measurements show that the average spacing between graphene layers in such particles is 0.355–0.365 nm.

The initial CB particles consist of two types of hybridized carbon atoms. According to the results reported by Muller [7], the primary CB particles contain 77 % of sp^2-bonded and 23 % of sp^3-bound carbon atoms. In this case, the sp^3-bonded carbon can be localized in those places in which there are graphene layer structures similar to the letters 'Y', 'T' or 'H' (Fig. 21.4b). During the irradiation, fast heating to a temperature of 4500 °C occurs, which is typical of carbon plasma. This causes transformation of sp^3-bonded carbon into sp^2-bonded carbon and results in particles with a graphite-like structure (Fig. 21.4c).

Thus, it is shown that pulse electron beam and laser irradiations of carbon black cause structural transformations similar to both types of exposure with the formation of either nanocapsules or rose-like particles depending on the particle diameter and structural organization of graphene layers in the initial carbon nanoglobules.

Acknowledgments The work was partially supported by RFBR grant No. 12-08-31478-mol_a.

References

1. H. Bladh, J. Johnsson, P.E. Bengtsson, On the dependence of the laser-induced incandescence (LII) signal on soot volume fraction for variations in particle size. Appl. Phys. B **90**, 109–125 (2008)
2. L.R. Vander Wal, M.Y. Choi, Pulsed laser heating of soot: morphological changes. Carbon **37**, 231–239 (1999)
3. S. Hu, P. Bai, F. Tian, S. Cao, J. Sun, Hydrophilic carbon onions synthesized by millisecond pulsed laser irradiation. Carbon **47**, 876–883 (2009)
4. F. Banhart, Irradiation of carbon nanotubes with a focused electron beam in the electron microscope. J. Mater. Sci. **41**, 4505–4511 (2006)
5. L. Dai (ed.), *Carbon Nanotechnology: Recent Developments in Chemistry, Physics, Materials Science and Device Applications* (Elsevier, Dayton, 2006), p. 733
6. YuG Kryazhev, N.N. Koval', V.A. Likholobov, A.D. Teresov, V.A. Drozdov, M.V. Trenikhin, Tech. Phys. Lett. **38**(4), 301–303 (2012)
7. J.O. Müller, D.S. Su, U. Wild, R. Schlögl, Bulk and surface structural investigations of diesel engine soot and carbon black. Phys. Chem. Chem. Phys. **9**, 4018–4025 (2007)
8. M.V. Trenikhin, O.V. Protasova, G.M. Seropyan, V.A. Drozdov, Structural transformations of technical carbon under the action of nanosecond laser radiation. Chem. Sustain. Dev. **1**, 109–114 (2013)
9. K. Oshida, T. Nakazawa, T. Miyazaki, M. Endo, Application of image processing techniques for analysis of nano- and micro-spaces in carbon materials. Synth. Met. **125**, 223–230 (2002)

Chapter 22
Investigation of Microstructural Properties of $Sm_{0.1}Ce_{0.9}O_{1.95}$ Powders

Aliye Arabacı

Abstract Ceria doped with rare earth oxides such as Gd_2O_3, Sm_2O_3 has been considered as one of the most promising candidate materials as electrolyte materials for intermediate temperature solid oxide fuel cells. Samarium doped ceria, $Sm_{0.1}Ce_{0.9}O_{1.95}$ (SDC), powders are successfully prepared by Pechini method using ethylene glycol, citric acid, cerium nitrate and samarium nitrate as the starting materials. Crystal structure and microstructure were characterized by means of X-ray diffraction (XRD) and scanning electron microscopy (SEM). The sintered SDC samples at 1400 °C had dense structure (98 % of the theoretical density).

22.1 Introduction

Compare to the traditional stabilized zirconia, ceria doped with aliovalent cations have been one of the most preferred materials for intermediate temperature solid oxide fuel cells (SOFCs) due to their higher ionic conductivity at lower temperatures. For example, rare-earth additions such as Gd_2O_3 or Sm_2O_3-doped ceria was found to have the highest conductivity [8, 14]. Samarium doped ceria (SDC) and Gadolinium doped ceria (GDC) have higher ionic conductivity than yttria-stabilised zirconia (YSZ) when compared at the same operating temperature. However, SDC and GDC are less redox stable than YSZ and begin to show significant electronic conductivity due to the variable oxidation state of cerium at higher temperatures, especially under reducing conditions. This creates a problem when operating at temperatures of 800–1000 °C. However, ceria-based ceramics are proposed as promising electrolytes for intermediate-temperature SOFCs (IT-SOFCs), working at low temperature (around 500–700 °C). Ceria and doped ceria materials are used in other

A. Arabacı (✉)
Faculty of Engineering, Department of Metallurgical Engineering, Istanbul University, 34320 Avcilar, Istanbul, Turkey
e-mail: aliye@istanbul.edu.tr

E. K. Polychroniadis et al. (eds.), *International Multidisciplinary Microscopy Congress*, Springer Proceedings in Physics 154, DOI: 10.1007/978-3-319-04639-6_22,

high-temperature applications, such as three-way automotive catalysts, gas sensors and ceramic oxygen generators, where their electronic conductivity is less problematic and even advantageous [5, 6, 11].

To synthesize ceria-based electrolytes with the desired properties, various methods such as hydrothermal synthesis [4, 15], homogeneous precipitation [3, 10], the sol-gel process [12], the glycine-nitrate process [13], combustion [7] and the reverse microemulsion process [2] have been employed.

In the present work, Pechini method was used to obtain homogeneous nanocrystalline $Ce_{(1-x)}Sm_xO_{2-\delta}$ (x = 0.10) precursor. And Crystal structure and microstructure were characterized by means of X-ray diffraction (XRD) and scanning electron microscopy (SEM). The relative densities of the pellets, which were sintered at 1400 °C, were 98 %. This result clearly indicates that SDC powder with adequate density can be prepared through the Pechini process at low temperatures.

22.2 Experimental

22.2.1 Powder Synthesis

Cerium nitrate ($Ce(NO_3)_3 \times 6H_2O$, 99 %, Aldrich) and samarium nitrate ($Sm(NO_3)_3 \times 6H_2O$, 99.9 %, Aldrich) were used as metal precursors and ethylene glycol (R.P. Normopur), citric acid (Boehringer Ingelheim) were selected for the polymerization treatment. Cerium nitrate and samarium nitrate salts were dissolved in de-ionized water individually and then the solutions were mixed in a beaker. Anhydrous citric acid was dissolved in de-ionized water and then was added to the cation solution. The molar ratio of total oxide (TO): citric acid (CA) and ethylene glycol: citric acid was selected as 2:1, 4:1, respectively. More detailed synthesis information were given our previous work [1].

22.2.2 Density Determination

The calcined powders were pressed with a cold isostatic press (CIP) at 95 MPa to obtain a disc-shaped test sample. The compact disc samples were then sintered at different temperatures (1200 and 1400 °C) for 6 h. The densities of the sintered discs were determined by Archimedes' method:

$$D_{pellet} = W.\rho / W_1 - W_2$$

where W is the dry weight, W_1 is the wet weight, W_2 is the body's submerged weight without fine wire and ρ is the density of water (water at 25 °C with a density of 0.997 $g{\cdot}cm^{-3}$). After determination of the bulk density, the relative densities of the samples were calculated by dividing the experimental bulk density values by the theoretical densities. The theoretical densities of the samples were taken as 7.21 $g{\cdot}cm^{-3}$ for SDC which were calculated by using the lattice parameters obtained from the XRD analysis.

22.2.3 X-ray Analysis

The XRD technique was used to examine the crystalline structure and phase purity. Calcined powders were investigated by XRD using a Rigaku D/max-2200 Ultima X-ray diffractometer with CuKα radiation. Data in an angular region of $2\theta = 10$–$90°$ were collected.

22.2.4 Scanning Electron Microscopy (SEM) Analysis

The scanning electron micrographs were obtained by using FEI Quanta FEG scanning electron microscopy to identify the morphology and microstructure of the calcined samples and sintered samples.

22.3 Results and Discussion

22.3.1 Powder Characterization

The XRD patterns for the SDC powders that were calcined at 400, 500 and 1000 °C are shown in Fig. 22.1.

As shown in Fig. 22.1 for powders obtained from gel, initial crystallization into fluorite structure have taken place for gel after heating at 400 °C for 2 h, which is the relatively low calcination temperature for the fluorite structure formation. The XRD patterns obtained after heating the gels sequentially to higher calcination temperatures are shown in Fig. 22.1. Heating the gels from 400 to 500 °C has little effect on peak width and intensity, indicating little change in crystallite size. However, after heating at 1000 °C, the peaks' widths become very narrow and intense, indicating rapid increase in the growth of the crystallite size.

The average crystallite sizes was calculated from the degree of line broadening in the main (111) diffraction peak using by the Scherrer formula from XRD data. The average particle size was about 10.8, 11.7 and 51 nm for powders calcined at 400, 500 and 1000 °C, respectively.

Electron microscopy was used to study the morphology of the SDC nanopowders. Figure 22.2 shows the microstructures of calcined powders at 400 and 1000 °C.

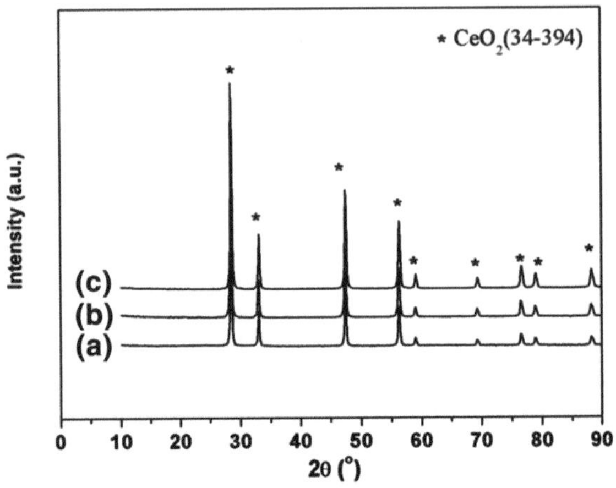

Fig. 22.1 XRD patterns of SDC powders calcined in air for 2 h at **a** 400 **b** 500 and **c** 1000 °C, respectively

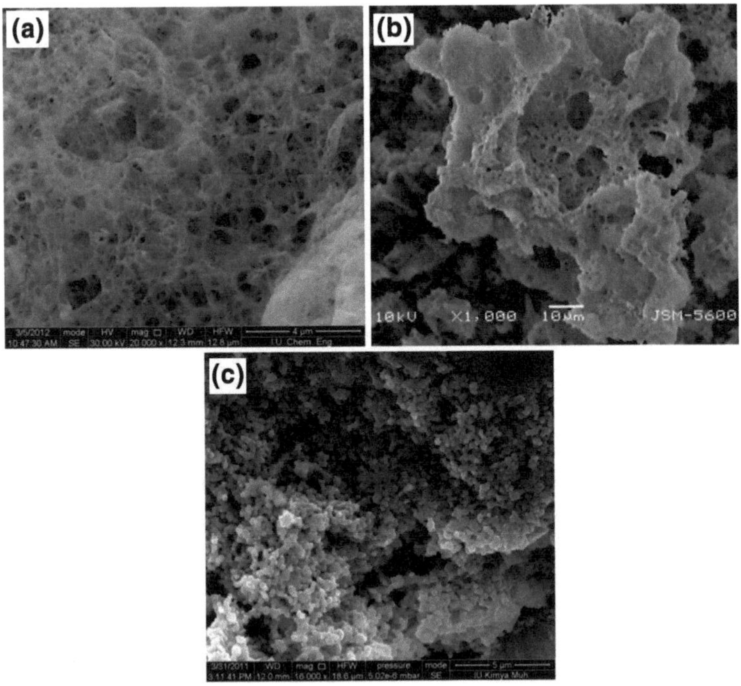

Fig. 22.2 Low and high magnification SEM photographs of SDC powders calcined at 400 °C (**a**), at 1000 °C (**b**) for 2 h and 1200 °C (**c**) for 6 h

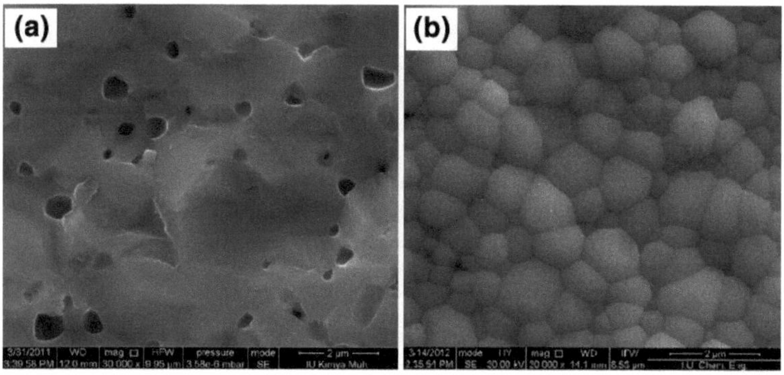

Fig. 22.3 3 SEM photographs of fracture surface (**a**) and the surface (**b**) of SDC pellet sintered at 1400 °C for 6 h

Sponge-like structures were observed for calcined SDC powders after calcination at 400, 1000 °C for 2 h. Similar structures were obtained by Baker [9] for the 10SDC composition. after calcination at 500 °C for 2 h, the highly porous morphologies of SDC samples calcined at 400 and 1000 °C is very likely to be caused by the large volumes of gas produced in the preparation method. During the calcination step, due to the the combustion process, a large amount of gaseous materials was evolved and hence the combustion product was highly porous as shown in the SEM micrographs (Fig. 22.2a, b).

From the SEM morphologies of SDC powders shown in Fig. 22.2c, It can be seen that the particles of the calcined sample at 1200 °C were agglomerated and irregular. The resultant SDC powders were pressed into a pellet of 8.6 mm diameter and 0.76 mm thickness under a cold isostatic press (CIP) a pressure of 200 MPa, and then sintered at 1400 °C for 6 h. The microstructures of the SDC ceramics are shown in Fig. 22.3. The Fig. 22.3a is the SEM image of the pellet fracture surface and the Fig. 22.3b is its pellet surface. It can be seen that the SDC pellet is nearly fully dense with very few residual pores. The well-developed grain boundary and the very low porosity are typical micro-structural features of a ceramic in the final stage of the sintering. The calculated relative density for sintered at 1400 °C samples is 98.0 % of the theoretical value.

22.4 Conclusions

CeO_2 powders doped with 10 mol % of samarium have been synthesized via Pechni process. This method is a low-cost and relatively simple method. Moreover It needs a relatively low calcination temperature compare to other hydrothermal treatments. By this method, the single-phase fluorite structure was obtained. The pechini processing of samarium and cerium nitrate precursors can be used for the production of nanocrystalline SDC powders after calcination at 400 °C. After sintering, the nano-sized samarium-doped ceria pellets achieved 98 % of the theoretical density at 1400 °C. the SDC pellet is nearly fully dense with very few residual pores Average grain size of SDC samples obtained from SEM images at sintered temperature of 1400 °C was found as 1.31 μm.

References

1. A. Arabacı, F.M. Oksüzomer, Preparation and characterization of 10 mol % Gd doped CeO_2 (GDC) electrolyte for SOFC applications. Ceram. Int. **38**, 6509 (2012)
2. J. Chandradass, B. Nam, K.H. Kim, Fine tuning of gadolinium doped ceria electrolyte nanoparticles via reverse microemulsion process. Colloids Surf. A **348**, 130–136 (2009)
3. H.I. Chen, H.Y. Chang, Homogeneous precipitation of cerium dioxide nanoparticles in alcohol/water mixed solvents. Colloids Surf. A **242**, 61–69 (2004)
4. M. Hirano, E. Kato, Hydrothermal synthesis of cerium (IV) oxide. J. Am. Ceram. Soc. **79**(3), 777–780 (1996)
5. J.-Y. Park, E.D. Wachsman, Lower temperature electrolytic reduction of CO_2 to O_2 and CO with high-conductivity solid oxide bilayer electrolytes. J. Electrochem. Soc. **152**, A1654–A1659 (2005)
6. C. Pijolat, G. Tournier, J.P. Viricelle, Detection of CO in H_2-rich gases with a samarium doped ceria (SDC) sensor for fuel cell applications. Sens. Actuators B: Chem. **141**(1), 7–12 (2009)
7. M.M.A. Sekar, S.S. Manoharan, K.C. Patil, Combustion synthesis of fine-particle ceria. J. Mater. Sci. Lett. **9**, 1205 (1990)
8. B.C.H. Steele, in *High Conductivity Solid Ionic Conductors: Recent 18 Trends and Applications*, ed. by T. Takahashi (World Scientific, New Jersey, 1989), p. 402
9. M.R. Kosinski, R.T. Baker, Preparation and property-performance relationships in samarium-doped ceria nanopowders for solid oxide fuel cell electrolytes. J. Power Sources **196**, 2498–2512 (2011)
10. N. Uekewa, M. Ueta, Y.J. Wu et al., Synthesis of CeO_2 spherical fine particles by homogeneous precipitation method with polyethylene glycol. Chem. Lett. **8**, 854–855 (2002)
11. J.B. Wang, W.-H. Shih, T.-J. Huang, Study of Sm_2O_3-doped CeO_2/Al_2O_3-supported copper catalyst for CO oxidation. Appl. Catal. A General **203**, 191–199 (2000)
12. G.S. Wu, T. Xie, X.Y. Yuan et al., An improved sol-gel template synthetic route to large-scale CeO_2 nanowires. Mater. Res. Bull. **39**, 1023–1028 (2004)
13. C.R. Xia, M.L. Liu, Microstructures, conductivities and electro-chemical properties of $Ce_{0.9}Gd_{0.1}O_2$ and GDC-Ni anodes for low- temperature SOFCs. Solid State Ionics **152**, 423 (2002)
14. H. Yahiro, Y. Eguchi, K. Eguchi et al., Oxygen ion conductivity of the ceria-samarium oxide system with fluorite structure. J. Appl. Electrochem. **18**, 527–531 (1988)
15. Y.C. Zhou, M.N. Rahman, Hydrothermal synthesis and sintering of ultrafine CeO_2 powders. J. Mater. Res. **8**, 1680 (1993)

Chapter 23
Alteration Phenomena on Surfaces of Stained Glasses by Masters From XIII to XV Century

E. L. Algarra Cruz, N. Barbier, S. Bruni, G. Maino, G. Marghella, E. Papalazarou and L. Volpe

23.1 Introduction

Stained glasses industries have shown a great diversity through the Europe and also through the times. Nevertheless all fabrication recipes share a common base: stained glasses result from a metastable state, the instantly cooling of a sand fusion liquid phase. In addition to silicate, that play the role of a network former, fluxes coming from vegetal ashes (K, Na, Mg) and traces element (Fe, Al, Mn, Pb) are added in order to modify texture, optic properties and mechanic response of the final product [1].

As far as stained glasses specificities, their behaviors in atmospheric environment, humidity, temperature variations, acid rain, etc. are concerned; stress exposition remains a complex issue because alteration phenomena are directly in relation with nature of materials used for the fabrication. Our study is devoted to the identification of alteration processes at the stained glasses' surface and to support conservation and restoration work that requires a higher visibility on factor which influences the state of conservation for such vulnerable systems [2]. We present in the following part the application of SEM/EDX for the characterization of stained glass alteration.

E. L. A. Cruz · S. Bruni · G. Maino · G. Marghella (✉)
ENEA-UTSISM and Laboratorio Laerte, Via Martiri di Montesole 4, 40128 Bologna, Italy
e-mail: giuseppe.marghella@enea.it

E. L. A. Cruz · G. Maino
Scuola di Lettere e Beni Culturali, Università di Bologna, office of Ravenna, Via Mariani 5, 48100 Ravenna, Italy

N. Barbier · E. Papalazarou
Département de Physique, Université Orsay Paris Sud 11, Paris, France

L. Volpe
Tecnopolo dell'Università di Ferrara, Dipartimento di Fisica e Scienze della terra, Università di Ferrara, Via Saragat 1, 44122 Ferrara, Italy

E. K. Polychroniadis et al. (eds.), *International Multidisciplinary Microscopy Congress*, Springer Proceedings in Physics 154, DOI: 10.1007/978-3-319-04639-6_23,

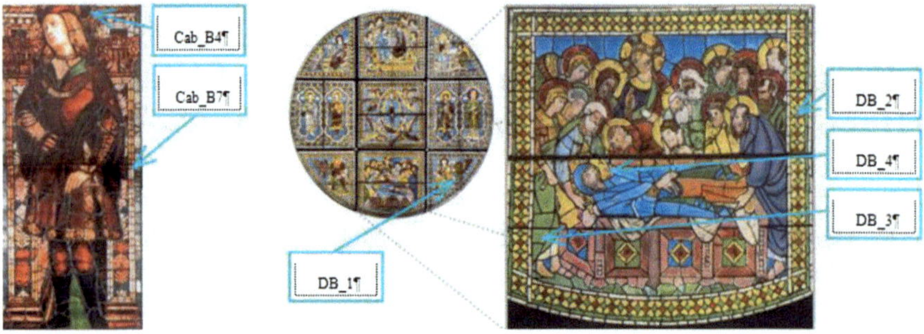

Fig. 23.1 Sampling points on Saint Petronio basilica window by the Cabrini (particular, on the *left*) and on the window by Duccio di Buoninsegna in Siena Cathedral (on the *right*)

23.2 Material and Methods

Broken fragments of glasses with different provenance, fabrication period, morphologies and state of preservation have been compared. Samples were obtained during the restoration campaigns of two famous Italian churches: Siena Cathedral (Fig. 23.1a) and St. Petronio Basilica in Bologna (Fig. 23.1b). Fours samples: DB_1, DB_2, DB_3 and DB_4 were taken from the rosette designed by the artist Duccio di Buoninsegna (1287). These samples present a wide range of different colors, and also an unusual good state of conservation [3, 4]. The second set of three samples, Cab_B4, Cab_B7_ext, Cab_B7_int, come from the St. Petronio window designed by a Master family: the Cabrini (XV). They present a great interest because of their limited color range which includes also samples with a grisaille layer; Cab_B4 sample shows an advanced corrosion state.

Cross sections of each sample were analyzed in laboratory UTSISM, ENEA, first by reflected light optical microscope REICHERT-JUNG-MF3 then by SEM Quanta Inspect S and for chemical composition by EDX Falcon. Complementary XRF chemical characterization, in order to improve heavy element detection on stained glasses surface, was realized with a PANanalytical Axios Fast instrument, in University Paris-Sud 11, Orsay. For these tests, samples' surfaces were exposed without any preliminary preparation, so that we obtain a spectra average on the whole sample surface.

SEM analyses on each cross-section give us some elements on structural modification due to alteration. Two samples coming from different sets have shown an evidence of degradation, namely Cab_B4 and DB_2. Altered cross-sections presented on Fig. 23.2a, b let us distinguish two processes. The first one, visible in Fig. 23.2a, induce localized loss of matter but apparently there is no modification of the glass layer. The other process is acting not only at the interface stained glass/athmosphere but modifies in depth the sane glass layer (modified layer thickness varies from 73 to 94 μm) inducing morphological changes and apparition of fractures in the network.

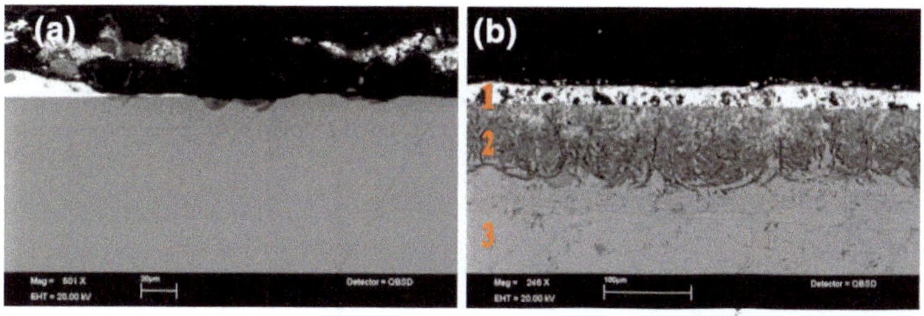

Fig. 23.2 SEM image of the Cabrini sample Cab_B4 (**a**) and of the Duccio di Boninsegna sample DB_2 (**b**)

Fig. 23.3 Porous structure of the modified layer of sample DB2 by SEM (magnification: 800x)

Thereafter, EDX and XRF analyses were performed on these samples in order to identify causes and chemical modifications related to the precedent observations.

DB2 sample (Fig. 23.2b) presents three layers: (1) heterogeneous and punctually lacunar grisaille, (2) morphologically modified layer, less dense in direct contact with the inner layer; (3) the sane glass. Higher magnification, on Fig. 23.3, underlines the porous structure of the second layer that is organized in adjoining circles with inclusion of sane glasses at the interface. Also optical microscope analysis has shown a fine, opaque modified layer. This morphology seems to be an evidence of a leaching process [5, 6]; named selective dissolution. Chemical characterization was performed on the modified layer and compared to the sane glasses composition.

EDX analysis confirmed that also chemical composition is really different in the two considered cases. Leached layer is impoverished in potassium and enriched in silicate; this depends on potassium cation diffusion toward the glass surface (Fig. 23.4).

This sample has a K-Ca high concentration, which explains why it is particularly subject to the alteration. Leaching is the result of a pH modification due probably to rain and condensed water forming at the surface. Hydrogen diffuses from the water into the glass matrix and substitutes to modifiers cations present in the silica

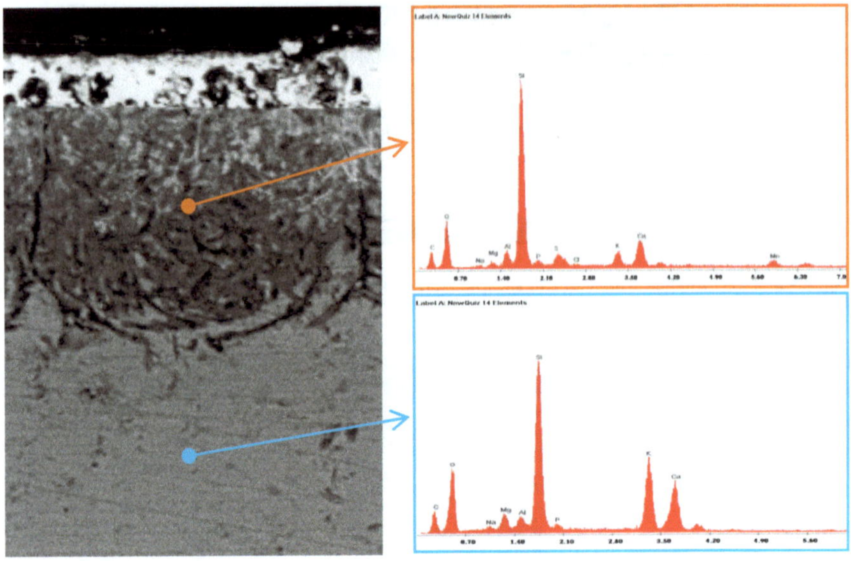

Fig. 23.4 EDX microanalysis on sample DB2 on modified layer (*orange*) and sane glass (*blue*)

network (Alkali and Alkaline-earth elements). These cations diffuse toward the glass surface and escape from this layer, which thus remains impoverished in modifiers and enriched in silicate [7]. This modified layer presents an arborescence aspect and a system of fractures parallel to the surface. This morphology reveals a radial alteration front [5, 6]. Leaching process begins in isolated points at the interface, and progresses on radial front which finally joined each other. This behavior depends on the presence of a homogeneous layer of grisaille at the interface. Water from rain or humidity interacts first with the grisaille layer and when punctually this layer is entirely degraded, leaching process begins. Grisaille layer acts as a protection layer which slowed the alteration process and can explain that it progresses heterogeneously. Porous structure and many fractures observed can be linked to the diffusion process that weakened the silicate networks and created surface tensions.

Sample Cab_B4 presented an advanced state of corrosion. Morphological analysis of Cab_B4 shows a crust on the glass surface that could be directly related to corroded zones; the surface reveals a high rugosity, where grisaille layer and glass are heterogeneously corroded, while sane glass presents wells dug (Fig. 23.2a). The XRF analyses on Cab_B4 surface deposits reveal an high concentration in potassium, calcium and iron. Trace elements as manganese, cobalt, copper, titanium and chrome are responsible for the deep blue color of this sample.

Iron and lead are principal component of grisaille, but in this case iron is present in a higher concentration compared to other samples with grisaille, a fact which can explain the red color of deposits. It is worth noticing also the Kα line of sulfur on Fig. 23.5: presence of sulfur, and a high level of calcium and potassium reveal probably neocrystallization products as gypsum ($CaSO_4{\cdot}2H_2O$) and potassium carbonate

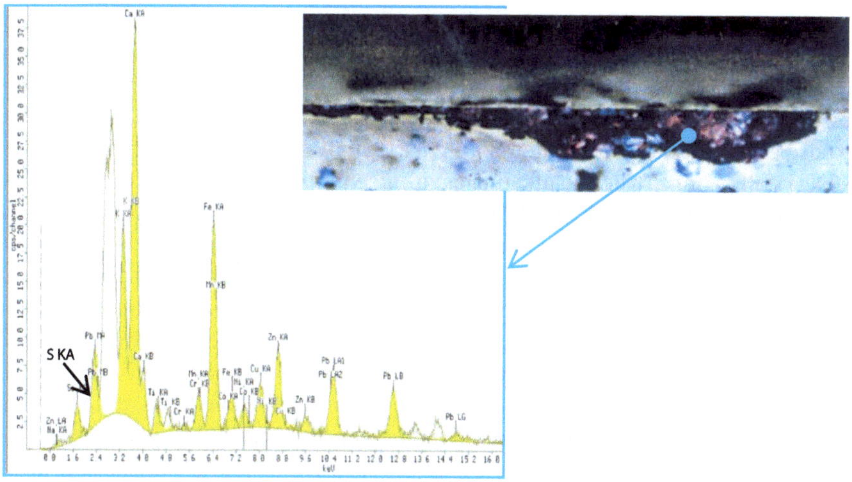

Fig. 23.5 XRF spectra acquired on the Cabrini Cab_B4 sample surface

(K_2CO_3), products due to atmospheric pollution. Previous studies [8] have shown that potassium carbonate has a dramatic impact on stained glasses surface since it induced corrosion process of the network in contact with an alkaline environment.

The other samples of the two sets, Cab_B7_ext, Cab_B7_in, DB_1, DB_3 and DB_4 present no evidence of alteration and a homogeneous, intact grisaille layer. XRF analyses on grisaille surface reveal a high level of lead, but also iron is found, as well as an important diversity of other trace elements as manganese, zinc and chrome. Glass layers on Cab_B7_ext and Cab_B7-in characterizations reveal high concentration of silica and high level of sodium and calcium. This composition characterizes high durability stained glasses. The silicate network is stabilized by a rich composition in network former and by the presence of calcium [5]. The sodium does not play the same role as potassium in the process described before; in fact the potassium is less linked to oxygen atoms in the glass network, so that it is more subject to diffusion in the leaching process. Same observations can be made for the three samples of Cabrini. EDX spectra have shown the same Si-Na-Ca rich composition, and observation at SEM and optical microscope confirm the absence of corrosion product or a modified layer. These samples with different colors but also different geographic and historic origin present a real proximity for their composition and for their morphologies.

23.3 Conclusion

Our study shows two main phenomena of surface alteration: neocrystallization and leaching process, known for being the principal cause of alteration for this glass system. These alterations reveal both different causes and morphological aspects.

Leaching process could depend on the cycle of rain and humid period following very dry ones, which are responsible for a variation of pH and for mechanic tension at stained glasses surface, as confirmed by optical and SEM analysis. Nevertheless, neocrystallization occurs more in an environment protected from humidity but exposed to atmospheric pollution, particularly high sulfur and CO_2 concentration. Neocrystallization creates an alkaline environment, favorable for corrosion process. A comparative study has shown the importance of considering stained glasses in their specificities. We saw that in the same set of samples, different behavior was observed when they are exposed to the same aggressive environment. Stained glasses diagnostic can be linked to many factors: a geographic area, exposition conditions, etc. An isolated preservation context cannot be considered as an exhaustive diagnostic process for preventive conservation and restoration project. Preventive conservation strategies need to integrate the complex response of stained glasses to specific environmental factors. Identification of nature and vulnerability of glasses used for windows fabrication and microscopic analysis of structural modification due to alteration help us to choose an appropriate restoration or conservation strategy adapted to each case of study.

References

1. E. Castelnuovo, *Vetrate Medievali* (Einaudi, Torino, 1994)
2. A. Coralli, V. Bertuzzi, *Il restauro delle vetrate* (Nardini Editore, Firenze, 1994)
3. A. Bagnoli, C. Tarozzi, *Duccio, la vetrata del Duomo di Siena e il suo restauro* (Silvana Editoriale, Milano, 2003)
4. D. Gimeno, M. Garcias-Valles, J.L. Fernandez-Turiel, F. Bazzocchi, M. Aulinas, M. Pugès, C. Tarozzi, B. Messiga, From Siena to Barcelona: deciphering colour recipes of Na-Rich Mediterranean stained glasses windows at the XIII-XIV century transition. J. Cult. Heritage **9**, 10–15 (2008)
5. L. Gentaz, Simulation et modélisation de l'altération des verres de composition médiévale dans l'athmosphère urbaine, Thèse de l'Université Paris-Est Créteil, (2011)
6. J. Sterpenich, Altération des vitraux médiévaux, contribution à l'étude du comportement à long terme des verres de confinement, Thèse de l'Université Nancy I, (1998)
7. J. Sterpenich, G. Libourel, Les vitraux médiévaux: caractérisation physic-chimique de l'altération. Techne **6**, 70–78 (1997)
8. L. Gentaz, T. Lombardo, Impact of neocrystallysation on the SiO_2-K_2O-CaO glass degradation in dry atmospheric conditions. Atmos. Environ. **55**, 459–466 (2012)

Chapter 24
Enhanced Optical Forces in Plasmonic Microstructures

Domna G. Kotsifaki, Dimitris Polyzos, Alexandros A. Serafetinides, Mersini Makropoulou, Giorgos Tsigaridas, R. J. Peláez and Carmen N. Afonso

Abstract Micromanipulation of dielectric objects, from polystyrene spheres to living cells, is achieved when radiation pressure forces create stable trapping by highly focused laser beams through microscopes. However, the impressive history of optical trapping is shadowed by the light diffraction limit, as research currently has focused on materials below the micron scale, requiring stronger optical confinement and higher intensities than can be provided by the conventional optical tweezers. Recently, plasmonic nanostructures have entered the field, either to assist or enhance it. In this study, we present experimental results on using localized fields of metallic structures for efficient trapping, with various patterns (dots, fringes and squares). The patterns were produced by laser interferometry on almost continuous Ag or Au films on glass and glass covered by an amorphous Al_2O_3 layer (10 nm

D. G. Kotsifaki · D. Polyzos · A. A. Serafetinides · M. Makropoulou (✉) · G. Tsigaridas
Department of Physics, School of Applied Mathematical and Physical Sciences,
National Technical University of Athens, Zografou Campus, Heroon Polytechniou 9,
15780 Athens, Greece
e-mail: mmakro@central.ntua.gr

D. G. Kotsifaki
e-mail: dkotsif@central.ntua.gr

D. Polyzos
e-mail: dpolyzos@mail.ntua.gr

A. A. Serafetinides
e-mail: aseraf@central.ntua.gr

G. Tsigaridas
e-mail: gtsigaridas@teilam.gr

R. J. Peláez · C. N. Afonso
Laser Processing Group, Instituto de Optica, CSIC, Serrano 121,
28006 Madrid, Spain
e-mail: ramon.j.pelaez@gmail.com

C. N. Afonso
e-mail: cnafonso@io.cfmac.csic.es

E. K. Polychroniadis et al. (eds.), *International Multidisciplinary Microscopy Congress*,
Springer Proceedings in Physics 154, DOI: 10.1007/978-3-319-04639-6_24,

thick) respectively. We have calculated the optical forces by measuring the particle's escape velocity. The results show that the effective quality factor Q in the patterned metal film is enhanced by a factor >10, with respect to the unpatterned metal film and a factor >100, with respect to an uncoated glass. In addition, mathematical simulation of plasmonic fields is investigated to confirm and explain theoretically, the experimentally observed plasmonic enhancement.

24.1 Introduction

The field of optical trapping has grown due to implementation in various arenas including physical chemistry, biology and medicine. This success lies in the non-invasive nature of optical forces. Recently, evanescent waves have been employed, instead of conventional propagating waves to form the optical trap. This is expected to allow development of optical traps with sub-wavelength volumes, which could enable direct manipulation of nano-scale objects [1–3]. Moreover, plasmonics involve a wealth of novel promising scientific applications such as the trapping of nanomaterials by an enhanced optical force. Optical trapping based on surface plasmons can potentially overcome the main disadvantage of the conventional optical trapping systems, i.e. due to the diffraction limit of the incident light the spatial resolution in trapping is ordinarily limited to more than several hundreds of nanometers. Nanometric optical tweezers with sub-diffraction-limited trapping volume have been realized by utilizing optical near field between gold nanodots fabricated by electron beam lithography [4, 5]. These studies indicate that such metallic nanostructures (nanogaps, nanoantennas etc.), can function as novel small light sources in the nanocosmos. Plasmonic traps may play a key role in the development of applications, such as lab-on-a-chip, with increased functionalities. Compared to conventional optical tweezers, plasmonic traps do not require elaborate optical setups. Lately, plasmonic traps have become an important tool for immobilizing nanoparticles or cells on a substrate.

In a previous work [6], we reported that the fabrication of an optical trap with enhanced efficiency could be based on localized plasmonic fields around sharp metallic features. The trap was created near the surface of a laser-structured silicon substrate with quasi-ordered microspikes, coated with a thin silver film.

In this work, we present experimental results on using localized fields of metallic structures for efficient trapping, with various patterns (dots, fringes and squares). The patterns were produced by laser interferometry on almost continuous Ag or Au films on glass or glass covered by an amorphous Al_2O_3 layer (10 nm thick) respectively. We have calculated the average of the optical forces by measuring the particle's escape velocity. An enhancement of the effective quality factor Q is observed for certain patterns with respect to unpatterned metal layers or uncoated glass. In addition, mathematical simulation of plasmonic fields is performed to confirm theoretically the experimentally observed enhancement.

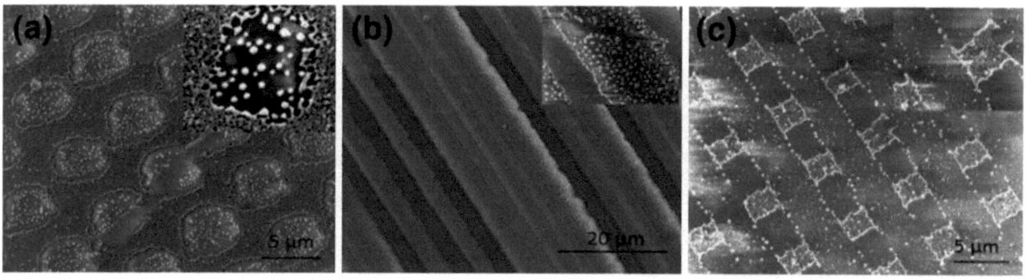

Fig. 24.1 SEM images of patterns produced in the Ag film: **a** dots, **b** fringes and **c** squares. The magnification of each image is noted. The insets show a magnification of the transformed areas where the morphology of the produced nanoparticles can be better appreciated

24.2 Materials and Methods

The experimental set-up has been described previously [3] and is only outlined here. The trap was created using a continuous wave Gaussian Nd:YAG laser beam at 1064 nm, focused through an objective onto the sample. The beam radius at the waist was $w_0 \sim 500$ nm, corresponding to the focusing lens with numerical aperture N.A. = 1.25, employed. The substrates were placed on a motorized x-y translation stage (Standa 8MT 167-2S) and the image of the trapped beads was captured on a CCD camera. As trapped objects we used polystyrene beads with mean diameter of 900 nm, suspended in deionized water of refractive index n = 1.33.

The metallic structures were produced by laser interferometry using phase masks at 193 nm on almost continuous Ag or Au films deposited on glass or glass covered by an amorphous Al_2O_3 layer (10 nm thick) respectively. Most results refer to patterns produced with a single pulse of energy density of 58 mJ/cm^2. Patterns as those shown in Fig. 24.1 were produced by exposing the films to 5 laser pulses (a, b) and 1 laser pulse (c), each with the phase mask rotated by 90° as described in details elsewhere [7, 8]. On the patterns nanoparticles with mean diameters in the range 70–100 nm were formed.

24.3 Results and Discussion

We determined the effective optical trap force, by performing escape velocity measurements. By translating the structured substrates at a constant velocity, under the focal point of the trapping laser beam, we were able to measure the escape velocity for which the polystyrene beads escaped the optical trap. At the escape velocity, the effective trapping force is considered to be equal to the viscous drag force, described by the modified Stokes law,

$$F = K \cdot 6\pi \eta r v_{esc} \quad (24.1)$$

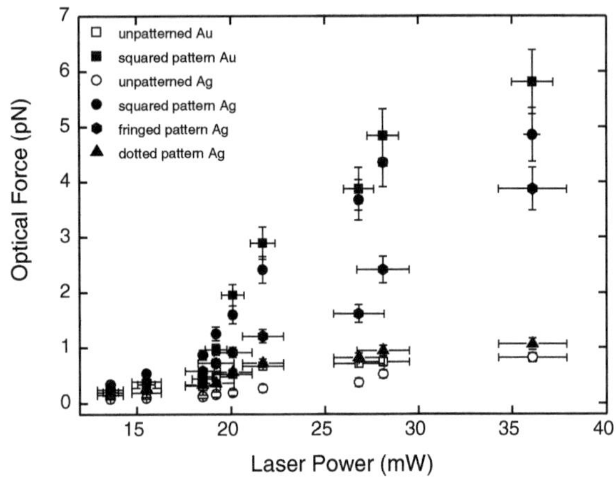

Fig. 24.2 Trapping force on 900 nm polystyrene beads as a function of laser power for different substrates. ■ squared pattern in Au film, □ unpatterned Au film, ▲ dotted pattern in Ag film (SEM image shown in Fig. 24.1a), • squared pattern in Ag film (SEM image shown in Fig. 24.1c), ♦ fringed pattern in Ag film (SEM image shown in Fig. 24.1b)

where η is the water viscosity, r the bead radius, v_{esc} the escape velocity of the bead and K (~1.33) a correction coefficient [9]. Figure 24.2 shows the effective force exerted on 900 nm polystyrene beads, optically trapped above the structures as a function of the power of the trapping laser beam. In this graph, we can observe that the trapping force changes with the substrate morphology. The squared and fringed patterns [Fig. 24.1b, c for Ag], show a significant enhancement of the trapping force, compared to that in the dotted pattern or the unpatterned films.

Knowing the trapping force, F, we can evaluate the effective quality factor of the optical trap, Q. The effective quality factor is defined as,

$$Q = Fc/nP \tag{24.2}$$

where c is the speed of light, n is the refractive index of the surrounding medium and P is the power of the trapping laser beam. The quality factors for the samples used in this study are given in the following Tables. We note that these are average of the optical force values and should be regarded as approximations, since for some samples the optical force is a nonlinear function of the trapping laser power, as shown in Fig. 24.2. In previous studies [9], it was observed that the increased trapping laser power changes the axial trapping point; hence a single value for the quality factor could not be derived for a given particle over all laser powers.

The calculated quality factors in Table 24.1, for the squared patterns in Au and Ag films, are one or two orders of magnitude larger than the ones of the other patterned substrates and uncoated glass respectively. The evaluated structures of the Table 24.1 have been developed with one laser pulse of energy density 58 mJ/cm^2.

Table 24.2 shows the dependence of the quality factor on the average energy density used to produce fringed patterns in Ag films using 5 laser pulses. The patterns produced with the highest energy density show more than twice enhancement of Q,

Table 24.1 Effective trapping quality factor, Q, for structures with squared patterns [similar to that shown in Fig. 24.1c] both in Ag and Au films together with the respective values for the unpatterned films and the uncoated glass substrate

Metal	Type of structure	Quality factor Q
Au	Squared pattern	$(9.20 \pm 0.60) \times 10^{-3}$
Ag	Squared pattern	$(6.40 \pm 0.80) \times 10^{-3}$
Au	Unpatterned	$(9.00 \pm 0.36) \times 10^{-4}$
Ag	Unpatterned	$(3.24 \pm 0.20) \times 10^{-4}$
–	Uncoated glass	$(3.80 \pm 0.50) \times 10^{-5}$

Table 24.2 Effective trapping quality factor, Q, for fringed patterns in Ag film [as that shown in Fig. 24.1b] for various laser energy densities and five laser pulses

Energy density	Quality factor Q
58 mJ/cm^2	$(1.24 \pm 0.10) \times 10^{-3}$
44 mJ/cm^2	$(5.04 \pm 0.54) \times 10^{-4}$
29 mJ/cm^2	$(4.85 \pm 0.42) \times 10^{-4}$

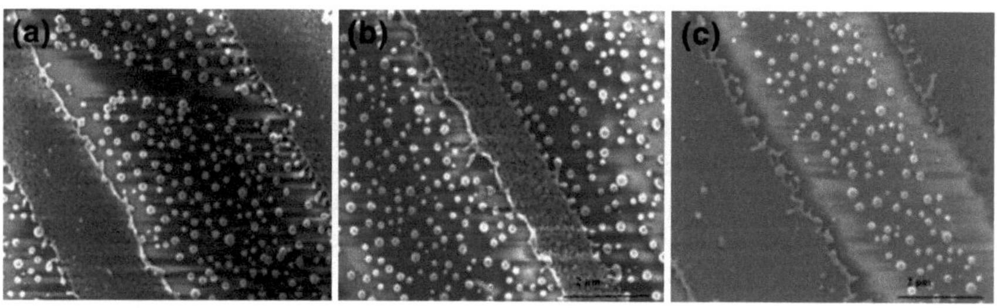

Fig. 24.3 SEM images of the fringed patterns in Ag, with 5 pulses and energy densities **a** 58 mJ/cm^2, **b** 44 mJ/cm^2 and **c** 29 mJ/cm^2

when compared to similar patterns produced with lower energy densities. Generally, lower energies produce less nanoparticles as expected and shown in Fig. 24.3.

A possible explanation for the enhanced efficiency of the described optical trap is based on the action of localized plasmonic fields, created around the patterns covered with metallic nanoparticles [10, 11]. However, the field enhancement is not an inherent property of plasmonics, but rather that of cavity effects as seen in Fabry-Perot interferometers and laser resonances [12]. For this reason more investigation is needed. However, we can conclude that the combination of metallic structures with metallic nanoparticles results in electromagnetic field enhancement, during optical trapping and increases the quality factor of the trap.

As far as the theoretical analysis is concerned, the trapping force shows an enhancement when the substrate is covered with a metal layer, and even more in the presence of nanostructures. The electromagnetic force can be calculated through the gradient of the square of the electric field norm, as described by the formula,

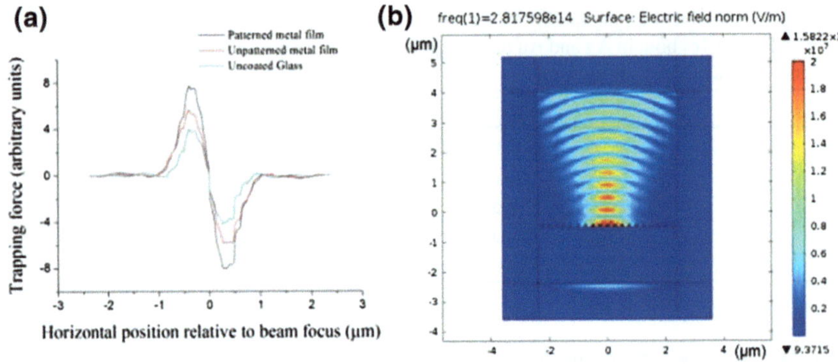

Fig. 24.4 **a** The trapping force in arbitrary units, as a function of the *horizontal position* relative to the beam focus, in μm. **b** Schematic representation of the electric field distribution over a metal substrate with nanospheres

$$F = -\frac{n_{env}}{2} a \nabla E^2 \tag{24.3}$$

where a is the polarizability of the particles and n_{env} is the refractive index of the medium, water in our case ($n_{env} = 1.33$) [13]. Specifically and in coherence with the experimental parameters, a Gaussian laser beam with wavelength $\lambda_0 = 1064$ nm was assumed to illuminate the samples, namely an uncoated glass, an unpatterned metal film and a patterned metal film. The beam radius, at the waist was $w_0 \sim 500$ nm corresponding to a focusing lens with numerical aperture N.A. = 1.25, as the one used in the experiments. The diameter of the nanospheres was chosen to be 100 nm, and the distance between them 250 nm, in order to fit best the experimental conditions. The material of the nanospheres was assumed to be gold, although similar results are produced for silver or other metals. The background environment was water, as in the experiments. Then, the Maxwell equation in the form,

$$\nabla \times \nabla \times \mathbf{E} - k_0^2 \varepsilon_r \mathbf{E} = 0 \tag{24.4}$$

was solved through finite element numerical analysis [14], using the COMSOL Multiphysics software package [15]. Here k_0 is the free space wavenumber and ε_r the relative permittivity of the material, given by the formula,

$$\varepsilon_r = n - i\kappa^2 \tag{24.5}$$

where n, κ are the real and the imaginary part of the complex refractive index, respectively. In this way, via (24.4) and (24.5), the electric field distribution in the focal region of the beam was determined, and then (24.3) was applied in order to calculate the trapping force.

The results are shown in Fig. 24.4a. It is clear that the trapping force reaches a maximum when patterned metal film is employed. Figure 24.4b shows the electric field distribution over a metal substrate with nanospheres. It is evident that the nanostructures strongly interact with the electromagnetic field, enhancing its gradient, and consequently the trapping force.

24.4 Concluding Remarks

In conclusion, we have demonstrated that the measured trapping force exerted on polystyrene beads of 900 nm diameter changes with the substrate morphology. Periodically patterned metal films create a plasmonic optical trap with a larger trapping efficiency, than that of a conventional system, and an additional potential for applications in nanoengineering and nanobiology. The calculated Q factor was almost one or two orders of magnitude larger than the one of a conventional optical trap, with the same objective lens. The maximum trapping force was achieved in squared patterns covered by metal nanoparticles, with gold leading to slightly higher enhancement than silver. The theoretical simulations confirm an optimization in optical trapping when metallic nanospheres are employed. However, as the field enhancement is not an exclusive property of plasmonics, more investigation is under way to establish the contribution of the localized plasmonic field to the total trapping field. In addition, as 1064 nm wavelength was not the optimum for the specific combination of substrate and nanostructures employed, a more favorable combination of all three is studied.

Acknowledgments The authors would like to thank Prof. A. Boudouvis for access to the COMSOL Multiphysics software package. The authors would also like to thank Mr. Sp. Zidropoulos for his help with the SEM imaging. RP acknowledges a grant from the JAE-doc program co-funded by European Social Fund. This work has partially been supported by MAT2011-28345-C02-02 (Spain) projects.

References

1. M. Gu, J.-B. Haumonte, Y. Micheau, J.W.M. Chon, X. Gan, Laser trapping and manipulation under focused evanescent wave illumination. Appl. Phys. Lett. **84**, 4236–4238 (2004)
2. K. Wang, E. Schonbrun, P. Steinvurzel, K.B. Crozier, Trapping and rotating nanoparticles using a plasmonic nano-tweezer with an integrated heat sink. Nat. Comm. **2**, 469-1–469-4 (2011)
3. D.G. Kotsifaki, M. Makropoulou, A.A. Serafetinides, Efficient and low cost multiple optical trap, based on interference. Optik **124**, 617–621 (2013)
4. A.N. Grigorenko, N.W. Roberts, M.R. Dickinson, Y. Zhang, Nanometric optical tweezers based on nanostructured substrates. Nat. Phot. **2**, 365–370 (2008)
5. M. Righini, G. Volpe, C. Girard, D. Petrov, R. Quidant, Surface plasmon optical tweezers: tunable optical manipulation in the femtonewton range. Phys. Rev. Lett. **100**, 186804-1–186804-4 (2008)
6. D.G. Kotsifaki, M. Kandyla, I. Zergioti, M. Makropoulou, E. Chatzitheodoridis, A.A. Serafetinides, Optical tweezers with enhanced efficiency based on laser structured substrates. Appl. Phys. Lett. **101**, 011102-1–011102-3 (2012)
7. R.J. Peláez, C.N. Afonso, J. Bulíř, J. Lančok, M. Novotný, Diffractive and coloured films by laser interferometry patterning. Surf. Coat. Techn. **211**, 205–208 (2012)

8. R.J. Peláez, G. Baraldi, C.N. Afonso, S. Riedel, J. Boneberg, P. Leiderer, Selective gold nanoparticles formation by pulsed laser interference. Appl. Surf. Sc. **258**, 9223–9227 (2012)
9. W.H. Wright, G.J. Sonek, M.W. Berns, Parametric study of the forces on microspheres held by optical tweezers. Appl. Opt. **33**, 1735–1748 (1994)
10. A. Kinkhabwala, Z. Yu, S. Fan, Y. Avlasevich, K. Müllen, W.E. Moerner, Large single-molecule fluorescence enhancements produced by a bowtie nanoantenna. Nat. Phot. **3**, 654–657 (2009)
11. W. Zhang, L. Huang, C. Santschi, O.J.F. Martin, Trapping and sensing 10 nm metal nanoparticles using plasmonic dipole antennas. Nano. Lett. **10**, 1006–1011 (2010)
12. S. Kawata, Plasmonics: future outlook. Jap. J. Appl. Phys. **52**, 010001-1–010001-10 (2013)
13. J. Wu, X. Gan, Optimization of plasmonic nanostructure for nanoparticle trapping. Opt. Exp. **20**, 14879–14890 (2012)
14. J.-M. Jin, The finite element method in electromagnetics. Wiley-IEEE Press, New York (2002)
15. http://www.comsol.com

Chapter 25
Photoemission of Ternary Sodium-Cesium-Potassium Alloys

Ruslan Arkhestov, Zareta Kegadueva, Barasbi Karamurzov and Boris Alchagirov

Abstract The photoemission properties and the temperature dependences of the electron work function (EWF) of the ternary Na-K-Cs alloys have been studied by means of the Fawler photoelectric method in conditions of high vacuum and of thermodynamic equilibrium in temperature range from 290 to 420 K. The additions of potassium to the binary Na-Cs alloy have been shown to increase the spectral sensitivity and to shift the maxima to the side of greater by 40 nm wavelengths. In ternary alloys, depending on the cesium content, the sign of the surface activity changes for EWF and the location of the isolines $\varphi(x)$ alters in respect to the cesium corner of the concentration triangle while sodium is being replaced with cesium.

25.1 Introduction

The study of the photoemission properties of the multicomponent systems of alkali metals and alloys involving them is of great importance due to the use of such in production of novel high-power consuming chemical sources of electric current, multialkali photocathodes and so on [1–3]. The reference review reveals [4–6] that the systematic studies of EWF, determining the emission properties of materials, are completed only for pure metals and their binary mixes by now. The electron work function of ternary alkali metal alloys is studied poorly [4].

R. Arkhestov (✉) · Z. Kegadueva · B. Karamurzov · B. Alchagirov
Kh.M. Berbekov Kabardino-Balkarian State University, Nalchik, Russia 360004
e-mail: ruslan.arhestov@mail.ru

B. Alchagirov
e-mail: alchg@kbsu.ru

E. K. Polychroniadis et al. (eds.), *International Multidisciplinary Microscopy Congress*, Springer Proceedings in Physics 154, DOI: 10.1007/978-3-319-04639-6_25,

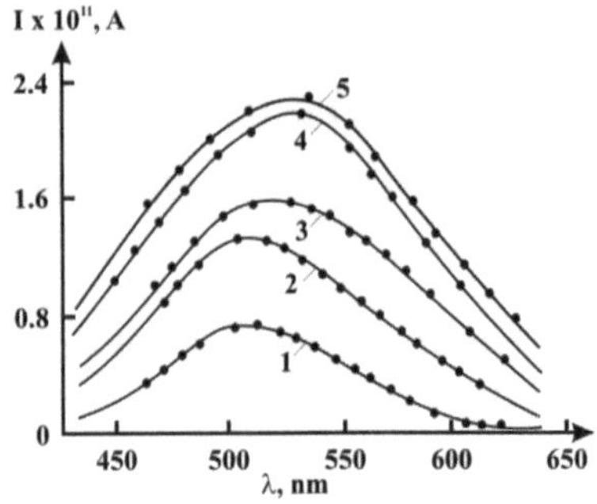

Fig. 25.1 The changes of the spectral characteristics of the photoemission current from the surfaces of the photocathode from the alloys: 1—Na + 77.7 at.% Cs; 2—Na + 76.6 Cs + 1.5 K; 3—Na + 74.4 Cs + 4.1 K; 4—Na + 66.2 Cs + 14.8 K; 5—Na + 46.3 Cs + 40.4 at.% K at 293K

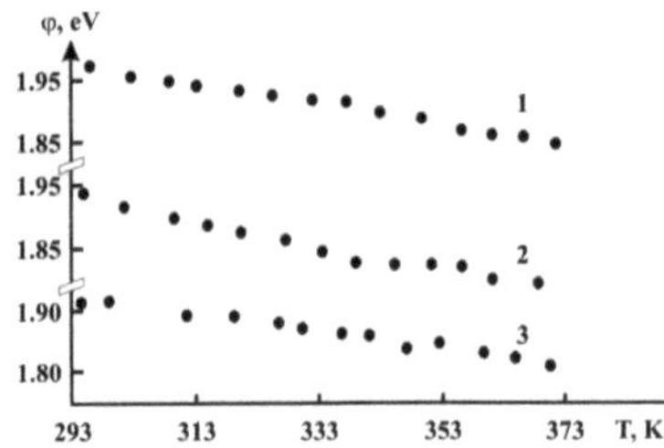

Fig. 25.2 The temperature dependence of EWF of the alloy Na + 77.7 at.% Cs in respect to the amount of potassium added: 1—1.5; 2—14.8; 3—40.4 at.% K

25.2 Discussion

In this research we present the results of the measurements of the emission properties and of the electron work function of ternary Na-K-Cs alloys the compositions of which lie along the radial sections originating from the potassium's corner. The experiments were carried out in a glass measuring cell [7] by means of the Fawler photoelectric method [8] in temperature range from 290 to 420 K. The setup and the method of determining of EWF allowed us to carry on the measurements in conditions of thermodynamical equilibrium and in high static vacuum (10^{-8} Pa) [9, 10] with the error about the 1.5 %. The metals used were those with a concentration of the basic element not less than 99.995 %.

Figures 25.1 and 25.2 represent graphically the data of the measurements of the spectral dependences of the photoemission currents at temperature 293 K and the calculated by them temperature dependences of EWF for a number of Na-Cs-K alloys of constant ratio of concentrations of sodium to cesium ($X_{Na} : X_{Cs} = \text{const}$). One can see from the Fig. 25.1 that the photoemission current from the surface of the binary alloy Na + 77.7 at.% Cs is 2–3 times less than that of ternary alloys (curves 2–5) that is the photoemission increases with the growing extent of the potassium added. The spectral dependences of the photocurrents from the samples transform their shape: the maxima of the spectral dependences of the photoemission currents of ternary alloys shift to the side of the greater by 40 nm wavelengths.

The dependences of EWF on temperature of selected ternary alloys of the given section are shown in Fig. 25.2 from which is seen that the polytherms $\varphi(T)$ of alloys are linear until 373K and that EWF decreases with rising temperature.

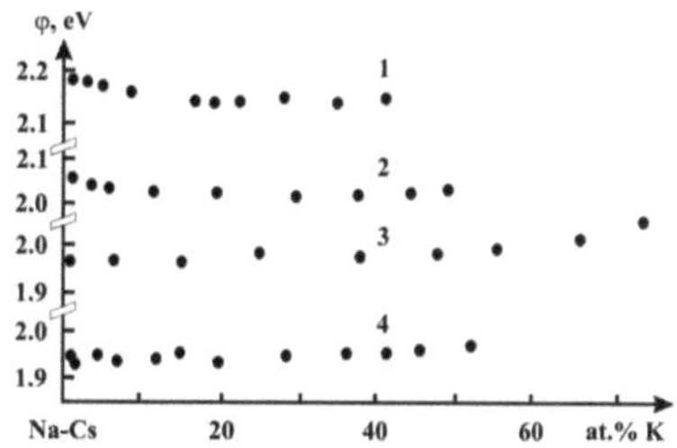

Fig. 25.3 The isotherms of EWF of Na-Cs alloys in respect to the amount of potassium: 1—Na + 1.7 at.% Cs; 2—Na + 6.5 at.% Cs; 3—Na + 36.6 at.% Cs; 4—Na + 77.7 at.% Cs

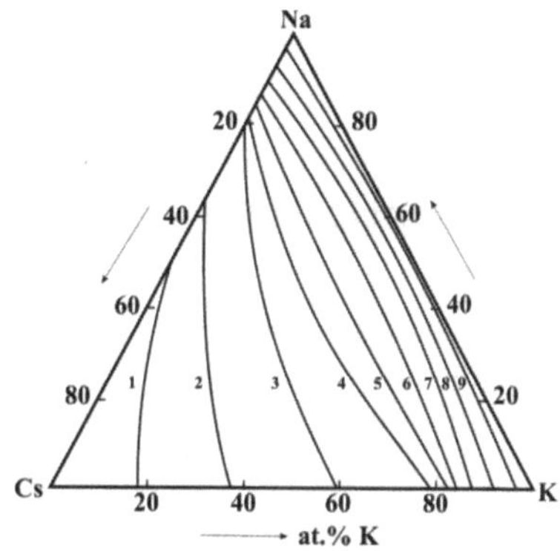

Fig. 25.4 The isolines of EWF of Na-Cs-K alloys at T = 293 κ: 1—2.15; 2—2.04; 3—2.02; 4—2.00; 5—1.99; 6—1.98; 7—1.97; 8—1.96; 9—1.95 eV

The temperature dependence φ(T) for the alloy 13.3 Na + 46.3 Cs + 40.4 at.% K, the closest to the eutectic (13.9 Na + 43.5 Cs + 42.6 at.% K) among the studied ones, is also linear and no peculiarities have been observed.

The concentration dependences φ(x) of a number of isotherms of EWF of Na-Cs-K alloys at temperature 293 K are given in Fig. 25.3. The addition of potassium up to 20 at.% to the binary sodium-based alloys results (curves 1 and 2 in Fig. 25.3) to the slight decrease of EWF. The further enrichment of ternary alloys in potassium up to the concentrations of 50 at.% K does not noticeably influence on the electron work function.

The increasing content of potassium in alloys, with the decreasing ratio of concentration of sodium to cesium ($X_{Na} : X_{Cs}$), results in increasing EWF: that is if sodium is replaced with cesium in ternary alloys then potassium switches from the surface-active addition to the inactive one (Fig. 25.3, curves 3 and 4).

The features of the influence of the potassium on EWF of the ternary Na-Cs-K alloys can be seen within the concentration triangle of the mentioned system either (Fig. 25.4).

Figure 25.4 demonstrates the compositions of ternary Na-Cs-K alloys having equal magnitudes of EWF φ(x) (isolines of EWF) on the concentration triangle. The isolines φ(x) represent the smooth curves without extremums. EWF increases with decreasing concentration of cesium in ternary alloys, especially sharply in area of the least contribution of cesium (the isolines lie closer).

The feature of the potassium influence on EWF of ternary alloys is that with the increasing concentration of Cs in Na up to 13 at.% Cs the addition of potassium results in isolines of EWF directed to the cesium corner by the concavity (curves 1–4 in Fig. 25.4) while in case of sodium replaced with cesium the addition of potassium alters the nature and degree of non-linearity of the isolines $\varphi(x)$—the latter become directed to the cesium corner of the concentration triangle by their convexity (curves 5–9 in Fig. 25.4).

The measurements have revealed that, according to the data on EWF, the isotherms $\varphi(x)$ of Na-Cs-K alloys possess, within the studied temperature range, the same features known for the isotherm of EWF for binary lateral systems.

25.3 Conclusions

- The addition of potassium to the binary Na-Cs alloy in ternary alkali photocathodes results in both increasing spectral sensitivity in area of greater wavelengths and shifting of the maxima of spectral sensitivity by 40 nm.
- It is shown that the sign of the surface activity of potassium alters in ternary systems in respect to the cesium content in them.
- The location of the isolines of EWF in respect to the cesium corner changes within the concentration triangle if sodium is being replaced by cesium.

References

1. M.E. Drits, L.L. Zusman, *Alloys of Alkali and Alkali-Earth Metals*. Handbook (Moscow, Metallurgia, 1986). (in Russian)
2. L.N. Dobretsov, M.V. Gomoyounova, *Emission Electronics* (Nauka, Moscow, 1966). (in Russian)
3. B.K. Petrin, Chemical sources of electric current of high-power consumption. in *Results of Science and Technology*. Series of Generators of Direct Transformation of Thermal and Chemical Energy into the Electricity (Moscow, Mir, 1972). (in Russian)
4. B.B. Alchagirov, V.B. Lazarev, Kh.B. Khokonov. Reviews of thermophysical properties of materials. TPhC, IVTAN USSR, Moscow, 1989, Issue 5(79), pp. 76–146. (in Russian)
5. Yu.I. Malov, V.B. Lazarev, M.D. Shebzukhov, in *Surface Phenomena in Semiconductors* (Moscow, MISIS, 1976), Issue 1, pp. 15–23 (in Russian)
6. V.S. Fomenko, Emission Properties of Materials. Handbook, 4-th edn. (Kiev, Naukova Dumka, 1981). (in Russian)
7. B.B. Alchagirov, Setup for the determination of the electron work function of metals and alloys versus the temperature. RF Patent #2008659, Bull. No. 4, 1994. No 20086595005115/25, p. 5 (in Russian)
8. R.H. Fawler, Phys. Rev. **38**(2), pp. 45–46 (1931)
9. B.B. Alchagirov, Kh.B. Khokonov, B.S. Karamurzov, *Modern Methods of Investigations of the Solid Surfaces* (KBSU, Nalchik, 1986). (in Russian)
10. B.B. Alchagirov, Kh.B. Khokonov, R.Kh. Arkhestov, Doklady AN SSSR, **326**(1), pp. 121–125 (1992)

Chapter 26
Structure of the Carbon Matrix Made by Pyrolyzed Yttrium Bis-Phtalocyanine by Atomic Force Microscopy

V. Y. Bairamukov, D. V. Lebedev and V. I. Tikhonov

Abstract Evolution of the structure of yttrium bis-phtalocyanine (Pc_2Y) during high-temperature pyrolysis was investigated by atomic force microscopy (AFM). A phase transition of the crystal structure to an amorphous structure was observed in the temperature range from 750 to 850 °C and the further evolution of the surface topology during pyrolysis at higher temperatures into a porous carbon-based matrix was observed.

26.1 Introduction

The problem of conditioning, storage and transmutation of radioactive waste (RW), containing long-lived isotopes, including minor actinides, calls for creation of new chemically and thermally stable matrices resistant to ionizing radiation [1–5]. Carbon-based matrix appears to be most suitable in meeting such hard requirements. Carbon is chemically resistant, can withstand high thermal loads, and both of its stable isotopes (C^{12} and C^{13}) have low neutron capture cross-sections (3.4 and 1.3 mb) that is especially important for the process of transmutation. Particularly, the method of manufacturing matrices for storage and transmutation of RW containing minor actinides that is based on pyrolysis of their bis-phtalocyanines has shown very promising results during initial tests of the chemical, thermo and radiation stability [6, 7].

Phtalocyanines—organometallic complex compounds that relate to a large class of porphyrins. Their mono- and bis-complexes are common for all transition and rare earth elements (REE), as well as for the actinides, which radioactive isotopes represent the main problem in the conditioning of RW. The structure of bis-phtalocyanines

V. Y. Bairamukov (✉) · D. V. Lebedev · V. I. Tikhonov
National Research Centre "Kurchatov Institute" B.P. Konstantinov Petersburg Nuclear Physics Institute, Orlova Roshcha, Gatchina, Leningrad district, Russia 188300
e-mail: vbayramukov@gmail.com

E. K. Polychroniadis et al. (eds.), *International Multidisciplinary Microscopy Congress*, Springer Proceedings in Physics 154, DOI: 10.1007/978-3-319-04639-6_26,

(Pc_2Me), their properties and methods of preparation are described in [8, 9]. Mono-phtalocyanine is a planar molecule, while the bis-complex, $[C_{64}H_{32}N_{16}Me]^-$, is a sandwich of two planar complexes with the metal atom between them. Pyrolysis of Pc_2Me in argon atmosphere at temperatures above 800 °C results in the metal-carbon composite with highly porous structure MeC_x (x = 30–35), where the metal atoms are distributed isotropically in the cavities of the carbon matrix. Incorporation of radionuclides into these microcavities provides a unique opportunity for their storage and transmutation. There are three temperature stages of pyrolysis concerned with substantial mass loss: 750−800 °C, 1200−1250 °C and approximately 1600−1700 °C. In the first stage of pyrolysis ca. 90 % of hydrogen and approximately half of nitrogen and carbon are lost, during the second stage the matrix loses the rest of hydrogen and some additional nitrogen, and after heating to 1700 °C the matrix consists of carbon and metal only. Further temperature increase (up to 2200 °C) results in only a small additional loss of mass.

To optimize the conditions of the RW incorporation into carbon matrices made of the pyrolyzed bis-phtalocianine, their structure was investigated at different temperatures of pyrolysis process by the small-angle neutron scattering (SANS) [10], X-ray diffraction [11] and AFM methods. The data obtained by AFM are presented in this paper.

26.2 Materials and Methods

Yttrium bis-phtalocianines were synthesized as described in [8] and deposited by vacuum sputtering on sapphire glass (Lighten Optics, China). The process was carried out by heating the powder of electrochemically purified Pc_2Y in tantalum oven at 450–500 °C in a vacuum of 10^{-2} Pa. Sapphire glass substrates were washed in a piranha solution (3 V H_2SO_4: 1 V H_2O_2) and then twice in distilled water. Sapphire glass substrate had atomically flat surface (roughness of about 0.3 nm) and high melting point (about 2000 °C), allowing us to study structure of the carbon matrix at different temperatures of pyrolysis (up to 1700 °C).

Surface topology measurements were performed in tapping mode in air at room temperature on Solver atomic force microscope (NT-MDT, Russia). Cantilevers NSG03 (NT-MDT, Russia) with a constant stiffness 1.74 N/m and the radius of curvature of the needle of 10 nm were used, the scan speed was 1 Hz. Autocorrelation was calculated using ImageAnalysis software (NT-MDT, Russia) and Radial Profile Plot plugin of ImageJ package. Fractal analysis was performed in ImageJ software using Box Count Fractal plugin.

26.3 Results and Discussion

The initial scan of topology of Pc_2Y on sapphire glass shows polydisperse crystals that were resistant to pyrolysis up to 500 °C (Fig. 26.1a). Noticeable destruction of Pc_2Y with the formation of amorphous phase starts at temperatures higher than

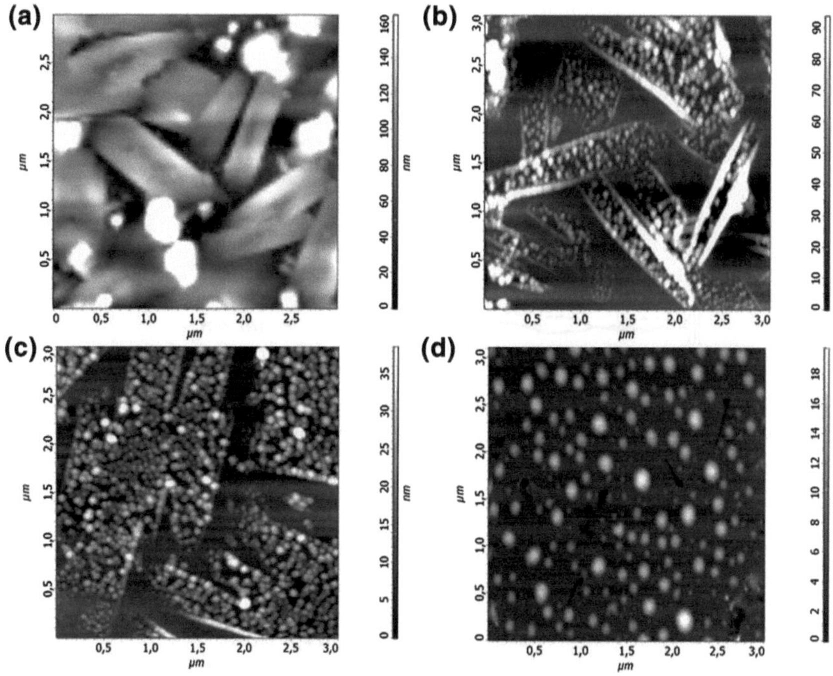

Fig. 26.1 Pc_2Y at different temperatures of pyrolysis, scan range $3 \times 3\,\mu m$. **a** initial Pc_2Y crystals; **b** Pc_2Y pyrolyzed at 790 °C; **c** Pc_2Y pyrolyzed at 850 °C; **d** Pc_2Y pyrolyzed at 1040 °C

750 °C. In the AFM images the destruction process manifests itself as a transition of crystalline particles to the amorphous phase with a variety of round shaped aggregates (Fig. 26.1). At 790 °C the boundaries of the crystal of the initial Pc_2Y are still visible, but the internal structure of the crystal is changed dramatically by formation of disordered particles close to spherical shape, with sizes ranging from 10 to 60 nm (Fig. 26.1b), and their height, compared with the original crystal, decreases almost three-fold, from 60 to 15–20 nm. At 850 °C the boundaries vanish and the round shaped particles tens of nanometers in diameter and under 15–20 nm in height tend to fill the entire field of scan (Fig. 26.1c). Further increase of the temperature to 1040 °C results in formation of the bigger spherical clusters with the diameter of up to 150–200 nm (Fig. 26.1d). An increase in the cluster size with the temperature of the pyrolysis is clearly seen in the autocorrelation curves calculated for the corresponding AFM images (Fig. 26.2). Growing of these clusters appears to be limited to ca. 250 nm. The sample surface formed by pyrolysis at 1300 °C did not show individual particles but rather a continuous porous matrix (Fig. 26.3).

Fractal analysis of the surface of Pc_2Y based on AFM images performed by box counting method have revealed a power law dependency, indicating the scaling law in the aggregate sizes characteristic for the fractal structures (Fig. 26.4). The power law exponent was close to 1.5 for the spherical aggregates during the initial stages

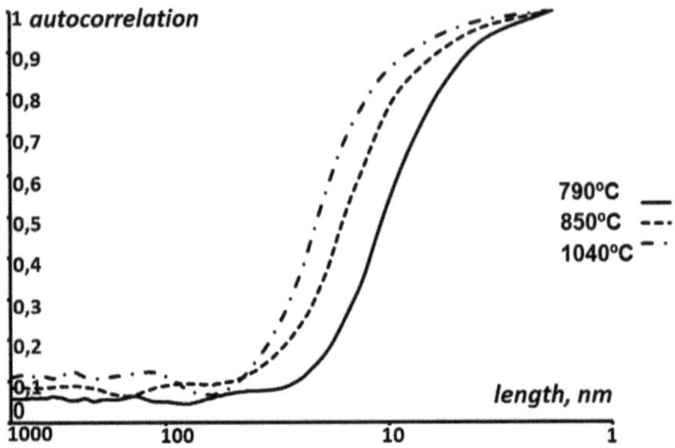

Fig. 26.2 Autocorrelation curves for Pc_2Y surface after pyrolysis at different temperatures; *solid line* corresponds to 790 °C; *dash line*—850 °C; *dash line* with dots—1040 °C

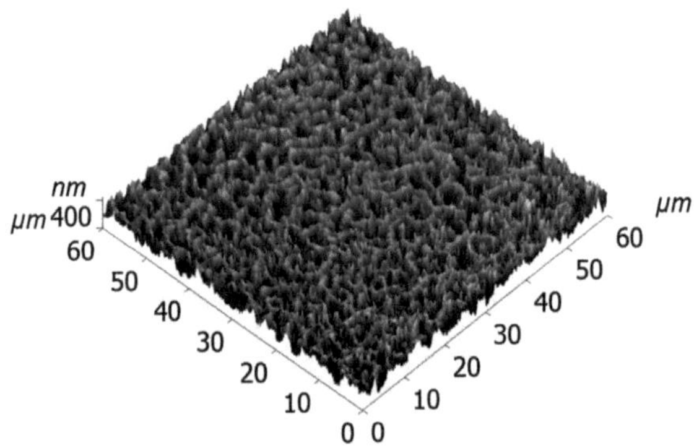

Fig. 26.3 A three-dimensional view of Pc_2Y pyrolyzed at 1300 °C; scan range 60 × 60 μm

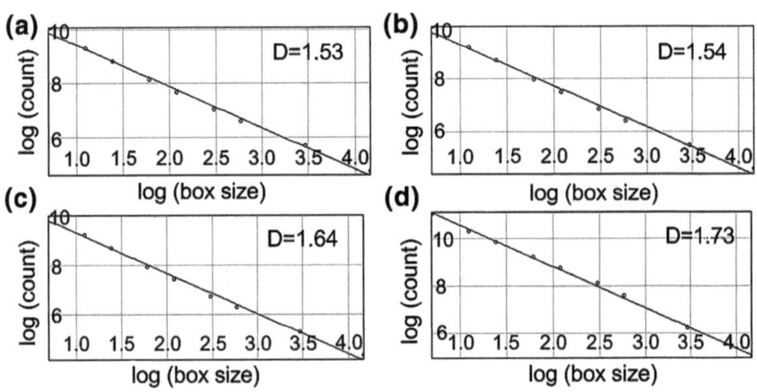

Fig. 26.4 Fractal analysis of Pc_2Y surface after pyrolysis at different temperatures. **a** 790 °C; **b** 850 °C; **c** 1040 °C; **d** 1300 °C

of pyrolysis and rose to 1.65 for the larger clusters formed at 1040 °C. The surface of the carbon matrix formed at 1300 °C has also exhibited fractal properties with the fractal dimension around 1.73.

The results of AFM study of Pc_2Y surface are in a good agreement with those obtained on bulk powdered samples by the SANS and X-ray diffraction methods [10, 11]. SANS results have shown that the pyrolysis process leaded to formation of two types of carbon clusters with typical sizes of 3–7 and 30–50 nm. The latter clusters exhibited properties of fractal aggregates. Similar sized particles were observed in this work by AFM at temperatures up to 850 °C.

X-ray diffraction measurements have shown that starting from 1200 °C the formation of crystal phase of inorganic salts (nitrides or carbides in accordance with properties of metal) begins, however, the dimensions of these crystallites do not growing above 130–150 nm. One could suppose from these data that a diffusion of the incorporated atoms is possible inside cavities only and is limited by the barriers of carbon atoms.

It appears likely that the formation of fine-dispersed carbon cluster during pyrolysis of Pc_2Me is the main basis for effective retention of atoms incorporated in such kind matrix, as it was shown in [6]. One could suppose that there are two different factors responsible for retention of metal atoms. In the temperature range 850–1200 °C it is the formation of closed carbon cells with the size of "windows" of 0.12–0.14 nm that keep the atoms fixed inside the carbon cells. At the higher temperatures (above 1200 °C), where a tendency of growing carbon clusters and evolution to the large-scale high porous structure is observed (Fig. 26.3), the decisive cause of high retention becomes a very slow diffusion. One could make an assumption that if on the first stage (up to approximately 1200 °C) the release of the incorporated atoms has a threshold property, while at the temperatures higher 1200 °C this process submit to common diffusion law in highly porous materials.

26.4 Conclusion

High-temperature pyrolysis of yttrium bis-phtalocyanine (Pc_2Y) results in a phase transition of the crystal structure to an amorphous spherical aggregates in the temperature range from 750 to 850 °C. Further pyrolysis leads to an increase in the size of the aggregates up to ca. 250 nm which at temperatures above 1200 °C transform to a highly porous surface that exhibits fractal properties.

Acknowledgments This work was funded by RFBR grant 12-026-12021/13021-ofi-m. Authors extend their gratitude to Igor Yung for his help in AFM image processing.

References

1. W. Donald, Waste immobilization in glass and ceramic based hosts: radioactive, toxic and hazardous wastes. *Materials Science Research Division, Atomic Weapons Establishment* (Wiley, UK, 2010) p. 526
2. L.L. Hench et al., Nucl. Chem. Waste Manag. **5**, 149 (1984)
3. M. Minatom, Strategy of nuclear power development in Russia in the first half of the 21st Century, Ministry of Russian Federation for Atomic Energy, Summary, Moscow (2000)
4. A.V. Baluev et al., Radiokhimija **42**, 295 (2000). (In Russian)
5. A.S. Gerasimov, G.V. Kiselev, Phys. Elementary Particles At. Nucl. **32**, 143 (2001). (In Russian)
6. V.I. Tikhonov et al., in *Proceedings of the 11th International Conference on Environmental Remediation and Radioactive Waste Management ICEM-2007*, Belgium, Report 7084 (2008)
7. V.I. Tikhonov et al., Method of fixation of the long-living radionuclides for storage and transmutation, Russian patent#2343575 (2009)
8. P.N. Moskalev, Koordinationnaja khimija **16**, 148 (1990)
9. A. Darovsky, Polymorphism of the lanthanide bis-phtalocyanines (Universiteit van Amsterdam, Amsterdam, 1992) p. 114
10. V.T. Lebedev et al., Study of structure of pyrolyzed bis-phtalocyanines of rare-earth elements by small-angle neutron scattering. Report PNPI, 2921, Gatchina, Russia, 13 (2013). (In Russian)
11. A.E. Sovestnov et al., Research of evolution of crystal structure and electron structure of pyrolyzed yttrium bis-phtalocianine. Report PNPI, 2921, Gatchina, Russia, 23 (2013). (In Russian)

Chapter 27
Microscopic Analysis Using Gaze-Based Interaction

Peter Frühberger, Edmund Klaus and Jürgen Beyerer

Abstract Fraunhofer IOSB is currently constructing an automated microscopic laboratory. Different optical microscopes will be used to analyze specimen and gain new information out of the combination of the acquired sensor data. In order to narrow down specific patterns or tune the subsequent automated microscopy process, specimen have to be examined before the actual analysis process. To keep this step less time consuming and fatiguing, we designed an intuitive human-machine interface consisting of automated focusing algorithms and gaze-based interaction. The most important task in microscopy is focusing a desired region of a specimen. Considering microscopic analysis as a visual search task [1], this region of interest is exactly the region the operator is looking at. We use this information to automatically detect the focus plane in this region, by utilizing the microscope's z-axis, and present the operator a focused image. The user does not need to know how to operate the microscope in detail, but however is successful by using the encapsulated sophisticated operational sequences like initial, continuous focusing or synthetic image enhancements. This paper presents results of realizations for guided and fully automated microscopic analysis incorporating intuitive interaction and eased operation while gaining high quality results.

27.1 Introduction

Industrial quality assurance especially in semiconductor industry uses microscopy devices to be able to guarantee a high level of quality. When analyzing different parts of variable height, width or other physical properties a fixed unit assembly line must continuously be adjusted to match those different components. Therefore an operator

P. Frühberger (✉) · E. Klaus · J. Beyerer
Fraunhofer Institute of Optronics, System Technologies and Image Exploitation IOSB, Fraunhoferstr. 1, 76131 Karlsruhe, Germany
e-mail: peter.fruehberger@iosb.fraunhofer.de

E. K. Polychroniadis et al. (eds.), *International Multidisciplinary Microscopy Congress*, Springer Proceedings in Physics 154, DOI: 10.1007/978-3-319-04639-6_27,

has to gain preliminary information about the series to properly setup the sensors of this assembly line. The operator has to use a microscopic device himself to define focal planes that are to be examined later on.

Successfully using a microscope is an ambitious task. Navigating too far into one direction may severely harm equipment or specimen. In contrast to e.g. consumer cameras microscopes are equipped with expert software, which is usually not to be used entirely in an intuitive way. Furthermore every vendor uses its own software components and therefore using different hardware also involves learning new technology and operation principals. Out of this reason microscope operators need to have expert knowledge of the microscopic sensor itself and also attended a professional training in order to get familiar with the microscope. In a semi or fully automated industrial process an operator should be able to concentrate on the results of his work, rather than on quirks of the concrete inspection hardware used. Most of the time the results of these operational sequences can be achieved by only a subset of the shipped software. It is therefore quite usual, when doing quality assurance on an industrial level, that software gets adjusted and often heavily changed to better integrate in the given process. Our approach goes one step further. On the one hand a highly sophisticated software encapsulates the functionality of the microscope software into an easy to use application programming interface (API) that simplifies the interaction with given standard software. On the other hand we instrument a state of the art human-machine interface to realize the operators' tasks like e.g. focusing, analysis via gaze-based interaction.

27.2 Hardware and Software Setup

In our example setup we use a macroscope of Leica company, Z16 APO A which is equipped with a movable z-axis and continuous motorized magnification between 7.1× and 115×. Pictures and video sequences are captured by a high resolution camera with more than 2 MP resolution. The macroscope is instrumented through the Leica software development kit (SDK), which is a low level API to control the specific features of the device through C++. For this example realization we used the light dome, the z-axis and the zoom control. For eye gaze measurement, a Tobii X60 remote eye tracker was used. It was combined with a 24″ monitor (1920 × 1080 pixel) using a Tobii Monitor Mount. The eye tracker provides gaze data up to 60 times a second. According to the manufacturer, the eye tracker features an accuracy of 0.5°, which corresponds to a "region of uncertainty" of about 1.13 cm (43 pixels) at a viewing distance of approximately 65 cm [2]. The eye tracker was calibrated for each user prior to usage (Fig. 27.1a).

They eye tracker data is received via gigabit ethernet and packaged into JSON packages for further transmitting. This fully transparent protocol for serializing and transmitting data over a network connection is used to separate the eye tracker from the actual computer system which controls the microscope. On the side of the control system a separate server thread is used to communicate over this network. Button

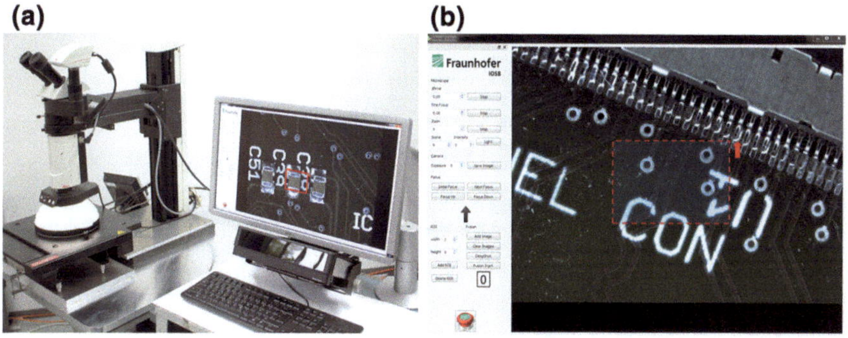

Fig. 27.1 Picture of setup with microscope, eyetracker and GUI realisation

presses and gaze data is used to control the region of interest (ROI) at the graphical UI. Furthermore the focusing direction of the microscope can be set. The same computer system instruments the high res camera to constantly display images on screen. In our implementation we use 30 fps which fits perfectly into the 60 hz refresh rate of the screen. Advanced software techniques are used in order to guarantee a direct visual feedback for the user and on the same time displaying images without any visual stutter. Image processing, like focusing, maximum contrast detection or image fusion is done in real-time, while the display is still updated. We used advanced threading concepts and modern CPU extensions, e.g. SSE2 to implement this experience. A great benefit of this loosely coupled design, is that other components, like a positioning stage or a robot can be integrated by simply reusing those components. The core was implemented with C++, the uncoupled graphical user interface (GUI) used QT4 (Fig. 27.1b).

27.3 Algorithm and Implementation

In this prototype the implementation was used to fulfill several use cases. The first use case was focusing the upper most focal plane of a specimen in order to get a visual pleasing start for the operator. Therefore the microscope z-axis was driven all the way up, as navigating away from the specimen is always safe concerning damage of the objective or specimen. From this upper point the first focal plane was searched by asynchronously driving the microscope down in the direction of the minimum position, while at the same time constantly measuring the focus value of the current camera image. This approach appears as fluent as possible to the operator, because the microscope does not have to stop while image processing is taking place. It is therefore very important that the queue of the images to be examined does not grow too large. The algorithm must be fast enough to handle the speed of the microscope's z-axis. This prototype uses a variance based contrast measure. Starting with a 3×3

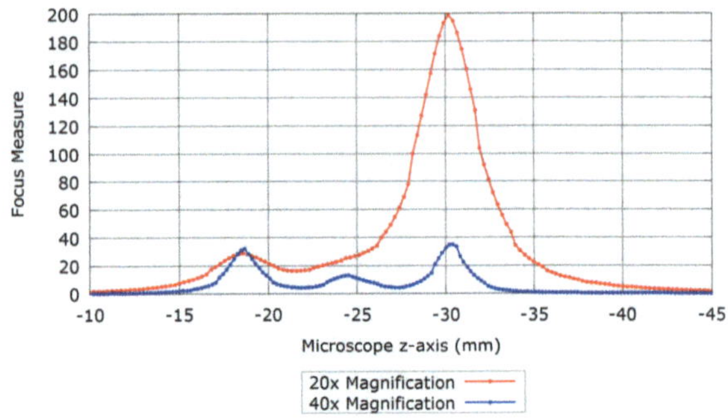

Fig. 27.2 Contrast values depending on the height of the z-axis

Median Filter, that was applied on the region of interest $R(h, w)$ with height h and width w to reduce noise effects, the local mean $\mu(x, y)$ at pixel position (x,y) of the image R was computed. For this lowpass operation a Gaussian filter with a mask of size 16×16 which corresponds to $\sigma \sim 5.0$ was used. The variance $\sigma(x, y)$ can be determined by calculating the square root of the low pass filtered absolute difference of $\mu(x, y)$ and the original image values at $R(x, y)$. The contrast value (c_R) was then computed as the sum of all intensity values of the resulting image. In order to compare that value between differently sized regions it was scaled with the product of the lateral dimensions of the Region of Interest $h \cdot w$. Figure 27.2 illustrates the contrast value when investigating a specific specimen region from top to bottom, the computations needed are summarized as follows [3]:

$$\mu(x, y) = LP\{R(x, y)\}$$

$$\sigma(x, y) = \sqrt{LP\{(R(x, y) - \mu(x, y,))^2\}}$$

$$c_r = \frac{1}{h \cdot w} \cdot \sum_{x=0}^{h} \sum_{y=0}^{w} \sigma(x, y)$$

Figure 27.2 makes clear, that changing the magnification will also alter the field of vision, e.g. the lateral dimension examined. The contrast measure must therefore cope with such different situations. It can also be seen, that the influence of the global maximum with low magnification is quite strong, as a small magnification also results in a larger depth of view, which results in local maximums that can easily be overseen when not adjusting the ROI. As the depth of view gets a lot smaller when increasing magnification, these regions get visible in the plot when increasing magnification. Here the eye gazed approach is of high value, as the ROI can be changed during run time and therefore relevant regions can be examined. On the other hand the contrast measure is helping the user to filter out false positives by not stopping too often and therefore giving the possibility for a smooth region driven operation. Another use case is to combine the region of interests throughout the different focal planes to create a synthetic enhanced image [4]. When measuring height differences between

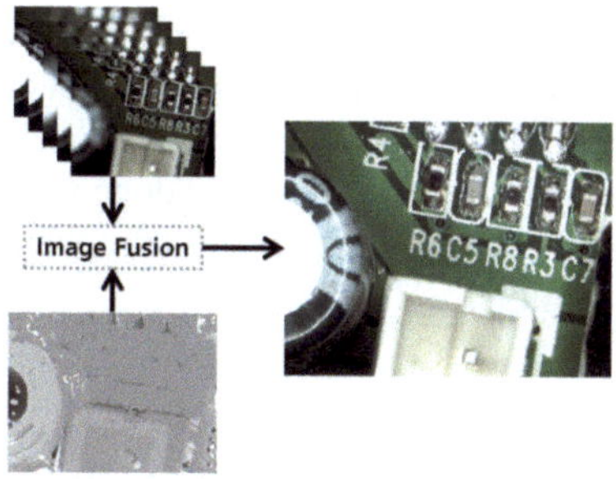

Fig. 27.3 Complete image fusion process

different focal planes, it is important to choose an appropriate contrast measure. In our prototypical realization the depth map was not used for further measurement, so the requirement for the contrast measure was not that demanding. But this depth map could be used as basis for the synthetic enhanced result image. It's goal was to present the specimen quality as an overview for the operator to see all relevant parts at once. Therefore the fusion process should not introduce additional artifacts either. The focal planes that were incorporated into the depth map were chosen by the operator himself when navigating through the relevant ROIs. The software remembered those positions and saved the relevant image data with the associated contrast measure and z-axis value on an image stack for later computation. Calculating the depth map is done by evaluating the contrast value at each pixel for each image on this stack. The resulting depth map might contain local artifacts because of noise or sensitivity of the used contrast measure. We therefore used an additional Gaussian filter to smooth this fusion map. The mask of this filter was chosen linearly scaled concerning the pixel resolution of the camera. Our tests have shown that a filter mask for an image $I(h, w)$ with the dimensions h, w of the size $\frac{1}{64} \cdot \max(h, w)$ was sufficient to compute the blurred depth image for the later image fusion process.

As the input images are RGB color images, great care needs to be taken when assembling the resulting fusion image [5]. Color images are quite sensitive when changing specific color channels. Those minimal changes are easily visible for human eyes. The fusion map was therefore calculated with fractional values, e.g. the values of the fusion map donated a weighting factor, which can automatically be used to scale the values of the specific channels between the channel value of two nearby planes. This approach greatly improved the resulting image quality. The complete fusion process can be seen in Fig. 27.3. Concerning the implementation, the fusion process has been multi-threaded, as it is quite easy to compute multiple depth maps from a selected range of input layers and merge them together later. In a normal scenario, when less than ten focal planes are of importance, it makes no sense of doing multi-threaded computation. For calculating a fine granular depth map with more than two hundred full resolution images on the stack, it is quite useful though. The image stack is sorted concerning the z-axis position of the images. Each thread t_i is computing a number of $\frac{\#images}{\#threads}$ input images, despite the last one, which could get the reminder. All threads store its resulting depth map and corresponding focus value image. After all threads have finished, it is enough to compute those resulting images once. As the input images are not changed at all, they even don't need to be copied and can be kept as is. This approach can become useful, when high resolution depth map estimation should be done with direct interaction of a user. With that technique and other optimization in the image computing code soft real-time image fusion is possible.

27.4 Summary and Outlook

We have shown a prototype implementation of a gaze based driven approach for controlling and using a microscope. It emphasized the fact that complex algorithms can be encapsulated into software, while making their usage quite intuitive. Visual inspection has been formulated as a search task, by directly using gaze based interaction to solve this task. For the future we plan on integrating a positioning stage into the proposed setup to also allow lateral navigation.

In order to validate this novel approach, an evaluation based on established statistic methods needs to be done. Therefore we will incorporate a user study that compares traditional usage of this microscopic setup in comparison to our new method of interaction. Furthermore it could be examined, if those results can be easily transferred onto other microscopic devices, where contrasting methodology cannot be used, e.g. White Light Interferometry. These devices are focusing differently, therefore the visual feedback has to be adjusted in order to present meaningful information.

The presented setup should be integrated into a productive environment in order to evaluate the possibilities of such a new approach under industrial conditions.

References

1. J. Hild, E. Müller, E. Klaus, E. Peinsipp-Byma, J. Beyerer, Evaluating multi-modal eye gaze interaction for moving object selection, in *International Conference on Advances in Computer-Human Interactions (ACHI)* (2013), pp. 454–459
2. J. Hild, E. Peinsipp-Byma, E. Klaus, Improving usability for video analysis using gaze-based interaction, in *Full Motion Video (FMV) Workflows and Technologies for Intelligence, Surveillance, and Reconnaissance (ISR) and Situational Awareness* (2012)
3. J. Beyerer, F. Puente León, C. Frese, *Automatische Sichtprüfung: Grundlagen, Methoden und Praxis der Bildgewinnung und Bildauswertung* (Springer Vieweg, Stroud, 2012).
4. M. Heizmann, F. Puente León, Fusion von bildsignalen. Tech. Mess. **74**, 130–138 (2007)
5. R.C. Gonzalez, R.E. Woods, *Digital Image Processing* (Pearson Education, Upper Saddle River, 2008)

Chapter 28
Epithelial Cell Detection in Endomicroscopy Images of the Vocal Folds

Firas Mualla, Simon Schöll, Christopher Bohr, Helmut Neumann and Andreas Maier

Abstract The reasons for a hoarse voice are manifold. Besides structural changes such as additional masses on the vocal folds, changes in the layers of the vocal fold mucus influence the acoustic properties of the voice signal [1]. In our research, we aim to examine this in vivo. One suitable technique for this purpose is the use of micro endoscopes. In contrast to traditional microscopes, the micro endoscopes have a reduced image quality and exhibit strong noise artifacts. Furthermore, images are affected by inhomogeneous illumination. All of the mentioned effects pose a challenge to automatic cell detection and segmentation methods. In this paper, we investigate whether automatic cell detection methods are also suitable for the cells of the epithelium of the vocal folds. Based on band-pass filtering, we could successfully reduce noise and emphasize cell boundaries at the same time. The pass-band was experimentally chosen to emphasize the regular structure of the epithelial cells which can be observed in the frequency domain of the cell image. Subsequently, we applied a watershed segmentation to identify the cell borders. Cell centers were located using a local minima search in the band-pass filtered image. First results indicate that

F. Mualla (✉) · S. Schöll · A. Maier
Pattern Recognition Lab, Department of Computer Science, Friedrich-Alexander University Erlangen-Nuremberg, Martensstr. 3, 91058 Erlangen, Germany
e-mail: firas.mualla@cs.fau.de

S. Schöll
ASTRUM IT GmbH, Am Wolfsmantel 2, 91058 Erlangen, Germany

S. Schöll · A. Maier
Erlangen Graduate School in Advanced Optical Technologies (SAOT), Friedrich-Alexander University Erlangen-Nuremberg, Erlangen, Germany

C. Bohr
Department of Otorhinolaryngology, Head and Neck Surgery, Erlangen University Hospital, Erlangen, Germany

H. Neumann
Department of Medicine I, Friedrich-Alexander University Erlangen-Nuremberg, Erlangen, Germany

E. K. Polychroniadis et al. (eds.), *International Multidisciplinary Microscopy Congress*, Springer Proceedings in Physics 154, DOI: 10.1007/978-3-319-04639-6_28,

the method is able to locate and outline epithelial cells with high accuracy. Future research will focus on the relation between such quantitative measures in cell images to acoustic properties of the voice signal and the mechanical properties of the vocal folds such as the synchrony of their vibration.

28.1 Introduction

Voice hoarseness can be caused by several reasons including laryngitis, cancer of larynx, and structural changes in the vocal folds such as nodules and polyps. Recently, it was shown that changes in the mucus of the vocal folds can be related to acoustic properties of the voice signal [1]. We aim to investigate the vocal fold mucus in vivo using a micro endoscope. An essential step towards this goal is the detection of epithelial cells in the mucus layer which is the topic of this paper.

A plenty of cell detection approaches are available in literature [2–6]. Images of the epithelial cells exhibit two important properties. Firstly, due to physiological reasons, the epithelial cells cover the whole scene. Therefore, the separation between cells and background, which is necessary in several proposed approaches [4, 6, 7], is not required. Secondly, there is a repetitive pattern. The latter was exploited in [8] and [9] for cell density estimation in the corneal endothelium. The purpose of this paper is to investigate whether it is possible to utilize the two aforementioned facts so that basic image processing algorithms can be applied in order to detect epithelial cells in endomicroscopy images of the vocal folds.

28.2 Materials and Methods

28.2.1 Materials

A sample of nine images of the epithelium of the vocal folds were acquired using a micro endoscope of a Cellvizio probe-based confocal laser endomicroscopy (pCLE) system. Figure 28.1 shows two examples.

28.2.2 Detection Pipeline

We apply a band-pass filter on the input image. Cell centers are then found using a minima search procedure. Watershed algorithm is utilized in order to delineate the cell borders. The pipeline is demonstrated in Fig. 28.2. Minima search and watershed in this pipeline are parameterless. On the other hand, the pass-band of the filter must

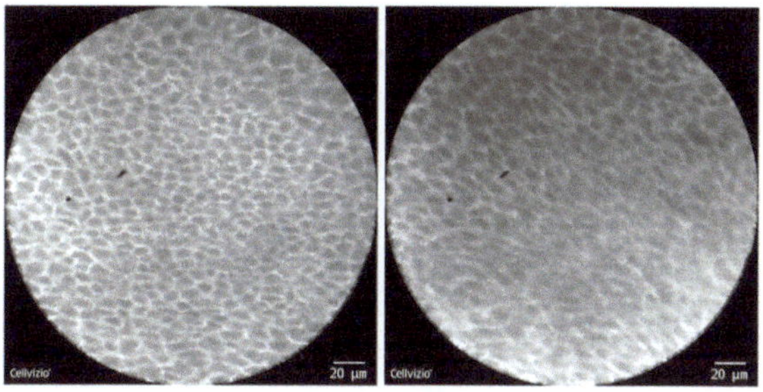

Fig. 28.1 Endomicroscopy images of the vocal fold epithelium

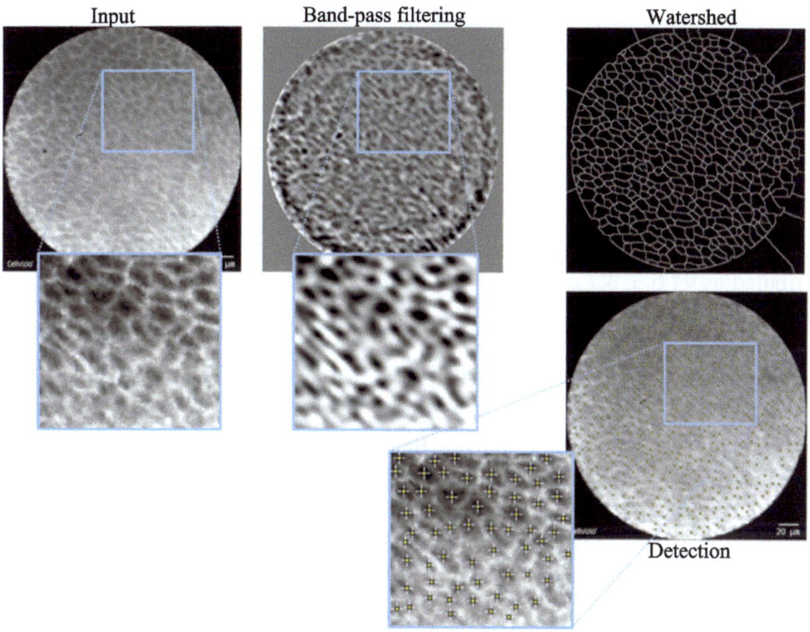

Fig. 28.2 Cell detection pipeline

be tuned. The goal of the tuning is emphasizing the regular pattern of the epithelial cells and at the same time reducing noise and smoothing cellular details.

28.3 Evaluation

A band-pass filter was manually designed in Fourier domain for each image and the pipeline described above was applied. Figure 28.3 exemplifies the results. The

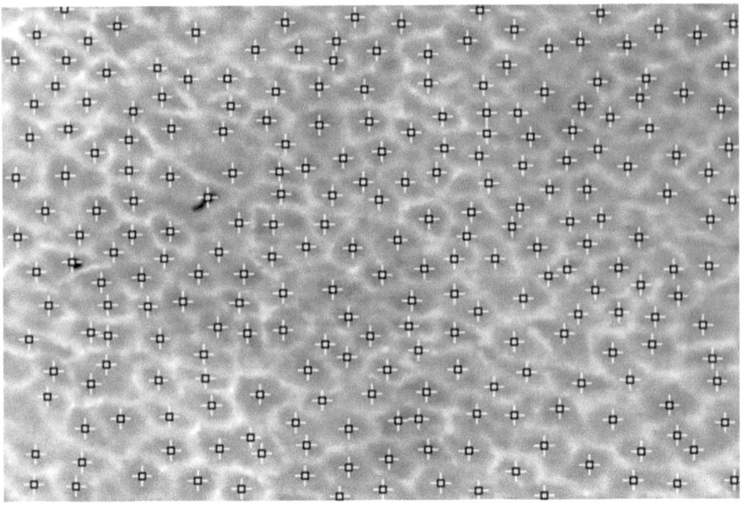

Fig. 28.3 An example of detection results

obtained F-measure of cell detection, averaged over the nine images, was 80.2 ± 4.7 distributed as 94.6 ± 3.7 recall and 70.0 ± 7.3 precision.

28.4 Conclusion and Discussion

It is well known from Fourier analysis that periodicity in space manifests itself in Fourier domain as a peak at the fundamental frequency of the signal. In the case of the 2D Fourier transform, a frequency component along a direction ϕ in space conforms to a peak at the corresponding frequency in the same direction ϕ in the 2D Fourier plane. This fact was exploited in [8] for cell density estimation in the corneal endothelium. Moreover, it was shown that the repetitive pattern information exists inside a ring in the Fourier domain. The radius of the ring is a measure of the endothelial cell density.

We noticed in preliminary experiments (data not shown) that the aforementioned ring is more apparent in the images of the corneal endothelium compared to our images. Therefore, estimating cell density by measuring the ring's radius is a harder task in our case. Nevertheless, the frequency domain is likely to have a distinguishable band. The question which naturally arose was whether there exists a band-pass filter for each image which makes cell detection possible using basic image processing techniques. Our results show that this filter exists. In future work, we plan to design the filter automatically based on the frequency content of the image. In addition, we want to investigate the relation between the quantitative image processing results, the mechanical characteristics of the vocal folds, and acoustic properties of the voice signal.

Acknowledgments The authors would like to thank the Bavarian Research Foundation BFS for funding the project COSIR under contract number AZ-917-10. In addition, we gratefully acknowledge funding of the Erlangen Graduate School in Advanced Optical Technologies (SAOT) by the German Research Foundation (DFG) in the framework of the German excellence initiative. Special thanks go to Bastian Bier for labeling the images.

References

1. S.A. Klemuk, T. Riede, E.J. Walsh, I.R. Titze, Adapted to roar: functional morphology of tiger and lion vocal folds. PLoS ONE **6**(11), e27,029 (2011)
2. G. Becattini, L. Mattos, D. Caldwell, A novel framework for automated targeting of unstained living cells in bright field microscopy. In: Proceedings of the IEEE International Symposium on Biomedical Imaging: From Nano to Macro, pp. 195–198 (2011)
3. X. Long, W. Cleveland, Y. Yao, A new preprocessing approach for cell recognition. IEEE Trans. Inf. Technol. Biomed. **9**(3), 407–412 (2005)
4. F. Mualla, S. Schöll, B. Sommerfeldt, A. Maier, J. Hornegger, Automatic cell detection in bright-field microscope images using SIFT, random forests, and hierarchical clustering. IEEE Trans. Med. Imag. **32**(12), 2274–2286 (2013). doi:10.1109/TMI.2013.2280380
5. T. Nattkemper, H. Ritter, W. Schubert, Extracting patterns of lymphocyte fluorescence from digital microscope images. Int. Data Anal. Med. Pharmacol. **99**, 79–88 (1999)
6. J. Pan, T. Kanade, M. Chen, Heterogeneous conditional random field: realizing joint detection and segmentation of cell regions in microscopic images. In: CVPR (2010)
7. F. Mualla, S. Schöll, B. Sommerfeldt, J. Hornegger, Using the monogenic signal for cell-background classification in bright-field microscope images. Proc. des Workshops Bildverarbeitung für die Med. **2013**, 170–174 (2013)
8. M. Foracchia, A. Ruggeri, Estimating cell density in corneal endothelium by means of fourier analysis. In: Engineering in Medicine and Biology, 2002. 24th Annual Conference and the Annual Fall Meeting of the Biomedical Engineering Society EMBS/BMES Conference, 2002. Proceedings of the Second Joint, vol. 2, pp. 1097–1098. IEEE (2002)
9. A. Ruggeri, E. Grisan, J. Jaroszewski, A new system for the automatic estimation of endothelial cell density in donor corneas. Br. j. Ophthalmol. **89**(3), 306 (2005)

Part II
Applications of Microscopy in the Biological Sciences

Chapter 29
The Elemental Composition of Cod and Salmon Bones Derived Powder Using SEM-EDX and ICP-OES

Krzysztof Marycz, Zbigniew Dobrzański, Fabiola Bubel, Agnieszka Śmieszek and Jakub Grzesiak

Abstract The study aimed at comparison of the results concerning elemental composition of cod and salmon fishbone powders determined using two methods, i.e. scanning electron microscopy with an energy dispersive X-ray analytical system (SEM-EDX) and inductively coupled plasma—optical emission spectroscopy (ICP-OES). The research material were fishbone powders of marine fish: cod (*Gadus morhua*) from Baltic Sea and salmon (*Salmo salar*) from Atlantic Ocean obtained using the method of chemical extraction. The analysis included main mineral elements: calcium, magnesium, sodium, phosphorus. The results obtained from SEM-EDX appeared to be compliant with the results obtained based on ICP-OES analysis. It was demonstrated that combination of two advanced techniques for fishbone powders characterization is essential for suitable analysis of its quality as well as quantity, especially for subsequent determination of supplements composition. It may be concluded that SEM-EDX method may be successfully used for preliminary analysis of macroelements content in biological material.

K. Marycz (✉) · A. Śmieszek · J. Grzesiak
Electron Microscopy Laboratory, Wroclaw University of Environmental and Life Sciences, Kożuchowska 5b, 50-631 Wroclaw, Poland
e-mail: krzysztof.marycz@up.wroc.pl

K. Marycz · Z. Dobrzański · A. Śmieszek · J. Grzesiak
Department of Animal Hygiene and Animal Welfare, Wroclaw University of Environmental and Life Sciences, Chełmońskiego 38C, 50-631 Wroclaw, Poland
e-mail: zbigniew.dobrzanski@up.wroc.pl

F. Bubel
Laboratory of Biotechnology, Institute of Opencast Mining, Poltegor—Institute, Parkowa 25, 51-625 Wroclaw, Poland
e-mail: fabiola.bubel@igo.wroc.pl

A. Śmieszek
e-mail: agnieszka.smieszek@up.wroc.pl

J. Grzesiak
e-mail: jakub.grzesiak@up.wroc.pl

E. K. Polychroniadis et al. (eds.), *International Multidisciplinary Microscopy Congress*, Springer Proceedings in Physics 154, DOI: 10.1007/978-3-319-04639-6_29,

29.1 Introduction

The organic trace minerals, such as calcium, phosphorus, magnesium and sodium, play a key role in animal nutrition, both in case of pets as well as in breeding animals. As macroelements, they also are significant in human metabolism [1].

Wastes from the fish industry are easy available and cheap source of organic and inorganic elements [2]. They may be used in bone meals production as mineral additives for the animals, or for manufacturing of special preparations rich in calcium—diet supplements for human [3, 4]. Numerous methods, mainly spectrometric ones, are used in laboratory analytics for an assessment of mineral composition of biological material [5, 6]. The methods of roentgen analysis are used more rarely [7, 8].

The aim of this study was to compare the elemental composition of cod and salmon fishbone powders with respect to macroelements content using two analytical methods.

29.2 Material and Methods

The research material were fishbone powders of marine fish from Baltic Sea (*Gadus morhua*) and Atlantic Ocean (*Salmo salar*) obtained using the method of chemical extraction described in Polish Patent Application (no P.403123). This material, as cod bone powder (CBP) and salmon bone powder (SBP) was in a dry state, minced to granulation of 0.1 mm. Analysis of chemical compositions (main mineral elements: calcium, magnesium, sodium, phosphorus) was performed using two advanced techniques: (I) scanning electron microscopy with an energy dispersive X-ray analytical system (SEM-EDX) and (II) inductively coupled plasma—optical emission spectroscopy (ICP-OES). Five samples from CBP and SBP were collected for analysis.

For scanning electron microscopy, cod and salmon fishbone powders were dried, placed on stages and coated with gold (Edwards, Scxancoat six). Specimen was observed using SE1 detector, at 10 kV of filament tension (SEM, Zeiss Evo LS15). Additionally, the distribution of elementals was performed at magnifications of 1000× using EDX (Energy-Dispersive X-ray) detector (Quantax, Bruker).

The content of macroelements was evaluated using spectrometric method with an application of ICP-OES Vista MPX analyzer (Varian, Australia). Previously, the material samples (bone powders) were mineralized with spectrally pure nitric acid (Merck) under the pressure of 100 atm. in microwave device of Milestone type (Italy).

The results were elaborated statistically calculating the mean values and standard deviations. The differences between the groups (CBP and SBP) were determined using students t-test at a significance level of $p < 0.05$.

Table 29.1 The content of main elements in fishbone powders [%]—SEM-EDX method

Element	CBP		SBP	
	$\overline{x}$	SD	$\overline{x}$	SD
Ca	37.96*	2.51	24.53	5.45
P	15.88*	0.73	9.71	1.74
Na	1.31	0.22	1.12	0.20
Mg	0.82*	0.06	0.60	0.05

*$P < 0.05$ (between CPB and SBP)

Table 29.2 The content of main elements in fishbone powders [%]—ICP-OES method

Element	CBP		SBP	
	$\overline{x}$	SD	$\overline{x}$	SD
Ca	27.39	5.48	24.86	4.97
P	13.49	2.70	12.41	2.48
Na	1.23	0.25	1.00	0.20
Mg	0.66*	0.13	0.46	0.09

*$P < 0.05$ (between CPB and SBP)

29.3 Results and Discussion

The results obtained using SEM-EDX method (Table 29.1) showed that CBP contained higher concentrations of calcium, phosphorus and magnesium compared to SBP. The differences observed were statistically significant. Concentration of sodium in both powders was similar. The results obtained from SEM-EDX were compliant with the results obtained from ICP-OES analysis (Table 29.2). Concentration of calcium, phosphorus, magnesium and sodium in CBP was higher compared to SBP, the differences were on a level of 10, 8.7, 43.5 and 23 %, respectively, however statistically significant difference was only noted in case of magnesium.

The results of main macroelements determination in fishbone powders, performed using two methods, are difficult to compare due to different techniques of samples preparation and analysis itself, but the tendencies in particular elements content were similar. In case of SEM-EDX method they were as follows: Ca>P>Na>Mg, while in case of ICP-OES method it was: Ca>P>Mg>Na. Generally, the concentrations of Ca (CPB) and P (CBP and SBP) were higher in case of SEM-EDX method compared to ICP-OES.

The results of main macroelements content obtained in this study are quite different compared to the literature data. For example, Toppe et al. [2] noted the values of Ca: 19 and 13.5 %, P: 11.3 and 8.1 %, Mg: 0.3 and 0.22 %, Na: 0.77 and 0.57 % in dry matter for unprocessed (raw) cod and salmon fishbones, respectively. Other data were provided by Liset et al. [9] for an ash from Atlantic salmon fishbone after enzymatic hydrolysis. An average content of Ca was 16.7 %, P—10.5 %, while Na—0.45 % and Mg—0.36 %. The results of this study are considerably higher than the cited ones, especially in case of Ca and P, which may result from the fact of almost all

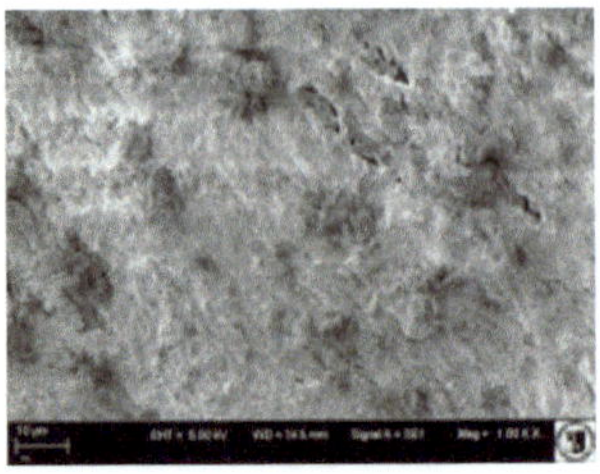

Fig. 29.1 SEM image of cod bone powder (mag. 1000×)

Fig. 29.2 SEM image of salmon bone powder from (mag. 1000×)

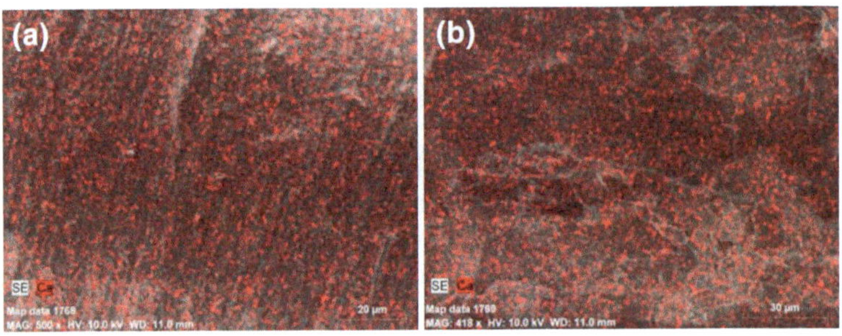

Fig. 29.3 Comparison of calcium (Ca) content in CBP (**a**) and SBP (**b**)

fat and protein removing from the fishbone during processing technology (alkaline extraction).

Figures 29.1 and 29.2 presenting morphological structure of cod and salmon bone powder are interesting, since they point insignificantly different histological structure which may affect chemical composition. It may be noted for example, that SBP contains higher number of extra cellular matrices compared to CBP, while CBP includes more bony (mineral) elements. This resulted from the fact that SBP contains slightly higher amount of organic matter (fat and protein residues) compared to CBP [4]. Figures 29.3, 29.4, 29.5 and 29.6 present the maps of the examined elements. It is difficult to conclude visually which powders contain higher amount of Ca, P, Mg or Na; colors intensity on the images is similar, however computer analysis points some values differentiation in relative units (percents), which is presented in Table 29.1.

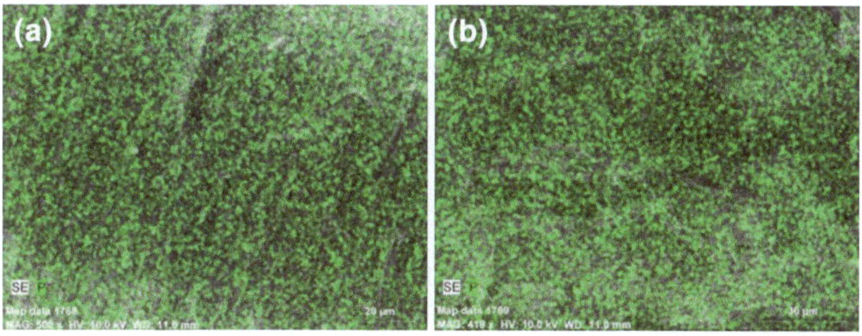

Fig. 29.4 Comparison of phosphorus (P) content in CBP (a) and SBP (b)

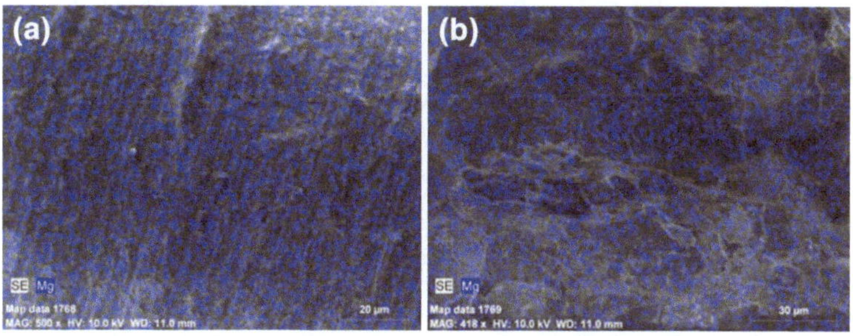

Fig. 29.5 Comparison of magnesium (Mg) content in CBP (a) and SBP (b)

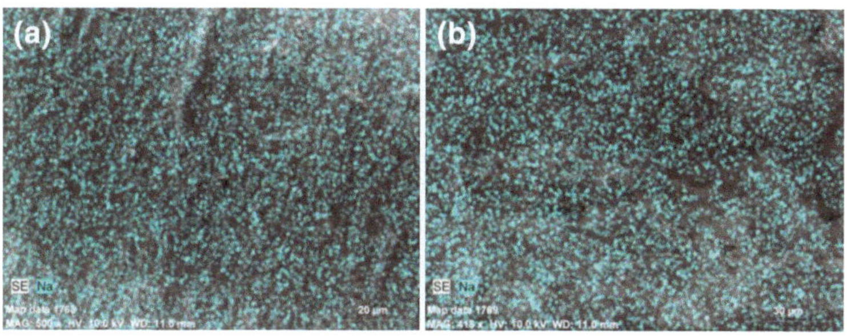

Fig. 29.6 Comparison of sodium (Na) content in CBP (a) and SBP (b)

29.4 Conclusion

We showed that combination of two advanced techniques for fishbone powders characterization is essential for suitable analysis of its quality as well as quantity, especially for subsequent determination of supplements composition.

It seems that SEM-EDX method may be used for preliminary analysis of macroelements content in biological material.

Acknowledgments The study was conducted within the research project No. NN209756840 financed by the National Science Center, Poland.

References

1. K.O. Soetan, C.O. Olaiya, O.E. Oyewole, The importance of mineral elements for humans, domestic animals and plants: a review. Afric. J. Food Sci. **4**(5), 200–222 (2010)
2. J. Toppe, S. Albrektsen, B. Hope, A. Aksnes, Chemical composition, mineral content and amino acid and lipid profiles in bones from various fish species. Com. Biochem. Physiol. Part B. **146**, 395–401 (2007)
3. J. Huo, S. Deng, Ch. Xie, G. Tong, Preparation and biological efficacy of haddock bone calcium tablets. Chin. J. Ocean. Limnol. **28**(2), 371–378 (2010)
4. Z. Dobrzański, F. Bubel, K. Chojnacka, P.J. Bykowski, J. Bolanowski, T. Trziszka, Physico-chemical properties of solid fish-processing waste and their uses in nutraceutical and biomedical productions. Przem. Chem. **91**(5), 733–738 (2012)
5. M. Baker, Mass spectrometry for biologists. Nature Meth. **7**, 157–161 (2010)
6. O.T. Butler, W.R.L. Cairns, J.M. Cook, C.M. Davidson, Atomic spectrometry update. Environmental analysis. J. Analyt. Atomic Spectr. **27**(2), 187–221 (2012)
7. K. Kaliński, K. Marycz, J. Czogała, E. Serwa, W. Janeczek, An application of scanning electron microscopy combined with roentgen microanalysis (SEM-EDS) in canine urolithiasis. J. Electron. Micros. **61**(1), 47–55 (2012)
8. E. Walkowicz, K. Czyż, B. Patkowska-Sokoła, Use of X-ray analysis of a biological material for an assessment of natural environment status. Przem. Chem. **92**(9), 1765–1767 (2013)
9. B. Liset, K. Julshamn, M. Espe, Chemical composition and theoretical nutritional evaluation of produced fractions from enzymic hydrolysis of salmon frames with Promamex™. Process Biochem. **38**, 1747–1759 (2003)

Chapter 30
Scanning X-Ray Electron Microscopy (SEM-EDX) as a Therapeutic Tool in the Diagnosis of Equine Metabolic Syndrome (EMS)

Krzysztof Marycz, Agnieszka Śmieszek and Jakub Nicpoń

Abstract Equine metabolic syndrome (EMS) is a disease entity characterized by a set of clinical sings such as significant obesity, increased insulin and glucose level and development of laminitis. This syndrome occurs more frequently in draft and/or pony horses. The diagnosis process of EMS is difficult and still not fully recognized. The screening tests, recommended for proper recognition of EMS, can be affected by external factors such as pain and/or stress. Moreover diagnosis poses life threatening situation for the patient. Searching for an appropriate and noninvasive method for the assessment of EMS symptoms, the authors decided to evaluate the usefulness of SEM-EDX technique. This analytical method was used to determinate chemical composition of hairs collected form healthy and EMS horses. Comparison of Pb, Cr, Ca and Fe content revealed statistically significant differences between investigated groups of animals. Basing on the obtained results authors recommend SEM-EDX method as an adjunctive diagnostics tool in equine metabolic syndrome (EMS) diagnosis.

K. Marycz (✉) · A. Śmieszek
Electron Microscopy Laboratory, Wroclaw University of Environmental and Life Sciences, Kożuchowska 5b, 50-631 Wroclaw, Poland
e-mail: krzysztof.marycz@up.wroc.pl

K. Marycz · A. Śmieszek
Department of Animal Hygiene and Animal Welfare, Wroclaw University of Environmental and Life Sciences, Chełmońskiego 38C, 50-631 Wroclaw, Poland
e-mail: agnieszka.smieszek@up.wroc.pl

J. Nicpoń
Department of Surgery, Faculty of Veterinary Medicine, Wroclaw University of Environmental and Life Sciences, pl. Grunwaldzki 51, 50-366 Wrocław, Poland
e-mail: jakub.nicpon@up.wroc.pl

E. K. Polychroniadis et al. (eds.), *International Multidisciplinary Microscopy Congress*, Springer Proceedings in Physics 154, DOI: 10.1007/978-3-319-04639-6_30,

30.1 Introduction

The metabolic disorders in horses are serious clinical issue not only due to high occurrence but also because of broad spectrum of accompanying symptoms, that are difficult to diagnose [1]. Genetic predispositions of animal as well as environmental factors play main role in the course of disease, especially the impaired metabolism of glucose and lipids but also improper diet. Equine metabolic syndrome (EMS) includes following characteristic features: (i) hyperinsulinemia, (ii) glucose intolerance, (iii) elevated plasma lipid concentrations, (iv) laminitis and obesity. To underline the relationship between obesity and development of laminitis, the metabolic syndrome is also defined as obesity dependent laminitis (ODL). This metabolic disorders most often concerns heavy horses, which are predisposed both genetically and environmentally, but also due to limited physical activity.

The diagnostic process of EMS is very complex and requires the identification of numerous and diverse clinical signs. Recommended procedures include dynamic metabolic tests for instance the Glucose Insulin Combine Test (GCIT). This approach is affected by various external factors, but what is more important is highly invasive and often induces hypoglycemia in the examined animals and causes life threading situations. The elemental hair analysis, as well as mechanical one, became efficient tool for adjunctive diagnostics of metabolic diseases, especially in humans [2, 6]. The mechanical analysis of horse hair was used to determination of the diet effectiveness [8]. Hair bulbs are metabolically active, and they require suitable nutrition, essential for their growth. Therefore the analysis of hairs' chemical composition brings many information about physiological state of patient. Scalanaya et al. performed multielemental hair analysis using atomic emission and mass spectroscopy with inductively coupled plasma. Results of Scalanaya's research group revealed that in type 2 diabetes numerous and serious changes in mineral metabolism of hair are observed.

We decided to determinate if this phenomenon is observed in horses with metabolic syndrome, and relate analysis of hair chemical composition to the diagnostic proceeding. The analysis of elemental composition of hairs was performed using scanning electron microscopy with energy-dispersive X-ray spectroscopy (SEM-EDX). Four essential chemical elements were detected: lead (Pb), chromium (Cr), calcium (Ca) and ferrum (Fe). To determine the influence of physiological state on hair chemical composition the comparative analysis was performed, the investigated groups of animals consisted from healthy horses and draft horses, coming from Silesia region, with diagnosed metabolic syndrome. The aim of this study was to determine the usefulness of SEM-EDX as a possible, convenient and noninvasive, diagnostic tool for in EMS recognitions. Such approach could serve as complementary survey, that would help veterinary clinicians in the process of complex diagnostics.

30.2 Material and Methods

The hair samples were collected from 20 draft heavy horses form the Silesia region (Poland). Investigated groups consisted from 10 horses with metabolic syndrome (EMS) and 10 healthy horses. Groups were divided taking into account the results of dynamic metabolic tests (Glucose Insulin Combie Test). Moreover, the animals qualified for the research were in similar age and clinical picture. Material collected from investigated horses derived from central part of the mane. Each sample consisted of 20 hairs. Preparation of samples for SEM-EDX analysis included: (i) washing hairs with demineralized and deionized water, (ii) completely drying of samples and (iii) fixing samples with carbon tape. Material was analyzed using SEM (Zeiss, EVO LS 15) appointed with Bruker Quantax ESPPT 1.8.2 roentgen microanalyser. The elemental concentration of Pb, Cr, Ca and Fe was evaluated in each sample. The analysis was performed in five various parts of the hair in follicle area under 20 kV voltage according to the procedure described previously [3].

Statistical analysis was performed using the Statistica 7.0 software (StatSoft, Inc., Statistica for Windows, Tulsa, OK). Evaluation of data normality was performed using Shapiro-Wilk test, whereas the Levene's test was used to assess the equality of variances. Differences between two variables were analyzed using t-Student test or non-parametric Mann U Whitney test depending on data validation. The values with p less than 0.05 were considered to be significant.

30.3 Results and Discussion

Metabolic disorders of animals individuals are reason for the development of various chronic diseases e.g. equine metabolic syndrome (EMS). The metabolic diseases requires applying of complicated and sophisticated diagnostic tools, which very often become life treating situations e.g hypoglycemia. That is why searching of novel, less invasive methods of EMS diagnosis is highly desirable. One of the most considerable method is evaluation of elemental composition of skin appendages which is less invasive procedure.

Examination of specified elements content in horse hairs might be adjunctive diagnostic tool in veterinary medicine [2, 5–7]. The particular elements e.g. lead (Pb), chromium (Cr), calcium (Ca) and ferrum (Fe) are known as a important factors in process of insulin metabolism regulation. In this study, we decided to evaluate the concentration of mentioned elements in horses' hairs. The draft heavy horses were divided into two groups, suffering from EMS and healthy horses. The comparison between investigated groups, allowed to determined the influence of animal health status on hair elemental composition.

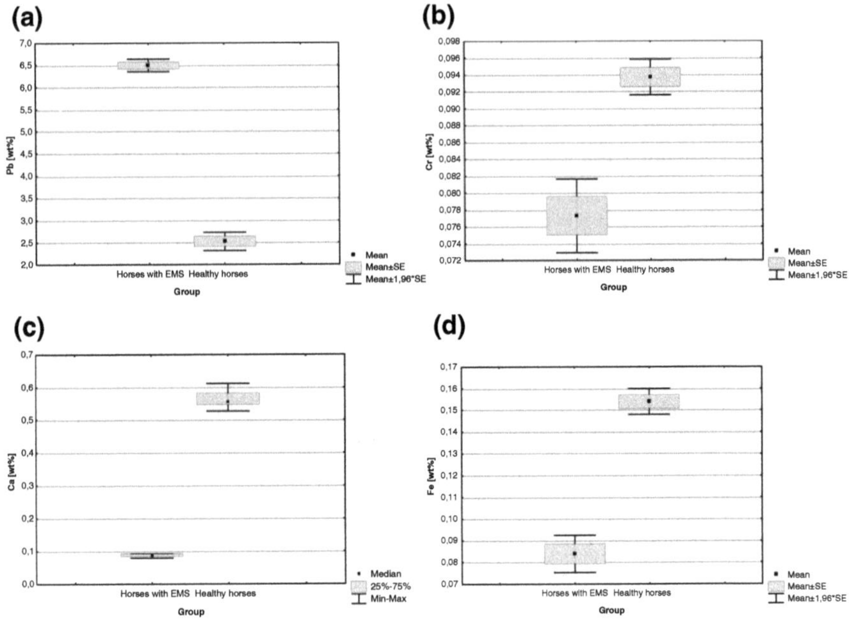

Fig. 30.1 The comparison of particular chemical elements content in the coat hair of EMS and healthy horses. Statistical analysis of the SEM-EDX results

The study performed with SEM-EDX demonstrated statistically significant higher level of lead in EMS horses when compared to the healthy individuals (Fig. 30.1a). This might result from an accumulation of lead ions in the hair coat which is a result of impaired metabolism. These results partially correlate with the findings published by Scalanaya et al. where high concentration of lead was noted in woman hairs with type 2 diabetes [6]. Another crucial chemical element is chromium, which play significant role in insulin action and its effects on carbohydrates. Our study demonstrated a significant decrease of this element in EMS horses' hairs, which confirms the abnormalities in metabolic regulations of insulin activity (Fig. 30.1b). Additionally the lower ions of calcium and ferrum levels were noticed in EMS group when compared to healthy one (Fig. 30.1c, d). Calcium plays an important role in maintaining of carbohydrates balance homeostasis and insulin regulation. Lowered level of Ca and Fe in animals with EMS may confirm metabolic clinical picture. However, the analysis of mentioned element in hair coat should be considered as an adjunctive diagnostic tool and still requires further research.

Figures presents the results of the comparative analysis performed within group of EMS and healthy horses, in terms of Pb (a), Cr (b), Ca (c) and Fe (d) content in coat hair. The results are statistically significant, $p < 0.05$.

30.4 Conclusion

In this research we demonstrated that SEM-EDX technique might be a helpful tool in equine metabolic syndrome (EMS) diagnosis. The SEM-EDX analysis of hair elemental composition is convenient and very reliable method, strongly correlating with physiological state of the organism.

Our data demonstrate that equine metabolic syndrome might affect on chemical composition of hairs, especially on Pb, Cr, Ca and Fe concentration.

In order to validate SEM-EDX technique as a diagnostic procedure, more groups of animals in different physiological condition is needed to research.

Acknowledgments The study was conducted within the research project No. NN 308266138 financed by the National Science Center, Poland.

References

1. R.A. Carter, L.J. McCutcheon, E. Valle, E.N. Meilahn, R.J. Geor, Effects of exercise training on adiposity, insulin sensitivity, and plasma hormone and lipid concentrations in overweight or obese, insulin-resistant horses. Am. J. Vet. Res. **71**(3), 314–332 (2010)
2. J.D. Campbell, Hair analysis: a diagnostic tool for measuring mineral status in humans journal of Orthomolecular. Psychiatry **14**(4), 276–280 (1985)
3. K. Kaliński, K. Marycz, J. Czogała, E. Serwa, W. Janeczek, An application of scanning electron microscopy combined with roentgen microanalysis (SEM-EDS) in canine urolithiasis . J. Electron. Micros. **61**(1), 47–55 (2012)
4. A.B. Chausmer, Zinc, insulin, and diabetes. J. Am. Coll. Nutr. **17**(2), 109–115 (1998)
5. K. Marycz, N.Y. Toker, J. Czogala, I. Michalak, J. Nicpon, J. Grzesiak, An investigation of the elemental composition of horse hair affected by equine metabolic syndrome (EMS) using SEM EDX and ICP-OES. J. Anim. Vet. Adv. **12**(2), 146–152 (2013)
6. M.G. Skalnaya, V.A. Demidov, Hair trace element contents in women with obesity and type 2 diabetes. J. Trace Elem. Med Biol. 21(SUPPL. 1), 59–61 (2007). doi:10.1016/j.jtemb.2007.09.019
7. M.D. Lakshmi Priya, A. Geetha, Level of trace elements (copper, zinc, magnesium and selenium) and toxic elements (lead and mercury) in the hair and nail of children with autism. Biol. Trace Elem. Res. **142**(2), 148–158 (2011)
8. M. Kania, D. Mikołajewska, K. Marycz, M. Kobielarz, Effect of diet on mechanical properties of horse's hair. Acta Bioeng. Biomech. **11**(3), 53–57 (2009)

Chapter 31
Effect of Different Forms of Hypokinesia on the Ultrastructure of Limbic, Extrapyramidal and Neocortical Areas of the Rat Brain: Electron Microscopic Study

Mzia G. Zhvania, Nadezhda J. Japaridze and Mariam G. Ksovreli

Abstract The effect of chronic restraint stress and chronic hypokinesia "without stress" on the ultrastructure of central and lateral nuclei of amygdala, CA1 and CA3 area of the hippocampus, cingular cortex, nucleus caudatus and motor cortex of adult male rats were elucidated. In some neurons and synapses of abovementioned regions pathological modifications were revealed. More significant alterations provokes chronic restraint stress. Alterations are mostly concentrated: first—in the nuclei of amygdala, then in the CA1 and CA3 areas. Moderate alterations were observed in cingular cortex and nucleus caudatus. In comparing with it, hypokinesia "without stress" provokes only moderate modifications: predominantly in the nucleus caudatus, in lesser degree—in the hippocampus and amygdalae.

The ability of restraint motor activity (hypokinesia) to produce alterations on the functioning of central nervous system has been studied extensively. Numerous articles and reviews have been written on the reasons, mechanisms and possible consequences of such alterations [1–5]. However several questions still demand further elucidation. Especially limited work has been accomplished investigating the effect of various forms of chronic hypokinesia on the structure of brain [2, 6, 7]. It is very likely that depending on the form of hypokinesia, different brain structures could be involved in pathology. In the present study, using transmission electron microscope (TEM) we elucidated how different forms of restraint motor activity are reflected

M. G. Zhvania (✉) · M. G. Ksovreli
Ilia State University, Tbilisi, Georgia, I. Beritashvili Center of Experimental Biomedicine, Tbilisi, Georgia
e-mail: mzia_zhvania@iliauni.edu.ge

M. G. Ksovreli
e-mail: mari.qsovreli@yahoo.com

M. G. Zhvania · N. J. Japaridze
I. Beritashvili Center of Experimental Biomedicine, Tbilisi, Georgia
e-mail: japaridze.nadia@gmail.com

E. K. Polychroniadis et al. (eds.), *International Multidisciplinary Microscopy Congress*, Springer Proceedings in Physics 154, DOI: 10.1007/978-3-319-04639-6_31,

on the ultrastructure of neurons and synapses of limbic, extrapyramidal and neocortical regions of rat brain. We are focused on two types of hypokinesia: (i) Chronic restraint stress—condition characterized with the increased blood plasma level of stress-hormone, corticosterone and (ii) Hypokinesia "without stress"—condition, characterized with normal corticosterone plasma level. The following structures were examined: central and lateral nuclei of amygdalae, CA1 and CA3 areas of the hippocampus, cingular cortex (limbic areas), nulceus caudatus (extrapyramidal area) and neocortical motor area.

31.1 Material and Methods

31.1.1 Model of Chronic Restraint Stress

Adult (85–90 days old) male Wistar rats (n = 7) were kept for 90 days (d) in individual Plexiglas cages, the dimesnions of which (35 cm × 35 cm × 30 cm) allowed movement to be strongly restricted. The cages were housed in normal controlled environment. Five age-matched control rats were kept in ordinary vivarial conditions. After 10 days the level of corticosterone in blood plasma was measured using corresponding ELISA kit.

31.1.2 Model of Chronic Hypokinesia "Without Stress"

The male Wistar rats were used. For the beginning of experiment the age of animals was 15–20 days. The features of such young rats are the high plasticity and reparative abilities of the central nervous system. Experimental rats (n = 5) were kept for 90 days in individual Plexiglas cages of special construction which gives the possibility to change their size. For the beginning of experiments the body weight of experimental rats varied from 120 to 150 g; the cage size was 35 cm × 35 cm × 30 cm. In parallel with the increasing of body size/weight, the size of cage was also gradulaly increased. So, for the end of experiments the body weight of rats was 220–240 g and the final size was 65 cm × 55 cm × 50 cm. After 10 days of experiment the level of corticosterone in blood plasma was measured using corresponding ELISA kit. Control rats (n = 3) were kept in ordinary vivarial conditions.

31.1.3 Perfusion and Material Processing

Experimental and control animals were transcardially perfused under pentobarbital injection (100 mg·kg) with heparinized 0.9 % NaCl followed by 500 mL of 2.5 % glutaraldehyde in 0.1 % phosphate buffer (PB) (pH—7.4, perfusion pressure—120 Hg). After removal from skull and washing in PB, the brains were postifxed in 0.1 %

osmium tetroxide. The tissue blocks containing areas of interest were processed using conventional procedures and embedded in Araldite. 70–75 nm tick sections were prepared with a Reichert ultramicrotome, pocked up on 200 mesh copper grids, double-stained with uranil-acetate and lead citrate and examined under TEMs, JEM 100 C (JEOL) and TESLA BS500.

31.2 Results and Discussion

31.2.1 Chronic Restraint Stress

As a result of 90 days of restraint stress, in experimental rats the blood plasma level of stress hormone, corticosterone, was elevated, indicating to the development of stress response. The most significant alterations were observed in limbic regions, first—in amygdala, then—in the CA1 and CA3 areas; small changes were seen in cingular cortex and nucleus caudatus; no changes were detected in motor cortex.

In the central nucleus of amygdala pathological alterations were detected in ∼40 % of neurons, and in the lateral nucleus of amygdala—in ∼30 % of neurons. In both structures the majority of altered cells have ultrastuctural peculiarities of pyramidal neurons. The mitochondria were affected most often. Thus, the disruption of several cristae, vacuolization of some forms or interruption of mitochondrial membrane (rarely) were observed. In other cells the moderate or significant swelling of cisterns of endoplasmic network, big cytoplasmic vacuoles (the result of vacuolization of cytoplasmic organelles or water balance disorder), chromatolysis, long and sometimes twisted invaginations of nuclear membrane or membrane-and/or myelin-like cytoplasm inclusions were seen. Some dendrites were swelled and dendritic tubules were disorganized. Rarely shrunk, high-osmiophilic shapless profiles (irreversibly degenerated presynaptic terminals or necrotic neurons), partially envoloped with astrocytic processes were observed. In parallel, the number of neurons with numerous "active" organelles were also observed. Among synapses more altered were axo-dendritic forms: in the central nucleus ∼15 %, in the lateral nucleus—∼13 %. Thus, agglutination of synaptic vesicles (Fig. 31.1a), the damaged presynaptic mitochondria, myelin-like presynaptic or postsynaptic (rarely) inclusions and few "dark", shapeless presynaptic terminals partially engulfed with astrocyte processed were observed. In other preterminals superfucial alterations—moderately destructive presynaptic mitochondriae or single small myelin-like/osmiophilic bodies—were seen. In comparing with control more numerous were presynaptic terminals with single synaptic vesicles located far from active zone, terminals with granular vesicles (Fig. 31.1b), almost symmetrical axo-dendritic synapses and stubby spines.

31.2.2 Hypokinesia "Without Stress"

Despite the fact that motor activity remained significantly reduced during all experiment, the situation was not stressful for animals from this group (due to high plasticity

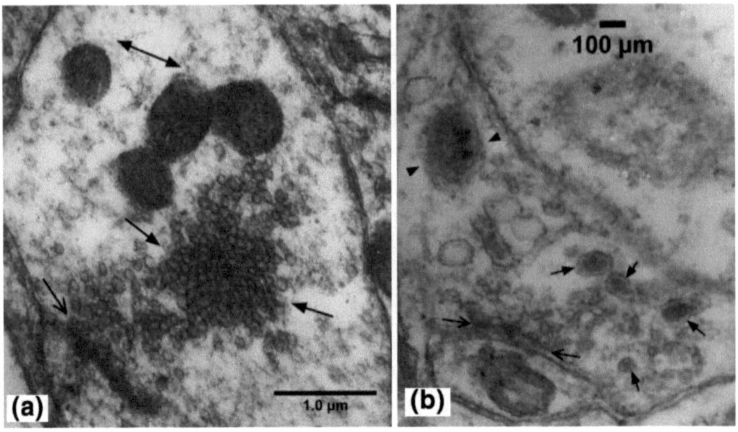

Fig. 31.1 Profiles of presynaptic terminals in rat brain after 90 days of hypokinetic stress; **a** *open arrows*—active zone, *filled arrows*—vesicle agglutination, *double head arrow*—mitochondria with moderately osmiophilic matrix, *bar*—1.0 μm; **b** *open arrows*—active zone, *filled arrows*—granular vesicles, *arrowheads*—mitochondria with osmiophilic matrix, *bar*—100 μm

and reparative abilities of the central nervous system of young animals and gradual increase of cage size): the level of corticosterone in blood plasma of such rats corresponded to blood plasma corticosterone level of control animals.

The brain ultrastructure of these rats was more preserved. Moderate (mostly) alterations were observed in the nucleus caudatus, then in amygdala and hippocampus while the cingular cortex and neocortex preserved alsmost normal structure. Among alterations, dominated moderate changes of mitochondria/endoplasmic network, and small cytoplasmic inclusions (myelin-like or small vacuoles). Moderately damaged mitochondria and/or small osmiophilic inclusions were observed in some altered presyanptic terminals also. Besides, like in the case of chronic restraint stress, the presynaptic terminals with single synaptic vesicles located predominantly at distance from active zone and almost symmetrical axo-dendritic (axo-spine) synapses were relatively numerous.

Thus, the main results of our study are: (i) different forms of hypokinesia provoke ultrastructural alterations in limbic and extrapyramidal regions of rat brain; (ii) hypokinetic stress provokes more sgnificant and numerous alterations then "hypokinesia without stress"; (iii) different forms of hypokinesia mostly affect different regions of brain: in the case of chronic hypokinetic stress more altered were limbic regions of the brain, in the case of hypokinesia "without stress"—extrapyramidal region—nucleus caudatus.

The nuclei of amygdala are known to play a crucial role in the orchestration and modulation of an organism's response to stressful events [8–11]. Some of pathological alterations observed in these regions as a result of hypokinetic stress could be related with the development of stress reaction. Thus, granular vesicles usually are considered as containing catecholamines. On the other hand, it is well-known that catecholaminergic systems are activated in different ways following exposure to distinct stressful stimuli [12–14]. Therefore, the increased number of synaptic terminals with granular vesicles could be provoked by stress response. Interestingly: such preterminals were rare in the brain of rats with normal level of corticosterone. The prevalence of stubby spines in comparing with other spine types could be one more feature of stress response: for unkown reasons prevalence of stubby spines was also described during other forms of stress [3, 15–17]. The "dark" neurons and presynaptic terminals, agglutination of synaptic vesicles or big myelin-, membrane- or vacuole like inclusions in the cytoplasm of some neurons and/or presynaptic terminals which are considered as irreversible, also could be the result of stress reaction: they were absent in the case of hypokinesia "without stress". Such pathological changes of neurons and synapses can ultimately significantly alter the connection profile for a neuron, providing the necessary anatomical foundation for changes in information processing and behavioral output. On the other hand: the occurence in some other neurons numerous and quite long nuclear membrane invaginations (that indicate to activation of nuclear and cytoplam interaction) or the presence of numerous "active" organelles, may be related with the development of some compensatory processes. It is notable that social partial isolation which often coexists with restraint stress, typically provokes moderate alterations not in amygdala but mostly in the hippocampus—brain region actively involved in the processes of learning and memory [18].

In contrast to restraint stress, the hypokinesia "without stress" provokes structural modifications mostly in extrapyramidal structure, which plays significant role in the modulation of motor acitivity and not in limbic areas. The alterations that are observed in these region (moderate changes of cytoplasmic organelles and some synapses, the increase of the number of preterminals with single vesicles and almost symmetrical axo-dendritic synapses) were common for both forms of hypokinesia. Therefore, such kind of changes could be considered as provoked by restraint motor activity.

In summary: according our results, the restraint motor activity does not provoke significant alterations in the ultrastructure of brain. However considerable and sometimes irreversible structural modifications could produce stress reaction, which is often developed in the organism subjected to hypokinesia.

References

1. S. Beretta, Cortico-amygdala circuits: role in the conditioned stress response. Stress **8**, 221–232 (2005)
2. C. Chotiwat, E. Kelso, R. Harris, The effects of repeated restraint stress on energy balance and behavior of mice with selective deletion of CRF receptors. Stress **13**, 203–213 (2010)
3. D. Christoffel, S. Golden, D. Dumitriu et al., IκB kinase regulates social defeat stress-induced synaptic and behavioral plasticity. J Neurosci **31**, 314–321 (2011)
4. G. Carrasco, L. van de Kar, Neuroendocrine pharmacology of stress. Eur J Pharmacol **453**, 235–272 (2003)
5. R. Carter, S. Pinnock, J. Herbert, Does the amygdala modulate adaptation to repeated stress? Neurosci **126**, 9–19 (2004)

6. E. Fuchs, G. Flugge, Chronic social stress: effects on limbic brain structures. Physiol. Behav. **79**, 417–427 (2003)
7. B. McEwen, Physiology and neurobiology of stress and adaptation: central role of the brain. Physiol. Rev. **87**, 873–904 (2007)
8. R. Admon, G. Lubin, O. Stern et al., Human vulnerability to stress depends on amygdala's predisposition and hippocampal plasticity. Proc Natl Acad Sci USA **106**, 14120–14125 (2009)
9. H. Cousijn, M. Rijpkema, S. Qin et al., PNAS **107**, 867–9887 (2010)
10. M. Zhvania, in *Neurochemistry: Cellular, Molecular and Clinical Aspect* (eds. A. Teelken and J. Korf) 1996, pp. 299–232
11. M. Zhvania, N. Japaridze, M. Ksovreli et al., The neuronal porosome complex in mammalian brain: a study using electron microscopy, in *NanoCell Biology—Multimodal Imaging in Biology and Medi cine*, ed. by B. Jena, D. Taatjes (Pan Stanford Publishing, Singapore, 2013)
12. R. Kvetnansky, E. Sabban, M. Palkovits, Catecholaminergic systems in stress: structural and molecular genetic approaches. M. Physiol. Rev. **89**, 535–606 (2009)
13. G. Rotoli, G. Grignol, W. Hu et al., Catecholaminergic axonal varicosities appear to innervate growth hormone-releasing hormone-immunoreactive neurons in the human hypothalamus: the possible morphological substrate of the stress-suppressed growth. J. Clin. Endocrinol. Metab. **96**, 606–611 (2011)
14. S. Spencer, T. Day, Role of catecholaminergic inputs to the medial prefrontal cortex in local and subcortical expression of Fos after psychological stress. J. Neurosci. Res. **78**, 279–288 (2004)
15. V. Beltran-Campos, R. Prado-Alcala, U. Leon-Jacinto et al., Increase of mushroom spine density in CA1 apical dendrites produced by water maze training is prevented by ovariectomy. Brain Res **1369**, 119–130 (2011)
16. M. Coulon, C. Wellman, I. Marjara et al., Early adverse experience alters dendritic spine density and gene expression in prefrontal cortex and hippocampus in lambs. J Psychoneuroendocrinology **38**, 1112–1121 (2013)
17. D. Diamond, A. Campbell, C. Park et al., Influence of predator stress on the consolidation versus retrieval of long-term spatial memory and hippocampal spinogenesis. Hippocampous **16**(7), 571–576 (2006)
18. M. Zhvaniia, Effect of social isolation on the ultrastructure of the dog brain. Arkh. Anat. Gistol. Embriol. **96**, 10–17 (1989)

Chapter 32
Diversity of Bacterial Spores from Brazilian Cerrado's Soil Strains by Transmission Electron Microscopy

Danilo A. Cavalcante, Juliana C. Orem and Marlene T. De-Souza

Abstract Transmission electron microscopy (TEM) studies in the 1960s started revealing a series of complex ultrastructural bacterial spore appendages that have, since then, been used as instruments for bypassing obstacles in finding answers about the appearance, distribution and possible function of spore structures. Our group's multi-phase taxonomy work, aims to obtain a consistent classification of diverse aerobic endospore-forming bacteria isolated from the Brazil's Cerrado (tropical Savanna) soil. In the present work, the spore's morphology of nineteen selected strains was resolved by TEM with the usage of two different staining techniques, a simple yet effective methodology, in order to determine characteristic surface features that could be used to distinguish and help to identify, also provide better understanding about our soil strains.

32.1 Introduction

Bacteria have many strategies for surviving environmental challenges. Formed in response to starvation, bacterial spores are the most resilient cell types known. Spores are frequently reported as being able to withstand extreme stress conditions [1–3] and as examples of long-term cell survival [4–6]. The spore formation, or sporulation, is an example of cell differentiation where a cell replicates its DNA, places a copy of its genome in a compartment called the forespore that further develops in a mature

D. A. Cavalcante · J. C. Orem · M. T. De-Souza (✉)
Cell Biology Department, University of Brasília, Brasilia, Brazil
e-mail: marlts@unb.br

J. C. Orem
e-mail: julianacapella@live.com

D. A. Cavalcante
e-mail: danilo.ac9@gmail.com

E. K. Polychroniadis et al. (eds.), *International Multidisciplinary Microscopy Congress*, Springer Proceedings in Physics 154, DOI: 10.1007/978-3-319-04639-6_32,

spore [7, 8]. Being released into the surrounding environment it can persist or quickly convert back to actively growing cells, a process called germination.

Spore-forming bacteria exhibit a gigantic array of environmental niches and habitats [9, 10], few of which have been explored in detail. A variety of spore-forming bacteria are routinely found in soil samples [11]. Recognition of the extent of soil bacterial diversity and environmental, clinical and industrial concerns has given rise to numerous studies that emphasize the important role of efficient phylogenetic diversity classification of soil spore forming bacteria. The diversity of soil bacteria is often quantified using determined indices [12, 13] such as richness, evenness or complementarity, with reference to a predefined taxonomic group. The current challenge, in bacterial taxonomy in general, is to find a systematic way to reconcile molecular data with ecology for a valid and biologically meaningful classification [14–17].

Our group's work aims to expand the fundamental knowledge of soil *Bacillus spp.* and related genera isolated from Brazil's Cerrado (tropical Savanna), the country's second-largest biome. We have previously generated a spore library containing 154 strains designated SDF001-154 presenting phenotypic diversity. This library consists on a part of a multi-phase taxonomy project, involving phylogenetic, genotypic and phenotypic characterization that seeks to obtain a consistent classification of diverse aerobic spore-forming bacteria from the Cerrado's soil.

32.2 Spore Layers and TEM

Spores are essentially constituted by concentrically arranged layers that can be built in a large number of ways, some of which are way too different from those presented on the model spore-forming Bacteria (Fig. 32.1). Most of these data are from electron microscopic analysis that reveals variations in the morphology of all classes of external spore structure [2, 8, 18, 19].

Spore structure of the model genus *Bacillus* consists, starting from the center, of an inner core, surrounded by a layer of peptidoglycan, the cortex, and a multilayered protein shell called the coat. By thin section electron microscopy, the coat frequently appears as a darkly staining layer, presenting or not, ridges and valleys that sometimes can be used as a tool for species differentiation through atomic force microscopy [20, 21]. Usually, two layers of the spore coat can be observed in thin section electron microscopy: a lamellar inner coat and a coarsely layered outer coat [7]. A recently identified outer-coat layer named crust can also be observed [22]. In some species, a distinct glycoprotein outermost layer, called the exosporium, can be also identified as a contiguous shell surrounding the coat and separated from it by a gap called the interspace [7, 23].

In the present work, the spore morphologies of *Bacillus spp.* and related genera were resolved by thin section transmission electron microscopy (Fig. 32.2), a simple and yet effective methodology in order to determine characteristic surface features that could be used to distinguish and help to identify, also provide better understanding about our soil isolates.

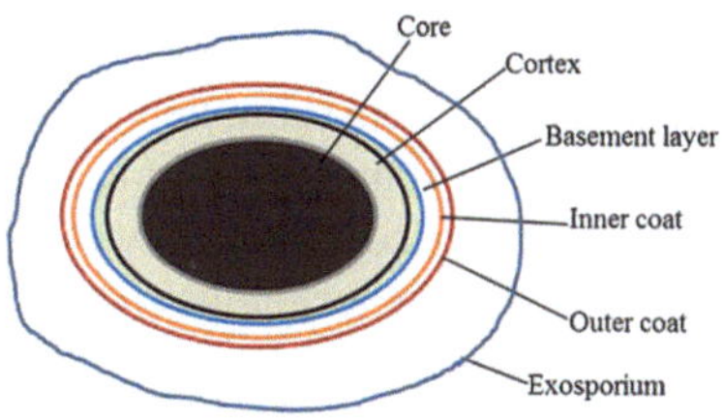

Fig. 32.1 Representative spore ultrastructure. Model of a typical *Bacillus spp*. spore the core (*black*) in the center is surrounded be the contex (*grey*) and by the multi-layer spore coat, which consists of the basement layer (*blue*), the inner coat (*orange*) and the outer coat (*red*) and sometimes presents an external layer called the crust. Some species exhibit an outermost layer called the exosporium (*deep blue*) that is separated from the rest of the spore by a large interspace

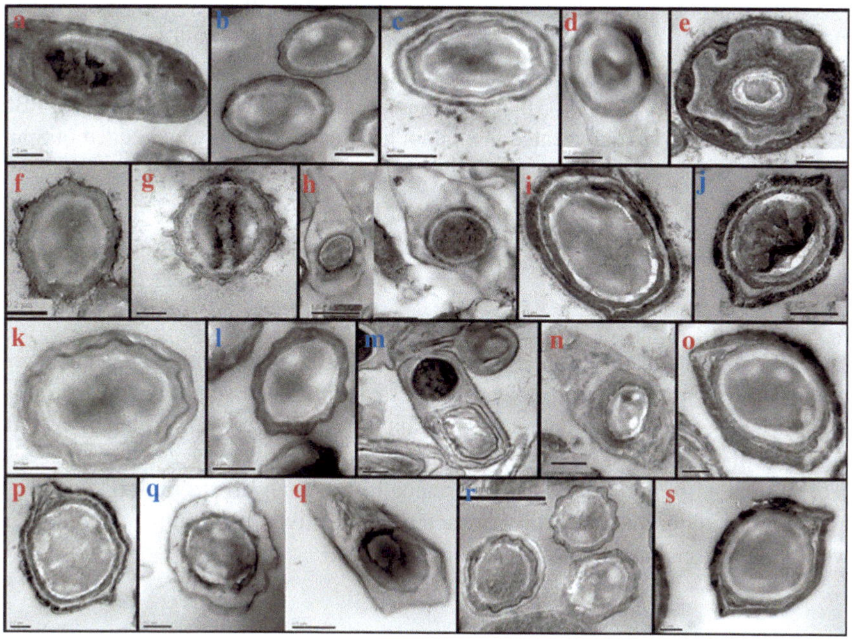

Fig. 32.2 Braxillian Cerrado's spore diversity. Thin section electron micrographs of spores isolated from Cerrado's soil were selected after analysis of the first and the second staining technique. The technique that provided better structure visualization was chosen as the representative strain model. **a** SDF01 sporangium; **b** SDF02 spores; **c** SDF12 spores; **d** SDF07 spores; **e** SDF08 sporangium presented a thick layer surrounding the coat; **f** SDF09 spore; **g** SDF29 spore presented a gap in the core and a spiked exosporium; **h** SDF05 spores presented a very long exosporium and a star-like structure surrounding the coat; **i** SDF15 spore; **j** SDF21 spore; **k** SDF24 spore; **l** SDF11 spore; **m** SDF30 sporangium presented a huge cristal-like structure; **n** SDF42 sporangium; **o** SDF47 spore; **p** SDF49 spore; **q** FT9 spores with comparative staining revealing exosporium appendages and fiber-like structure; **r** SDF51 spore; **s** SDF48 spore. *[Red] = Ruthenium red **[Blue] = Uranyl acetate

Nineteen different strains selected from our spore library (Fig. 32.2: a-s) for presenting diverse phenotypic properties were submitted to osmium tetroxide and either uranyl acetate or ruthenium tetroxide staining. Cultures of the strains were grown in LB medium for 48 h at 28 °C and washed 2 times in water.

Samples selected for the first staining technique (Fig. 32.2: marked in blue) were fixed for 4 h in Karnovsky (1 % paraformaldehyde; 3 % glutaraldehyde, and 0.07 M sodium cacodylate). Samples were washed in 0.1 M sodium cacodylate and post-fixed in 0.1 M potassium ferrocyanide and 1 % osmium tetroxide for 1 h followed by washing in 0.1 M sodium cacodylate. Samples post-fixed were then stained in 0.5 % uranyl acetate for 12 h, dehydrated in acetone gradient, and slowly embedded in Spurr resin at 60 °C for 48 h.

Samples selected for the second staining technique (Fig. 32.2: marked in red) were processed in a primary fixative-1 mg ruthenium red and 69 % (0.1 M) sodium cacodylate in 31 % (8 %) glutaraldehyde-at 37 °C for 1 h followed by washing in PBS (150 mM NaCl; 10 mM $NaHPO_4$; 0.005 % NaN_2; pH 7.4). Samples were processed in a secondary fixative-1 mg Ruthenium red and (0.05 M) sodium cacodylate in 50 % (4 %) Osmium Tetroxide-for 3 h at room temperature followed by washing in PBS. Samples stained were dehydrated in ethanol gradient and slowly embedded in Spurr resin at 60 °C for 48 h.

32.3 Conclusion

The data analysis of the spore's morphology, layers and external appendages represents an important first step towards understanding the assembly and function of these structures as well as generate significant information regarding phenotypic taxonomy of these species. Beyond providing information to our group spore database project, this work highlights the importance of the analysis of non-model systems both to expand our understanding of bacterial diversity and to complement ongoing work in traditional model systems.

References

1. A.I. Aronson, P. Fitz-James, Structure and morphogenesis of the bacterial spore coat. Bacteriol. Rev. **40**, 360–402 (1976)
2. A. Driks, Bacillus subtilis spore coat. Microbiol. Mol. Biol. Rev. **63**, 1–20 (1999)
3. P. Setlow, Spores of Bacillus subtilis: their resistance to and killing by radiation. J. Appl. Microbiol. **101**, 514–525 (2006)
4. M.J. Kennedy, S.L. Reader, L.M. Swierczynski, Preservation records of micro-organisms: evidence of the tenacity of life. Microbiology **140**, 2513–2529 (1994)
5. R.J. Cano, M.K. Borucki, M. Higby-Schweitzer, H.N. Poinar, G.O. Jr, Poinar, K.J. Pollard, Bacillus DNA in fossil bees: an ancient symbiosis? Appl. Environ. Microbiol. **60**, 2164–2167 (1994)

6. R.H. Vreeland, W.D. Rosenzweig, D.W. Powers, Isolation of a 250 million-year-old halotolerant bacterium from a primary salt crystal. Nature **407**, 897–900 (2000)
7. P.T. McKenney, A. Driks, P. Eichenberger, The Bacillus subtilis endospore: assembly and functions of the multilayered coat. Nat. Rev. Microbiol. **11**(1), 33–44 (2012)
8. A. Driks, Maximum shields: the assembly and function of the bacterial spore coat. Trends Microbiol. **10**(6), 251–254 (2002)
9. M. Pignatelli, A. Moya, J. Tamames, EnvDB, a database for describing the environmental distribution of prokaryotic taxa. Environ. Microbiol. Rep. **1**, 191–197 (2009)
10. J.G. Holt, Gram-Positive Bacteria Other than Actino-Mycetes, in *Bergey's Manual of Systematic Bacteriology*, ed. by P.H.A. Sneath, S. Mair, M.E. Sharpe, J.G. Holt (Williams & Williams, Baltimore, MD, 1986)
11. A.M. Earl, R. Losick, R. Kolter, Ecology and genomics of Bacillus subtilis. Trends Microbiol. **16**, 269–275 (2008)
12. G. Vilas Boas, V. Sanchis, D. Lereclus, M.V. Lemos, D. Bourguet, Genetic differentiation between sympatric populations of Bacillus cereus and Bacillus thuringiensis. Appl. Environ. Microbiol. **68**, 1414–1424 (2002)
13. L. Genevieve, Spatial scales of soil bacterial diversity–the size of a clone. FEMS Microbiol. Ecol. **48**, 119–127 (2004)
14. H. Maughan, G. Van der Auwera, Bacillus taxonomy in the genomic era finds phenotypes to be essential though often misleading Infection. Genet. Evol. **11**, 789–797 (2011)
15. E. Helgasin, N.J. Tourasse, R. Meisal, D.A. Caugant, A.B. Kolsto, Multilocus sequence typing scheme for bacteria of the Bacillus cereus group. Appl. Environ. Microbiol. **70**, 191–201 (2004)
16. K.K. Hill, L.O. Ticknor, R.T. Okinaka, M. Asay, H. Blair, K.A. Bliss, Fluorescent amplified fragment length polymorphism analysis of Bacillus anthracis, Bacillus cereus, and Bacillus thuringiensis isolates. Appl. Environ. Microbiol. **70**, 1068–1080 (2004)
17. F.G. Priest, M. Barker, L.W.J. Baillie, E.C. Holmes, M.C.J. Maiden, Population structure and evolution of the Bacillus cereus group. J Bacteriol **186**, 7959–7970 (2004)
18. A. Driks, Surface appendages of bacterial spores. Mol. Microbiol. **63**, 623–625 (2007)
19. J.R. Walker, A.J. Gnanam, A.L. Blinkova, M.J. Hermandson, M.A. Karymov, Y.L. Lyubchenko, P.R. Graves, T.A. Haystead, K.D. Linse, Clostridium taeniosporum spore ribbon-like appendage structure, composition and genes. Mol. Microbiol. **63**, 629–643 (2007)
20. M. Plomp, T. Leighton, K.E. Wheeler, A.J. Malkin, The high resolution architecture and structural dynamics of Bacillus spores. Biophys. J. **88**, 603–608 (2004)
21. B.J. Panessa-Warren, G.T. Tortora, J.B. Warren, High resolution FESEM and TEM reveal bacterial spore attachment. Microsc. Microanal. **13**, 251–266 (2007)
22. P.T. McKennedy et al., A distance-weighted interaction map reveals a previously uncharacterized layer of the Bacillus subtilis spore coat. Curr. Biol. **20**, 934–938 (2010)
23. A. Driks, The Bacillus anthracis spore. Mol. Aspects Med. **30**, 368–373 (2009)

Chapter 33
DNA Feulgen Cytophotometry and Chromatin Diminution

Irina G. Palchikova, Elena A. Ivankina, Valery F. Semeshin, Leonid V. Omelyanchuk, Igor F. Zhimulev and Eugeny S. Smirnov

Abstract The essential conditions for image acquisition and processing in the DNA image cytometry are revealed. To eliminate the specific cytometric errors we propose the calibration procedure for camera. Special test objects with different absorption are elaborated and made by photolithography. We first revealed chromatin diminution in two endemic Baikal species of *Cyclopoida*: *Acanthocyclops incolotaenia* and *Diacyclops galbinus*. We estimated the extent of chromatin diminution in *Diacyclops galbinus* as 95.5–96.2 %.

33.1 Introduction

The method of DNA image cytometry is based on the Bouguer–Lambert–Beer law. For this purpose the scanning technique [1] is accepted as correct and the most appropriate method. However, densitometry method [1], combined with the use of a Vickers M86 standard scanner, is very expensive and, therefore, its practical application is limited. An alternative and prospective approach in this field could consist of the use of digital equipment. Currently, the amount of DNA in the cell nucleus is evaluated numerically by the total optical density of the digital image of the Feulgen stained structures [2]. The obvious advantage of the image cytometry method is the

I. G. Palchikova (✉) · E. S. Smirnov
Technological Design Institute of Scientific Instrument Engineering,
Sib.Br. RAS, 41, Russkaya str, Novosibirsk, Russia 630058
e-mail: palchikova@tdisie.nsc.ru

E. A. Ivankina · V. F. Semeshin · L. V. Omelyanchuk · I. F. Zhimulev
Institute of Molecular and Cellular Biology, Sib.Br. RAS, 8, Acad. Lavrentiev ave,
Novosibirsk, Russia 630090

L. V. Omelyanchuk
e-mail: ome@mcb.nsc.ru

E. K. Polychroniadis et al. (eds.), *International Multidisciplinary Microscopy Congress*,
Springer Proceedings in Physics 154, DOI: 10.1007/978-3-319-04639-6_33,

possibility to assess groups with small numbers of cells detected in the preparation in situ.

In this paper we represent how the accurate monitoring of the instrumentation and calibration ones [2–4] allows getting the high validity and reproducibility of measurements with help of the quantitative analysis of digital images. In our studies, the presence of chromatin diminution in two Baikal Cyclops species—*Acantocyclops incolotaenia* and *Diacyclops galbinus*—was established for the first.

33.2 Materials and Methods

Material fixation and treatment. We used blood cells (smears) from several species (*Gallus domesticus, Danio rerio, Homo sapiens, Rana arvalis*) with the known DNA content per haploid chromosome set (1C) [2, 5], which are commonly applied as the standard for a comparative evaluation of the DNA content in other animal and plant species. Blood smears were fixed in a mixture of 96 % ethanol with glacial acetic acid (3:1) for 20 min, washed in three ethanol changes, and dried. DNA was hydrolyzed in 5 N HCl for 5 min at room temperature [6, 7]. Samples were stained according to Feulgen by the routine technique as follows: treated with Schiff's reagent for 2 h, washed with sulphur water for 15–40 min (three to eight changes), dehydrated with alcohols, and mounted in Canadian balsam.

Instrumentation. The experimental data process implies the handling of specimen, micro image acquisition and its digital processing. The inaccuracies are inserted into the measurement results at the each stage of the process.

Digital images [4] were registered with different cameras (AxioCam MRc, Canon EOS 500D, Leica DFC420) and different microscopes (AXIOVERT-200 (Carl Zeiss), DM-4000 (Leika), DIALUX 20 EB (Leitz) with objectives of 20×/0.40). The interference filter with average of 80 % transmission in the wavelength band of 551 ± 10 nm is used in experiments.

To avoid the optic blurs which can significantly alter the measurements we design [2] the special procedure for the nucleus image acquisition and processing. We produce the black control object which holds the China ink micro pieces. Intensity of the digital control object image is subtracted from the intensity of the nucleus image pixel by pixel. Then findings are normalized to the incident intensity. The resulting digital matrix is composed of values that proportional to the light transmission coefficient (τ) at the pixels of nuclear area. According to the Buger–Lambert–Ber law the optical density D for each pixel in the elementary area is calculated by formula

$$D = -\log_{10} \tau.$$

The reverse summarized matrix logarithm of transmission coefficients was used to determine the nuclear DNA content.

The validity and reproducibility of measurements in DNA image cytometry depend mainly on the transfer linearity of the densitometry device. We characterized

Fig. 33.1 Special test object

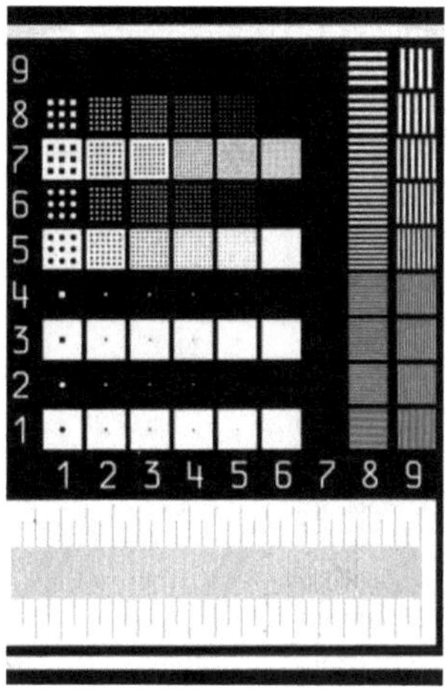

the properties of the photo matrix in the lighting mode applied in our densitometry device, based on transmission microscope. For this purpose, samples were replaced by a multistage spectrophotometric attenuator with a step transmission of 100, 70.79, 50.12, 31.62, 19.95, 14.8, 8.91, 6.31 and 100 %, respectively. The average pixel intensity was estimated in the photomatrix region used for densitometric measurements for each step images of the attenuator. In the areas of low transmission the diagram is described with linear dependence. Linear deviation in the area of high transmission is not exceeding 10 % from transmission value.

For purposes of both photogrammetric and densitometric calibration of camera we elaborated and made special test objects in chromium films by photolithography. The test object is shown in Fig. 33.1, it combines the operating characteristics of stage micrometer, bright point-object and linear gratings. Transmission coefficients of test objects vary within the limits of 0.1–0.9.

To perform the DNA calibration of our device, we carry out experiments on the DNA staining of blood cells from four organisms with known genome DNA content (*Gallus domesticus, Danio rerio, Homo sapiens, Rana arvalis*). The dependence between the measured mean DNA values (relative units) and values (pg/cell) known from the literature is linear. The calibration inclined line crosses the ordinate axis in the point less than 10 relative units. This value approximately corresponds to the signal variation in our samples and is the standard measurement error that arises due to residual cytoplasm staining.

33.3 Experimental Results

The results of measuring the DNA content in the nuclei of *Cyclops kolensis* cells are presented in Table 33.1. The most consistent calibration data on blood cells were obtained in the first three experiments (Moscow population of *Cyclopoida*). In these

Table 33.1 DNA content in *C. kolensis* and *C. insignis* cells [9]

Experiment number, population	Pre-diminution development			Post-diminution development		Eliminated DNA
	Anaphase (2C)	Metaphase (4C)	Polar body (1C)	Late embryos (2C)	Swimming legs, female (2C)	
1. *C. kolensis*, Moscow population (Axiovert-200)[a]	39.9 $\sigma = 3.5$; n = 28	80.8 n = 1	21.2 $\sigma = 1.27$; n = 23	–	1.82 $\sigma = 0.24$; n = 18	95.4–95.7
2. *C. kolensis*, Moscow population (Axiovert-200)	41.5 $\sigma = 3$; n = 69	84.8 $\sigma = 8.3$; n = 44	–	–	1.98 $\sigma = 0.18$; n = 14	95.2–95.3
3. *C. kolensis*, Moscow population (DM-4000)	40.6 $\sigma = 3.6$; n = 38	80.9 $\sigma = 6.9$; n = 177	20.6 $\sigma = 2.8$; n = 28	1.6 $\sigma = 0.12$; n = 13	1.77 $\sigma = 0.17$; n = 38	95.6–96.1
4. *C. kolensis*, Baikal population (Axiovert-200)	40.7 $\sigma = 3.4$; n = 3	85.6 $\sigma = 10.8$; n = 49	–	1.66 $\sigma = 0.36$; n = 24	1.52 $\sigma = 0.28$; n = 50	96.1–96.4
5. *C. kolensis*, Baikal population (Axiovert-200)	–	106.5 $\sigma = 15.1$; n = 8	18.9 $\sigma = 4.0$; n = 7	2.02 $\sigma = 0.45$; n = 43	2.08 $\sigma = 0.43$; n = 50	96.2
6. *C. insignis*, Moscow population (DM-4000)	7.5 $\sigma = 0.46$; n = 51	14.9 $\sigma = 1.04$; n = 120	3.7 $\sigma = 0.44$; n = 19	Anaphase 7.4 $\sigma = 0.34$; n = 30 Metaphase 14.2 $\sigma = 1.0$; n = 50	6.81 $\sigma = 0.36$; n = 74	Absent

Microscope type is indicated in brackets; n is the number of nuclei in the sample, σ is standard deviation of DNA content in the sample

experimental series 1 and 2 (Table 33.1), anaphases (A) and metaphases (M) were found for *C. kolensis* and measured; their DNA content may serve as the inner control of measurements. The results show that, in the first measurement series, the M/A ratio (80.8/39.9) was 2.03, instead of the expected 2. Thus, the difference between these two mean values of DNA content is $(2.03 - 2)/2 \times 100 = 1.5\,\%$. In the second experimental series M/A ratio was $84.8/41.5 = 2.04$. The difference was $(2.04 - 2)/2 \times 100 = 2\,\%$. The difference between the two experimental series in DNA content was $(41.5 - 39.9)/40.7 \times 100 = 4\,\%$ in anaphase and $(84.8 - 80.9)/82.8 \times 100 = 5\,\%$ in metaphase. So, the measurements were quite accurate. In the third experiment, the DNA content was measured under a DM-4000 microscope. The results are similar to those obtained in the previous two experiments and imply that DNA measurements performed under various systems are highly

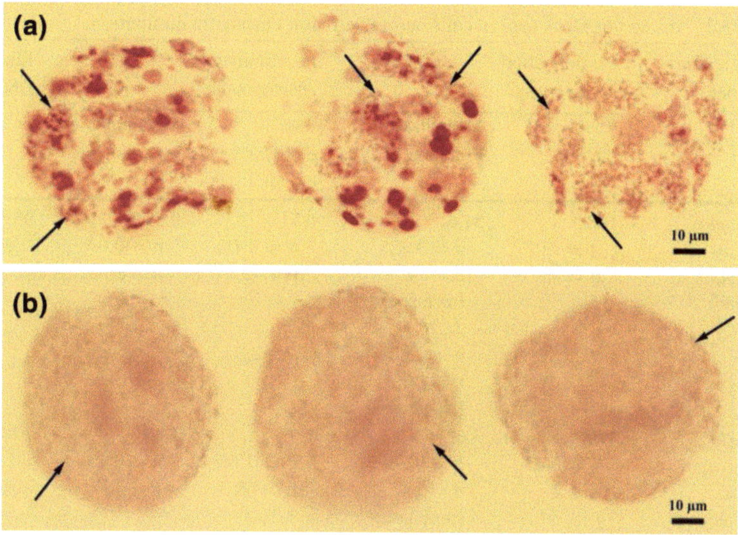

Fig. 33.2 Embryos of *A. incolotaenia* (Feulgen staining). **a** CD in *A. incolotaenia* embryos; **b** *A. incolotaenia* embryos at later stages of cleavage, after CD. The *arrows* indicate the granules of discarded DNA (**a**) and presomatic cells after CD (**b**). 40× objective [10]

reproducible. The data obtained for Baikal Cyclopoida populations (experiments 4 and 5, Table 33.1) are comparable to those obtained for the Moscow population. The results calculated relative to haploid DNA content (1C) for the cells of prediminution development (anaphase, metaphase, and polar body) coincide well. In cells after CD, the DNA content is more variable (from 0.83 to 1.04 pg). It cannot be excluded that the cells were at various stages of mitosis; therefore, these results are considered to be approximate values. The DNA content in two different populations of *C. kolensis* proved to be higher than that reported previously [8]. Thus, the authors estimated that the DNA content (1C) in *C. kolensis* was 2.3 ± 0.03 pg (n = 91) before CD and 0.14 ± 0.01 pg (n = 70) after CD; this is 8.6–11.5 and 5.7–7.4 times less, respectively, than the values presented in Table 33.1. On the other hand, our results are in a good agreement with the relative values of eliminated DNA in this species. These values vary in the range of 94.7–96.4 %, rather than the 93.5 % reported previously.

In our studies, the presence of chromatin diminution in two Baikal Cyclops species—*Acantocyclops incolotaenia* and *Diacyclops galbinus*—was established for the first. *Cyclops A. incolotaenia* at the stage of chromatin diminution (Fig. 33.2a) and at the stage of late embryonic development (Fig. 33.2b). In Fig. 33.2a, Feulgen positive granules of eliminated DNA can be observed, which serves as evidence of CD in this species. Figure 33.2b shows *A. incolotaenia* embryos at the late stage of embryonic development, probably after diminution. In the center of each embryo, there are two or three large nuclei not undergoing diminution; they are surrounded by smaller nuclei, in which some of the DNA has already been eliminated.

Table 33.2 Amount of DNA (pg) in cells before and after chromatin diminution

Cyclops species (population, microscope used	Prediminution development		Postdiminution development		Eliminated DNA
	Anaphase (2C)	Metaphase; Prophase (4C)	Late embryos (2C;4C)	Swimming legs, female (2C)	
Diacyclops galbinus (Baikal, Axiolab)	16.53 σ = 2.02; n = 33	31.6 σ = 3.22; n = 19 (metaphase); 31.76 σ = 2.00; n = 11 (prophase)	0.71 σ = 0.15; n = 15; –	0.63 σ = 0.09; n = 36	95.51–96.19
Mesocyclops leuckarti, (Baikal, Axiolab)	1.73 σ = 0.13; n = 4	3.62 σ = 0.26; n = 5 (prophase)	1.55 σ = 0.14; n = 73; 3.32 σ = 0.36; n = 39	1.60 σ = 0.15; n = 50	Absent
Mesocyclops leuckarti, (lake 1, Axiolab)	1.72 σ = 0.11; n = 34	3.72 σ = 0.19; n = 5 (metaphase)	1.51 σ = 0.08; n = 8; 3.38 σ = 0.11; n = 3	1.62 σ = 0.07; n = 47	Absent
Mesocyclops leuckarti, (lake 3, Axiolab)	1.74 σ = 0.09; n = 22	3.85 σ = 0.24; n = 3 (metaphase)	1.60 σ = 0.14; n = 7; 3.54 σ = 0.22; n = 4	1.56 σ = 0.31; n = 30	Absent

Arithmetic mean, σ (standard deviation) and number of measurements (n) are given

According to the results of cytophotometrical analysis, in *D. galbinus* the percent of eliminated DNA amounts to around 95.5–96.2 %, whereas the size of the pre- and post diminution genome differs significantly and amounts to 8.27 ± 1.01 pg and 0.36 ± 0.08 pg, respectively (Table 33.2). As a result of analysis of the DNA content in the *M. leuckarti* germ and somatic cells, we observed [9] an amount of DNA in somatic cells that was reduced by 7–10 % compared with the germ ones.

Such a low level of differences allows it to be suggested that there is an absence of CD in this species. Nevertheless, we cannot rule out as well the presence of a small DNA elimination. The good reproducibility of results for the cyclops collected in different habitats is a good corroboration of this (Table 33.2). At the same time, the DNA content in this species in somatic cells (2C) amounts to 1.56–1.60 pg and is almost twice the value (0.76 pg) established earlier [11, 12]. This discrepancy may be due to population differences in the cyclops used for analysis: Rasch and

Wyngaard used *M. leuckarti* of the European population (Germany). Elucidation of this problem requires carrying out additional comparative analysis.

Acknowledgments We are very grateful to A. L. Novitsky—head of the Biological Station of the Limnological Institute—for help in carrying out investigations, as well as to researchers of the Limnological Institute of the Siberian Branch of the Russian Academy of Sciences V. I. Teterina, A. V. Kupchinskii, and I. V. Khanaev and researchers from Institute of Molecular and Cellular Biology of the Siberian Branch of the Russian Academy of Sciences G. V. Pokholkova and A. V. Ivankin for valuable assistance in the work. This work was supported by Interdisciplinary Integrational Project of the Siberian Branch of the Russian Academy of Sciences no. 51M and by the Russian Foundation for Basic Research, projects no. 12-08-00396-*a*, 09-08-00651 and 09-04-00854.

References

1. E.M. Rasch, G.A. Wyngaard, B.A. Connelly, J. Morphol. **269**, 387 (2008)
2. L.V. Omelyanchuk, I.G. Palchikova, V.F. Semeshin, A.L. Alekseeva, I.F. Zhimulev, Tsitologiya **52**(4), 349–353 (2010)
3. M. Puech, F. Giroud, Cytometry **36**(1), 11–17 (1999)
4. I.G. Palchikova, L.V. Omelyanchuk, E.I. Palchikov, E.S. Smirnov, V.F. Semeshin, Sens. Syst. Issue **3**, 2–12 (2012)
5. http://www.genomesize.com
6. I.A. Magakian, E.M. Karalova, R.E. Khachikian, A.S. Avetisian, Tsitologiia **22**, 1054 (1980)
7. E.M. Rasch, *Methods in Molecular Biology*, vol. 247 (Clifton, NJ, 2004), p. 163
8. A.K. Grishanin, A.K. Shekhovtsov, T.V. Boĭkova, A.P. Akif'ev, I.F. Zhimulev, Tsitologiia **48**, 379 (2006)
9. V.F. Semeshin, L.V. Omel'ianchuk, A.L. Alekseeva, E.A. Ivankina, N.G. Sheveleva, I.F. Zhimulev, Tsitologiia **53**, 285 (2011)
10. E.A. Ivankina, A.L. Alekseeva, L.V. Omel'ianchuk, I.G. Pal'chikova, N.G. Sheveleva, S.V. Kiril'chik, I.F. Zhimulev, Tsitologiia **55**, 52 (2013)
11. E.M. Rasch, G.A. Wyngaard, Biol. J. Linn. Soc. **87**, 625 (2006)
12. G.A. Wyngaard, E.M. Rasch, Hydrobiologia **417**, 43 (2000)

Chapter 34
Comparison of Microstructural and Mechanical Properties of Hydroxyapatite-ZrO_2 Composites with Commercial Inert Glass Addition

Berrak Bulut, Nermin Demirkol, Ziya Engin Erkmen and Eyup Sabri Kayali

Abstract Hydroxyapatite(HA) is very popular bioceramic for hard tissue restorations. But, bulk HAs are not suitable for load bearing applications. So, HA material can be reinforced with a second phase (i.e polymers, metals and ceramics) to make a stronger composite material. The aim of this study is to compare the microstructural and mechanical properties of HA-ZrO_2 composites with the addition of 5 and 10 wt% commercial inert glass (CIG), separately. The green samples were produced according to BS 7253 at 350 MPa. Then, they were sintered at between 1000 and 1300 °C for 4 h. Microstructural properties of the obtained composites were characterized using SEM (Scanning electron microscopy) + EDS; phase analysis was done by X-ray diffractometer (XRD), mechanical properties were measured by compression and hardness tests. According to the results of ongoing bioactivity and biocompatibility studies, these two composites will possibly be good candidates for orthopedical applications.

B. Bulut (✉) · E. S. Kayali
Metallurgical and Materials Engineering Department, Faculty of Chemical and Metallurgical Engineering, Istanbul Technical University, Istanbul, Turkey
e-mail: berrakbulut@hotmail.com

E. S. Kayali
e-mail: kayali@itu.edu.tr

N. Demirkol
Ceramic, Glass and Tile Department, Vocational School of Degirmendere Ali Ozbay, Kocaeli University, Kocaeli, Turkey
e-mail: nermin.demirkol@kocaeli.edu.tr

Z. E. Erkmen
Metallurgical and Materials Engineering Department, Faculty of Chemical and Metallurgical Engineering, Marmara University, Istanbul, Turkey
e-mail: eerkmen@marmara.edu.tr

E. K. Polychroniadis et al. (eds.), *International Multidisciplinary Microscopy Congress*, Springer Proceedings in Physics 154, DOI: 10.1007/978-3-319-04639-6_34,

34.1 Introduction

Hydroxyapatite (HA, $Ca_{10}(PO_4)_6(OH)_2$) is one of the most widely used biomaterials for reconstruction of the skeleton and dental tissues. HA is a nontoxic and biocompatible material. This material improves biocompatibility and bonds well to growing bone [1, 2]. But it has poor mechanical properties especially in wet environments. There have been many investigations aimed in improving mechanical properties. The way is to improving mechanical properties to make composite materials. Addition of second phase to the HA matrix in order to obtain products with improved strength and toughness has been researched in recent years [3].

Zirconia has been one of the most important ceramic materials for well over a century. Zirconia is a biomaterial that has a bright future because of its high mechanical strength and fracture toughness. Zirconia ceramics have several advantages over other ceramic materials due to the transformation toughening mechanisms operating in their microstructure that can be manifested in components made out of them [4, 5]. Bioactive glasses and glass ceramics are highly bioactive biomaterials due to their osteoconductive and osteoinductive properties which result in strong bonding with the neighboring bone [5, 6]. The glass ceramic has superior mechanical properties, good biocompatibility, bioactivity and no toxicity making it useful as a biomaterial in artificial bone and dental implants [5].

There are few studies related to HA-ZrO_2 composites. Erkmen et al. [7] had studied microstructural and mechanical properties of HA-ZrO_2 composites. In this study, enamel-derived HA (EHA) and commercial HA (CHA) were chosen as the matrix. For EHA-PSZ (partially stabilized zirconia) composites, the density and mechanical properties were generally enhanced by adding 5 wt% PSZ, especially after sintering at 1200 °C, whereas CHA-PSZ composites showed lower strength values at sintering temperatures of 1200 and 1300 °C with respect to EHA-PSZ composites. Evis [2] had studied reactions in hydroxylapatite–zirconia composites. In this study, addition of zirconia caused increased decomposition of the HA in composites, forming tricalcium phosphate phases (TCP). The porosity of the sintered composites increased with increase in zirconia concentration in the composites, probably because of water formed by the decomposition of the HA. Demirkol et al. [3] had studied influence of commercial inert glass (CIG) addition (5 and 10 wt%) on the mechanical properties of commercial synthetic HA (CSHA). The mechanical properties of composites decreased with increasing CIG content. The highest mechanical properties and the highest density were obtained in CSHA-5 wt% CIG composites sintered at 1300 °C.

The aim of this study is to compare the microstructural and mechanical properties of HA-ZrO_2 composites with the addition of 5 and 10 wt% commercial inert glass (CIG), separately.

34.2 Materials and Methods

In this study, a commercial synthetic hydroxyapatite (Acros Org, BE) was used as hydroxyapatite. The CIG obtained from conventional window glass has the same chemical analysis with Demirkol et al.'s study [3]. Two batches were prepared: Batch 1 (B1) was composed of 90 wt% of HA, 10 wt% of ZrO_2 and 5 wt% of CIG, whereas Batch 2 (B2) was 90 wt% of HA, 10 wt% of ZrO_2 and 10 wt% of CIG. The HA-ZrO_2 powders were separately mixed with 5 and 10 wt% commercial inert glass (CIG) powder for 4 h using ballmilling (Ozmak Comp, TR) which were then pressed at 350 MPa. Pressed samples were sintered between 1000 and 1300 °C (+5 °C min^{-1}) for 4 h. Density, Vickers microhardness, and compression strength tests were performed consequently. Microstructural characterization was carried out by optical and scanning electron microscopy (SEM-Joel JSM-5910 LV-Low Vacuum Scanning) and energy dispersive spectroscopy (EDX-Oxford Inca Energy 200) and X-ray diffraction (XRD-Brucker D8-Advanced X-ray diffractometer using Cu K_α radiation) analysis. Microhardness values were determined under 200 g. loading. The compression tests were done with an universial testing machine (Shimadzu, JP) at a crosshead speed of 3 mm/min.

34.3 Results and Discussion

Table 34.1 summarizes density and porosity data of composites sintered at different temperatures.

Table 34.2 summarizes product phases obtained in XRD diagrams for HA-ZrO_2 with 5 and 10 wt% CIG addition composites sintered at 1000–1300 °C. Both composites sintered at 1000 °C include hydroxyapatite (HA), zirconia (ZrO_2), whitlockite (beta-tricalcium phosphate) (W), calcium zirconate (CZ). HA-ZrO_2-5 wt% CIG composite sintered at 1300 °C includes hydroxyapatite (HA), zirconia (ZrO_2), whitlockite (beta-tricalcium phosphate) (W), sodium calcium silicate (SC-Si) phases. However, HA-ZrO_2-10 wt% composite sintered at 1300 °C additionally showed calcium silicate (C-Si) phase to these phases in addition.

The highest density was obtained in HA-ZrO_2-5 wt% CIG composite sintered at 1300 °C.

Figures 34.1 and 34.2 present the micrographs of composites at 1000–1300 °C temperature.

When Fig. 34.1a is compared with Fig. 34.1c, grain growth occured with increasing sintering temperature. Increase of CIG content enhanced grain growth as clearly seen in Fig. 34.1a–d. Comparing Fig. 34.2a, b, grain growth and glassy phase formation along with the cracks were observed.

Table 34.3 summarizes the experimental results of Vickers microhardness and compression strength of the samples sintered at different temperatures.

Table 34.1 Density and porosity values of HA-ZrO_2 with 5 and 10 wt% CIG addition composites sintered at different temperatures

Temperature (°C)	Density (gr/cm^3)		Porosity (%)	
	5 wt%	10 wt%	5 wt%	10 wt%
1000	1.95 ± 0.036	2.02 ± 0.14	35.86 ± 0.66	42.48 ± 10.04
1100	2.12 ± 0.027	1.98 ± 0.09	31.67 ± 1.61	31.81 ± 8
1200	2.53 ± 0.095	2.34 ± 0.041	18.29 ± 3.56	22.71 ± 1.67
1300	2.74 ± 0.005	2.39 ± 0.25	8.44 ± 0.42	19.84 ± 8.57

Table 34.2 Phases obtained in XRD diagrams of HA-ZrO_2 with 5 and 10 wt% CIG addition composites sintered at different temperature

	1000 °C	1300 °C
HA-ZrO_2-5 wt% CIG	(1) $Ca_{10}(PO_4)_6(OH)_2$ (HA)	(1) $Ca_{10}(PO_4)_6(OH)$
	(2) ZrO_2(monoclinic)	(2) ZrO_2(tetragonal)
	(3) $Ca_3(PO_4)_2$(W)	(3) $Ca_3(PO_4)_2$
	(4) $CaZrO_3$(CZ)	(4) Na_2CaSiO_4 (SC-Si)
HA-ZrO_2-10 wt% CIG	(1) $Ca_{10}(PO_4)_6(OH)_2$	(1) $Ca_{10}(PO_4)_6(OH)_2$
	(2) ZrO_2(monoclinic)	(2) ZrO_2 (tetragonal)
	(3) $Ca_3(PO_4)_2$	(3) $Ca_3(PO_4)_2$
	(4) $CaZrO_3$	(4) Na_2CaSiO_4
		(5) Ca_2SiO_4(C-Si)

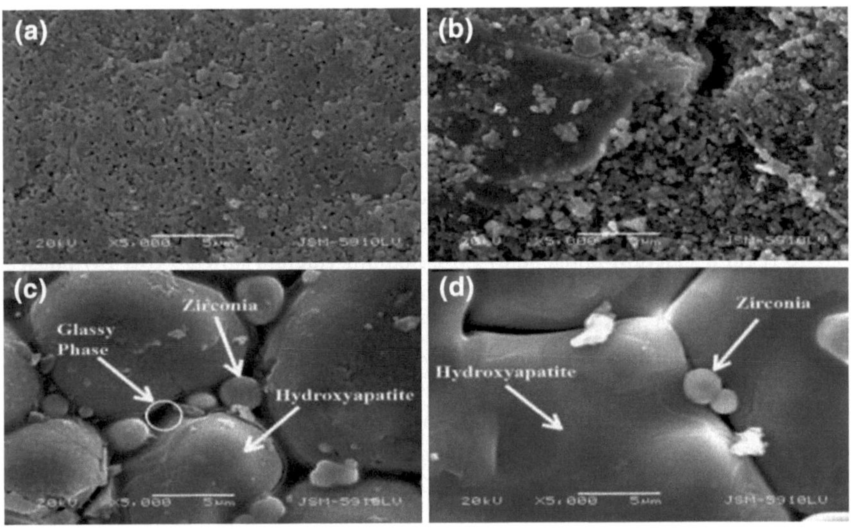

Fig. 34.1 Microstructures of composites sintered at different temperatures. **a** HA-ZrO_2-5 wt% CIG at 1000 °C, **b** HA-ZrO_2-10 wt% CIG at 1000°C, **c** HA-ZrO_2-5 wt% CIG at 1300 °C, **d** HA-ZrO_2-10 wt% CIG at 1300 °C

Fig. 34.2 Microstructures of HA-ZrO_2-10 wt% CIG composite sintered at 1300 °C temperature. **a** HA-ZrO_2-10 wt% CIG at 1300 °C backscattered image, **b** HA-ZrO_2-10 wt% CIG at 1300 °C

Table 34.3 Mechanical properties of HA-ZrO_2 with 5 and 10 wt% CIG addition composites sintered at different temperature

Temperature (°C)	Compression strength (MPa)		Vickers microhardness (HV)	
	5 wt%	10 wt%	5 wt%	10 wt%
1000	50.74 ± 14.16	30.19 ± 5.00	65,00 ± 2.20	32.77 ± 8.39
1100	50.31 ± 6.07	29.13 ± 9.20	57.82 ± 9.18	46.56 ± 5.55
1200	75.22 ± 16.43	47.54 ± 11.94	98.27 ± 3.89	53.19 ± 6.11
1300	46.08 ± 10.07	16.42 ± 1.68	382.49 ± 7.1	39.66 ± 3.93

The highest Vickers microhardness was obtained in HA-ZrO_2-5 wt% CIG composite sintered at 1300 °C. The highest compression strength was measured in HA-ZrO_2-5 wt% CIG composite sintered at 1200 °C.

HA-ZrO_2 composites with 5 and 10 wt% CIG addition reached high density and low porosity values with increasing sintering temperature. Grain size increased with increasing sintering temperature, glassy phase formed above 1200 °C with CIG addition. The latter caused an increase in hardness and a decrease in strength because of crack formation in HA-ZrO_2 composites with 5 wt% CIG addition. When the sintering temperature increased, the monoclinic ZrO_2 began to transform to tetragonal and cubic ZrO_2. At 1100 and 1170 °C sintering temperatures, calcium from HA diffused into the zirconia phase, forming CZ and the HA phases which decomposed to tricalcium phosphate and formed calcium zirconate phases. Addition of excess CIG (glass powders) enhanced the HA decomposition to TCP (tricalcium phosphate). Clearly, there were chemistry changes in the composites due to the addition of zirconia into HA, which might have influenced microstructural and mechanical properties. Both composites showed reduction in compression strength values at 1300 °C due to SCS (Sodium Calcium Silicate) glassy phase in which crack propagations were enhanced. Additionally, HA-ZrO_2-10 wt% CIG composite showed the presence of CS phase. In the latter, there was dramatically reduction of mechanical properties (16, 42 MPa; 39, 66 HV) due to brittle nature of CS and CSC phases.

34.4 Conclusions

The findings of this study are summarized as follows:

The highest density and the lowest porosity were obtained in HA-ZrO_2-5 wt% CIG composites at 1300 °C. The highest Vickers microhardness data were obtained in HA-ZrO_2-5 wt% CIG composites at 1300 °C. The highest compression strength was obtained in HA-ZrO_2-5 wt% CIG composites at 1200 °C. The mechanical properties of composites decreased with increasing CIG content. Biocompatibility studies are continuing on and results will be presented in near future.

Acknowledgments This work was a part of MSE Project of the corresponding author. Special thanks for Dr. Mustafa Ilhan for SEM analyses and Research Assist. Murat Alkan for XRD analyses.

References

1. N. Demirkol, F.N. Oktar, E.S. Kayali, Mechanical and microstructural properties of sheep Hydroxyapatite (SHA)-Niobium oxide composites. Acta Phys. Pol. A **121**, 274–276 (2012)
2. Z. Evis, Reactions in Hydroxylapatite-Zirconia composites. Ceram. Int. **33**, 987–991 (2007)
3. N. Demirkol, F.N. Oktar, E.S. Kayali, Influence of commercial inert glass addition on the mechanical properties of commercial synthetic Hydroxyapatite. Acta Phys. Pol. A **123**, 427–429 (2013)
4. I. Mobasherpour, M.S. Hashjin, S.S.R. Toosi, R.D. Kamachali, Effect of the addition ZrO_2-Al_2O_3 on Nanocrystalline Hydroxyapatite bending strength and fracture toughness. Ceram. Int. **35**, 1569–1574 (2009)
5. T.V. Thamaraiselvi, S. Rajeswari, Biological evaluation of bioceramic materials—a review. Trends Biomater. Artif. Organs. **18**, 9–17 (2004)
6. G. Goller, H. Demirkıran, F.N. Oktar, E. Demirkesen, Processing and characterization of bioglass reinforced Hydroxyapatite composites. Ceram. Int. **29**, 721–724 (2003)
7. Z.E. Erkmen, F.N. Oktar, Y. Genç, Microstructural and mechanical properties of Hydroxyapatite-Zirconia composites. J. Am. Ceram. Soc. **90**, 1–8 (2007)

Chapter 35
Superhelicity of Chiral Strings

Sergey V. Stovbun, Aleksey A. Skoblin and Iakov A. Litvin

Abstract A phenomenon of self-assembly of solitary rigid supramolecular strings with lengths up to $\sim$0.1 cm and anisometry ratios $\sim(10^2 - 10^4)$ was discovered in dilute ($\sim 10^{-3}$ M) solutions of homochiral trifluoroacetylated aminoalcohols (TFAAA). Macroscopic chirality (helicity) of strings with diameters from $\sim$2 nm to 20 μm and lengths from 10^2 nm to 0.1 cm was revealed as well as their chiral hierarchy: thicker strings consist of intertwisted thinner ones (superhelicity), with helix step lengths from $\sim$10 nm to several tens of μm. Empirical rule of switching of the macroscopic chirality sign between superhelicity levels was established, which fits the fundamental synergetic law of chirality sign switching between hierarchy levels of complexity in biological systems. It was shown that molecular chirality act as a macroscopic helicity structural factor.

The subject of investigation was a series of solutions of trifluoroacetylated aminoalcohols TFAAA-1–TFAAA-8 (Fig. 35.1) in different organic liquids (heptane, hexane, cyclohexane, isopropanol, cumene, benzene, carbon tetrachloride, chloroform, methanol, ethanol, acetone, dimethyl sulfoxide and others) and water [1–4]. Homochiral solutions got cured at concentrations of $(10^{-4} - 10^{-2})$ M which is much less than that of percolation threshold:

$$c_p = (0.1 - 0.2)\,\mathrm{M} \tag{35.1}$$

corresponding to an infinite cluster on the molecular grid (independently of the molecular design) [5]. Indeed, dispersive anisometric phase was observed on scales of 10 μm and more, establishing a visible connected grid (Fig. 35.2). The geometry of such grid with small own volume per solution volume corresponds to an infinite

S. V. Stovbun (✉) · A. A. Skoblin · I. A. Litvin
Department of Kinetics and Catalysis, Semenov Institute of Chemical Physics,
Kosigina Street 4, Moscow, Russian Federation 119991
e-mail: s.stovbun@chph.ras.ru

E. K. Polychroniadis et al. (eds.), *International Multidisciplinary Microscopy Congress*, Springer Proceedings in Physics 154, DOI: 10.1007/978-3-319-04639-6_35,

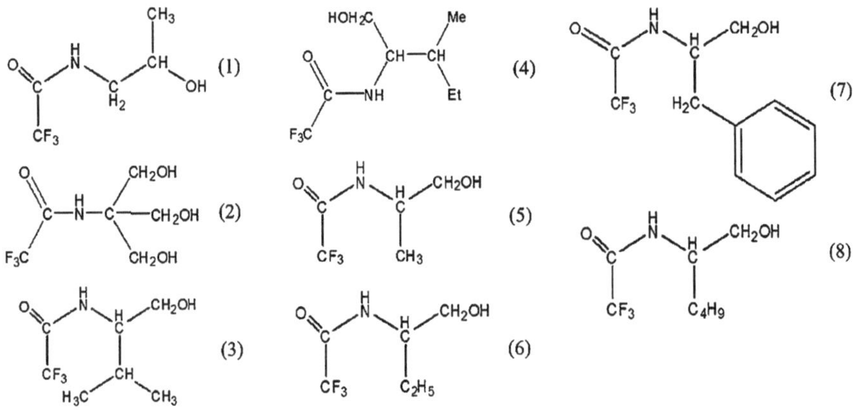

Fig. 35.1 Structural formulae of the TFAAA-1—TFAAA-8 molecules

Fig. 35.2 Solution of TFAAA-4 in cyclohexane, 1 mg/ml. System of "ideal" strings

cluster with the percolation threshold less than that of isometric gels: $c \ll c_p$. Such anisometric gel is isotropic and homogeneous on scales of $\gtrsim 0.1$ cm. Estimation of the activation energy of reversible gelation based on the assumption of kinetic equilibrium near the threshold gives [1, 6]:

$$W = kT \ln(2\, l^* / 3\, D\, n_p) = (0.2 - 0.5)\, eV \tag{35.2}$$

(k is the Boltzmann constant, T—temperature, D—diffusion constant, l^*—parameter of the cell model of liquid, namely the size of the cell, n_p—gelation concentration), which is a typical value for supramolecular low concentration gels.

In nonpolar solvents (cyclohexane, heptane, nonane, cumene, carbon tetrachloride) irregular grids of "ideal" strings are formed spontaneously. The lengths of the strings obtained experimentally are up to $l \sim 10^{-1}$ cm and more, diameters $d \sim 10^{-4}$ cm are constant along the lengths, and anisometry ratio is $A = l/d \sim 10^4$

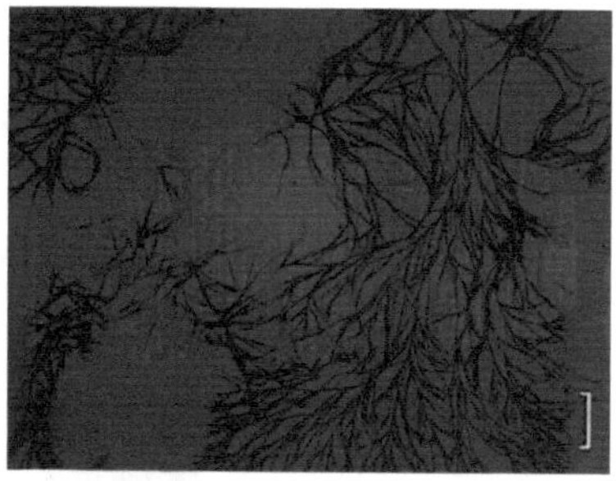

Fig. 35.3 Xerogel of a solution of TFAAA-5 in benzene, scalebar equals 100 μm. Breaks and branching

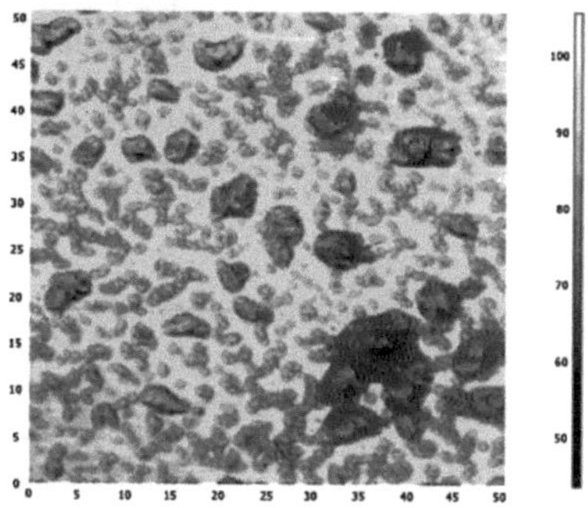

Fig. 35.4 Xerogel of TFAAA-2 in heptane, 1 mg/ml, AFM image, phase contrast. Isometric granules. Image size is 50 μm × 50 μm

or more. Strings are rectilinear on scales of 10^{-2} cm and greater, which points their mechanical elasticity (Fig. 35.2). In solvents which occupy the possibilities of creating weak intermolecular bonds more actively (benzene, methanol, ethanol, propanol, chloroform, acetone, dimethyl sulfoxide, water), breaks and branching are observed (Fig. 35.3), down to formations of dendrites, brushes etc, and anisometry ratio can be considerably less, $\mathbf{A} \sim 10^1 - 10^2$.

Gels are not formed within this concentration range in solutions of achiral TFAAA-2 which have electric dipole moment $d = 4\,\mathrm{D}$, their xerogels consist of faintly anisometric structures $\mathrm{A} \lesssim 5$ (Fig. 35.4). This means that no structures are formed on macroscopic scales in solutions of achiral TFAAA-2. Thus, chirality of the molecules is of principle importance for the growth of the strings in low concentration solutions. Racemic mixtures of chiral TFAAA-1 in heptane remain liquid at all concentrations, their xerogels consisting of isometric elements with $\mathbf{A} \sim 1 - 3$. To sum up, in racemic (TFAAA-1) and achiral (TFAAA-2) solutions in heptane strings are not formed at concentrations up to the TFAAA solubility limit, which is accompanied by isometric condensation and precipitation of TFAAA.

The absence of sorbed water on the surface of homochiral strings which was observed in xerogels (heptane) using AFM after a durational exposure to 10^{-4} Torr vacuum indicates low polarity and packing of hydrogen bonds of the strings under the surface, in contrast to the surface of racemic granules and granules of achiral TFAAA which is partially covered with water (Fig. 35.4) [7].

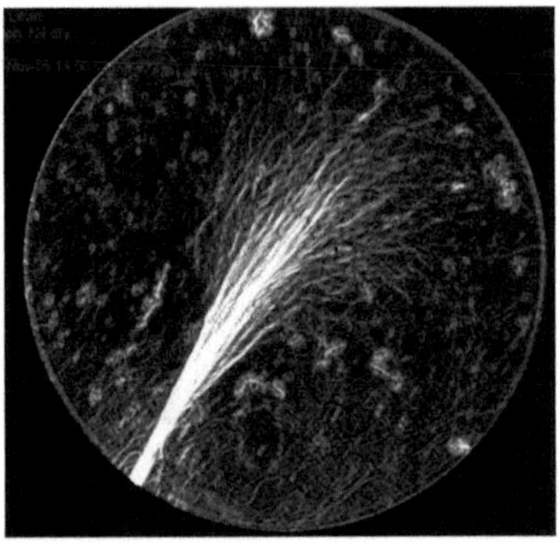

Fig. 35.5 Solution of TFAAA-4 in heptane, 1 mg/ml. Diameter of the field is 25 μm. Untwisting strings grade into disclinations

The phenomenon of solitary strings observed in capillaries indicates that a string is a separate (discrete) physical entity which is independent of the gelation [3].

In solutions at different concentrations, distances between strings amount tens of microns and more, i.e. there's only negligible interaction between strings aside the areas of direct mechanical contact. Thus, TFAAA solution curing is a trivial consequence of the assembly of rigid strings with macroscopic lengths.

Strings in heptane, hexane, benzene, isopropanol and water are surrounded with a mesophase which has a different optical density, while in cyclohexane and CCl_4 visible mesophase is not observed [3, 4]. Mesophase wets strings and presumably possesses orientational order, which is supported by the presence of Shlirren textures with characteristic lengths $l \sim 0.5\,\mu m$ and multiple disclinations ($l \sim 1\,\mu m$) observed in linearly polarized light. When untwisting in mesophase, thin strings continuously grade into disclinations (Fig. 35.5). Disclinations are one-dimensional and are formed in the frustration areas [8].

Two principle features of string formation were ascertained [4, 9, 10]: chirality in a form of spiral geometry of each string and hierarchy, or superhelicity, which means that thicker strings consist of spirally intertwisted thinner ones which, in turn, have their own spiral structure. Spiral structure of strings can be observed on scales between diameters of $d \sim 2\,nm$ and "gigantic" $d \sim 20\,\mu m$ and helix steps from $\sim 10\,nm$ to several tens of μm (Figs. 35.6, 35.7).

The sign of chirality (helicity) is switched between adjacent hierarchy levels [4, 11].

Mutual attraction of strings is due to their Van der Waals interaction. Two parallel strings with the radii R and lengths L on a distance $D \lesssim R$ are attracted with a force equal to [12]:

$$F = ALR^{1/2}/16D^{5/2} \qquad (35.3)$$

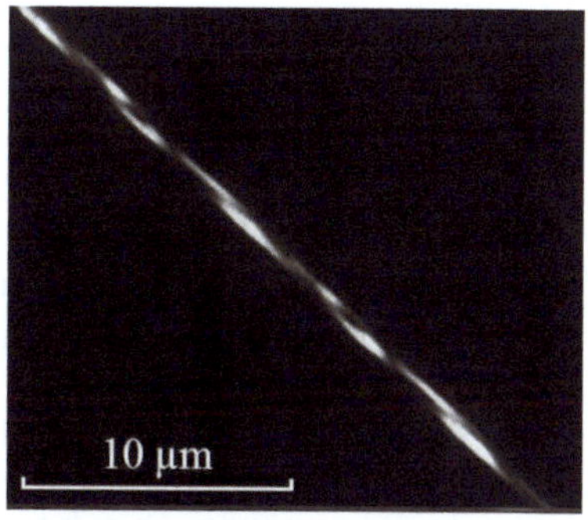

Fig. 35.6 Solution of TFAAA-4 in heptane, 0.4 mg/ml. Spiral structure of a thick string

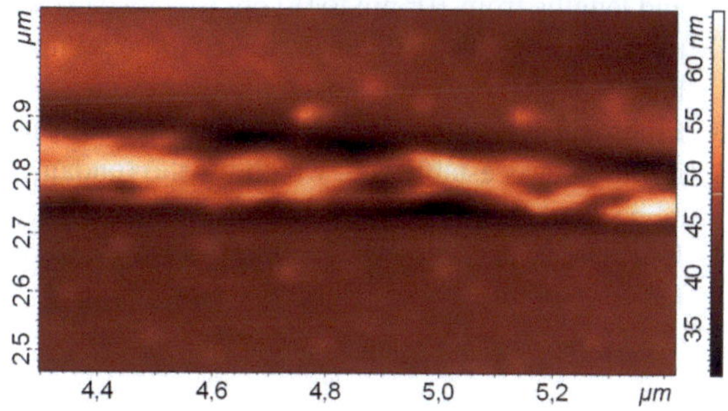

Fig. 35.7 Xerogel of TFAAA-5 in heptane, 1 mg/ml, AFM. A thin string consisting of intertwisted smaller ones

(A is the Hamaker constant of the interaction of strings). Characteristic time of the mutual attraction of the strings can be estimated as:

$$T \approx (128\pi/7)\,\eta\,D^{7/2}/AR^{1/2}\,\ln(740\eta^2 R/\rho\,A) \qquad (35.4)$$

(η is the viscosity of the solution, ρ—its density). Assuming $R = 1\,\mu m$, $D = 1\,\mu m$, $A = 0.5 \cdot 10^{-13}$ erg, $\rho = 0.8\,g/cm^3$, $\eta = 10^{-2}$ g/cm·s we deduce $T \approx 1$ s.

The presence of the Van der Waals attraction can provide the spiral intertwisting which explains the superhelicity phenomenon. Estimation of the requisite strain leads to the following value of the helix step h when two strings get intertwisted:

$$h^4 = \xi E R^{9/2} D^{5/2}/A \qquad (35.5)$$

($\xi \sim 10^1$ depends on the details of the string structure, E—elastic modulus of the strings). Assuming $E = 10^9$ dyne/cm^2 (this was estimated from other experimental data) and $D = 1$ nm, $h \approx 0.5\,\mu m$ if $R = 0.1\,\mu m$, $h \approx 10\,\mu m$ if $R = 1\,\mu m$, which is in a good agreement with experimental data.

The effect of chirality switching on transitions from one level of a hierarchy to the next [13] is present in all significant biological systems and provides their stratification [13–17]. The discovery of superhelicity of the strings opens up possibilities to simulate general synergic regularities of chirality sign switching in discrete hierarchical transitions.

35.1 Conclusion

In the present work:

1. In dilute $\sim 10^{-3}$ M solutions of homochiral TFAAA, a phenomenon of self-assembly of solitary supramolecular strings with lengths up to ~ 0.1 cm and anisometry ratios $\sim (10^2 - 10^4)$ was discovered. It was shown that in racemic and achiral solutions in heptane the strings are not formed at concentrations up to the TFAAA solubility limit which is accompanied by isometric condensation and precipitation.
2. Macroscopic chirality (helicity) of strings with diameters from ~ 2 nm to 20 μm and lengths from 10^2 nm to 0.1 cm was revealed as well as their chiral hierarchy: thicker strings consist of intertwisted thinner ones (superhelicity), with helix step lengths from ~ 10 nm to several tens of μm.
3. Empirical rule of switching of the macroscopic chirality sign between superhelicity levels was established, which fits the fundamental synergetic law of chirality sign switching between hierarchy levels of complexity in biological systems.
4. It was shown that molecular chirality act as a macroscopic helicity structural factor.

References

1. S.V. Stovbun, Formation of wirelike structures in dilute solution of chiral compounds. Russ. J. Phys. Chem. B **5**(4), 546–553 (2011)
2. S.V. Stovbun, O.N. Krutius, A.M. Zanin, D.S. Skorobogat'ko, R.G. Kostyanovskii, Experimental observation of anisometric structures in solutions with a low gelator content. Russ. J. Phys. Chem. B **5**(5), 846 (2011)
3. S.V. Stovbun, A.M. Zanin, A.A. Skoblin, A.I. Mikhailov, A.A. Berlin, Phenomenological description of the spontaneous formation of macroscopic strings in low-concentration chiral solutions and the formation of anisometric gels. Dokl. Phys. Chem. **442**(2), 36 (2012)
4. S.V. Stovbun, A.A. Skoblin, Chirotropical phenomena in biological fluids and their homochiral models. Mosc. Univ. Phys. Bull. **67**(3), 278–281 (2012)
5. P. de Gennes, *Ideas of Scaling in Polymer Physics* (Mir Publishers, Moscow, 1982)
6. S.V. Stovbun, A.I. Mikhailov, A.M. Zanin, R.G. Kostyanovskii, Chirality in self-organization of strings in liquid phase and principles of economy in nature. Vestnik Moskovskogo Oblastnogo Universiteta **3**, 92–97 (2011) [in Russian]
7. S.V. Stovbun, A.M. Zanin, A.A. Skoblin, D.P. Shashkin, A.I. Mikhailov, M.V. Grishin, B.R. Shub, Compaction of intermolecular bonds in the macroscopic chiral phase of strings. Russ. J. Phys. Chem. B **7**(1), 1 (2013)
8. M. Kleman, O.D. Lavrentovich, *Soft Matter Physics* (Springer, Berlin, 2003)
9. S.V. Stovbun, A.M. Zanin, A.A. Skoblin, A.I. Mikhailov, R.G. Kostyanovskii, M.V. Grishin, B.R. Shub, Macroscopic chirality of strings. Russ. J. Phys. Chem. B **5**(6), 1019 (2011)
10. S.V. Stovbun, A.A. Skoblin, A.M. Zanin, M.V. Grishin, B.R. Shub, Yu.M. Rybin, I.M. Ageev, G.G. Shishkin, V.A. Tverdislov, Superspiralization of chiral strings. Bull. Exp. Biol. Med. **154**(1), 34–36 (2012)
11. S.V. Stovbun, A.A. Skoblin, V.A. Tverdislov, Biological fluids as chiral anisometric media. Bull. Exp. Biol. Med. **152**(6), 703–706 (2012)

12. J.N. Israelachvili, *Intermolecular and Surface Forces* (Academic Press, New York, 1992)
13. V.A. Tverdislov, Chirality as a primary switch of hierarchical levels in molecular biological systems. Biophysics **58**(1), 128 (2013)
14. V.A. Tverdislov, L.V. Yakovenko, Physical aspects of the emergence of living cell precursors: the ion and chiral asymmetries as two fundamental asymmetry types. Mosc. Univ. Phys. Bull. **63**(3), 151–163 (2008)
15. V.A. Tverdislov, A.E. Sidorova, L.V. Yakovenko, From symmetries to the laws of evolution. I. chirality as a means of active media stratification. Biophysics **57**(1), 146 (2012)
16. G. Shultz, R. Shirmer, *Principles of Structural Organization of Proteins* (Mir Publishers, Moscow, 1982)
17. A.B. Rubin, *Biophysics, Biophysics of Cell Processes* (MSU Publishers, Moscow, 2004)

Chapter 36
Development of Bactericidal Ag/Chitosan Nanobiocomposites for Active Food Packaging

Amir Zarei, Saeideh Ebrahimiasl and Saeed Jafarirad

Abstract The use of nanomaterials as active fillers in polymeric nanocomposite is an innovative concept to prolong the shelf-life and safety of product for food packaging. This concept is of special importance in the area of fresh and high quality products. Recently new compounds based on the use of nanoparticles and biodegradable polymers developed. Among this material the antimicrobial properties of AgNPs have been exploited in consumer product. Also Chitosan (Cts) is a natural biopolymer with excellent biodegradability and nontoxicity, as well as its solubility in aqueous medium makes it good candidate for food packaging. In this research AgNPs were incorporated in the biodegradable chitosan matrix for applications as food packaging materials. The properties of Ag/Cts nanocomposites were studied as a function of the AgNPs concentration. Mechanical analysis and water vapor barrier properties of AgNPs/Cts nanocomposites were analyzed. It was observed that mechanical and water vapor barrier properties of the films were improved by the ratio of AgNPs. The antibacterial activity of AgNPs/Cts thin films were evaluated based on the diameter of inhibition zone in a disk diffusion test against Gram positive bacteria i.e., *Staphylococcus aureus* and *Bacillus cereus* and Gram-negative bacteria, i.e., *Escherichia coli* and *E. Faecalis* using Mueller Hinton agar at different ratio of AgNPs. The results revealed a greater bactericidal effectiveness for nanocomposite films containing 5 % w/w of AgNPs.

36.1 Introduction

Globalization of food trade and distribution from centralized processing poses major challenges for food safety and quality. One option is to use packaging to provide an increased margin of safety and quality. The new generation of food packaging

A. Zarei · S. Ebrahimiasl (✉)
Department of Chemistry, Islamic Azad University, Ahar, Iran
e-mail: ebrahimi.saeideh@yahoo.com

S. Jafarirad
Inorganic Chemistry Department, University of Tabriz, Tabriz, Iran

E. K. Polychroniadis et al. (eds.), *International Multidisciplinary Microscopy Congress*, Springer Proceedings in Physics 154, DOI: 10.1007/978-3-319-04639-6_36,

is including materials with antimicrobial properties. These packaging technologies could play a role in extending shelf-life of foods and reduce the risk from pathogens.

Metal nanocomposites are a new generation of nano food packaging which is made by incorporating metal nanoparticles into polymer films [3]. Cellulose-based natural polymers are being widely used as they offer advantages including edibility, biocompatibility, barrier properties, attractive appearance, non-toxicity, non-polluting and low cost [1]. Among the natural polymers, chitosan as a natural cationic biopolymer (Cts) has been extensively investigated for many different applications due to its excellent bioactivity, and multifunctional groups, as well as its solubility in aqueous medium and excellent barrier to oxygen. Consequently, they have a variety of current and potential applications in biomedical products [2] food packaging film, bone substitutes, and artificial skin [3].

Silver nanoparticles have been long known to have microbial inhibition [4]. The antimicrobial activity of these nanoparticles may be related to several mechanisms including, induction of oxidative stress due to generation of reactive oxygen species (ROS) which may cause the degradation of the membrane structure of the cell [5–7], release of ions from the surface of nanoparticles that has been reported to cause bacterial death due to binding to cell membrane [8, 9].

In this research, the spherical structure of AgNPs was synthesized incorporated in the chitosan matrix in aqueous solution by using $AgNO_3$, and $NaBH_4$ as the silver precursor, and the reduction agent, respectively, at room temperature. Then capability of the prepared Ag/CHI NBCs film was evaluated for antibacterial active packaging.

36.2 Material and Methods

$AgNO_3$ (99.98) used as the silver precursor, was obtained from Merck (Darmstadt, Germany) and used without purification. Deionized water was used to prepare all the solutions employed. $NaBH_4$ (98.5 %), low-molecular-weight Cts, and glacial acetic acid (99 %) used as a reduction agent, a stabilizer, and for solubilizing Cts, respectively, were obtained from Sigma-Aldrich (St Louis, MO). Gram positive: *S. aureus* (ATCC 25922), *B. cereus* (ATCC 11788), and Gram negative: *E. coli* (ATCC 25923), *E. faecalis* (ATCC 29212), used for the antibacterial assay were obtained from the BaharAfshan Ltd, Iran.

For the synthesis of Ag/Cts BNCs, soluble Cts (100 mL, 0.5 wt%) was prepared by solubilization in 1.0 wt% of acetic acid solution (pH $\sim$ 3.53) under constant stirring for 90 min, for each of the samples. Following the usual preparation method for AgNPs, $AgNO_3$ solutions were added to each soluble Cts sample under constant stirring for synthesis of the $AgNO_3$/Cts solutions. The aqueous Cts-acetic acid solution thickened after the addition of $AgNO_3$ solution. The $AgNO_3$ concentration of the samples was 0.015, 0.03, 0.06 and 0.15 (mol/L). Freshly prepared $NaBH_4$ (4×10^{-2} M) solution, was then added to the suspensions under continuous stirring to reach a constant $AgNO_3/NaBH_4$ molar ratio (1:4). After the addition of the reducing agent, stirring was continued for another hour. The suspensions of Ag/Cts BNCs obtained were then centrifuged, washed four times using double-distilled water to remove the silver ion residue, and dried at 40 °C under vacuum overnight. All the experiments were conducted at ambient temperature.

The in vitro antibacterial activity of the samples was evaluated by the well diffusion method using Mueller Hinton agar with determination of diameter of the inhibition zone formed around the well, which conformed to the recommended standards of the National Committee for Clinical Laboratory Standards (NCCLS 2000). Petriplates containing 20 ml Muller Hinton medium were seeded with 24 h culture of bacterial strains. Wells were cut and 20 μl of the bionanocomposite solution were added. The plates were then incubated at 37 °C for 24 h. Streptomycin antibiotic was used as a positive control and Dimethyl sulfoxide was used as a negative control.

The mechanical properties of the composites were evaluated in rectangular pieces of the film. An INSTRON INSTRUMENTS was used to determine the maximum tensile strength (TS), elongation at break. Films were stretched up using a speed of 50 mm min^{-1}. Tensile properties were calculated from the plot of stress versus strain. The mechanical properties were analyzed as a function of nanoparticles concentration in the film. The relative humidity (RH) at the film underside was determined using a Labthink WVP System according to Angles et al. [10].

36.3 Results and Discussion

36.3.1 Mechanical Analysis

Figure 36.1 shows the tensile strength of the AgNPs/Cts nanocomposites. The incorporation of Ag nanoparticles increased the mechanical resistance, which is due to the intermolecular bonding increase between chitosan and nanoparticles. The maximum toughness was observed for the film with 0.15 % (mol/L) AgNPs.

36.3.2 Water Vapor Permeability

Water is one of the undesirable and detrimental factors in the environment. Higher water activity in materials supports more microorganisms' growth. Bacteria usually require at least 0.91, and fungi at least 0.7 water activity. For many years, researchers tried to prevent bacterial growth with moisture content control. Consequently, a need developed for information on the water resistant characteristics of many packaging material in use. The hydrophilic natures of polysaccharide films make them poor water barriers.

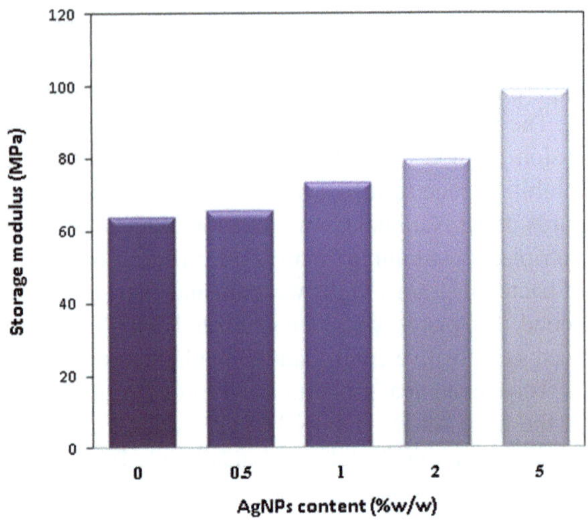

Fig. 36.1 Tensile strength for neat Chitosan films, and Chitosan films containing AgNPs with different ratio

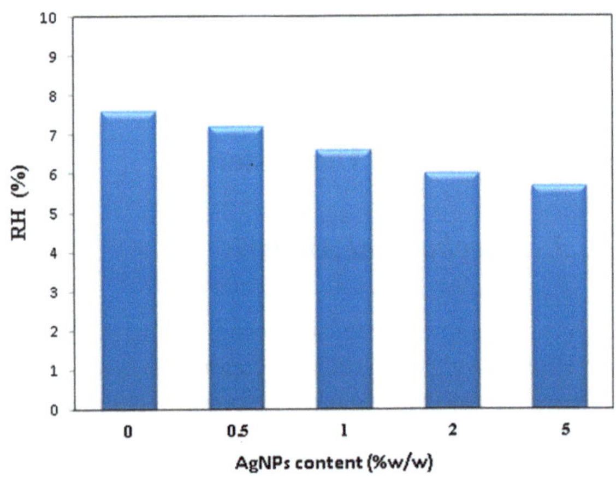

Fig. 36.2 Relative humidity at film underside of Cts/AgNPs BNCs film in different ratio of AgNPs

The silver nanoparticles effect on the relative humidity at film underside is shown in Fig. 36.2 for Cts films and Cts/AgNPs with different concentration of $AgNO_3$. Upon addition of nanoparticles in chitosan matrix, a decrease in relative humidity at film underside was observed. The RH at film underside value was 70 % for the native chitosan film. Addition of nanoparticles induced a decrease in RH at film underside. The RH values varied from 68 $\pm$ 0.9 % for chitosan films containing 0.5 % w/w AgNPs to 37 $\pm$ 0.9 % for films containing 5 % w/w nanoparticles respectively. The presence of nanoparticles increases in cohesion structures and reduce the water vapor permeability through of film.

Table 36.1 Average inhibition zone for chitosan/silver bionanocomposites at different concentration of AgNPs

Bacteria	Inhibition zone (mm)		Control positive (mm)
	Cts/AgNPs (3 %)	Cts/AgNPs (15 %)	SM
S. aureus	4.21	15	21.2
B. cereus	6.42	14.5	21.2
E. coli	5.5	15.23	21.2
E. Faecalis	6.9	16.3	29.1

36.3.3 Antibacterial Activity

Inhibition zone values were obtained for the synthesized Cts/AgNPs BNCs tested against Gram positive and Gram negative bacteria listed in Table 36.1. The tests were repeated three times for each treated sample, are presented as average values in Table 36.1. The result shows that the AgNPs in chitosan suspension had high antibacterial activity against Gram-negative bacteria and did not show any antibacterial activity against and Gram positive bacteria.

Several mechanisms argue for the antibacterial activity of nanoparticles, including generation of oxygen species to degradation of cell structure or release of ions from the surface of nanoparticles to binding cell membrane [11–13]. With the regard to the excessive antibacterial effect of Cts/AgNPs film against Gram positive bacteria which is responsible to deterioration of foods and clinical infections, it can be used as substitute for commercial chemical antibiotics.

36.4 Conclusion

AgNPs were successfully prepared from Cts/$AgNO_3$ suspension at different $AgNO_3$ concentrations by using $NaBH_4$ as a chemical reduction agent without any heat treatment or reducing agent. Tensile strength of the synthesized Cts/AgNPs BNCs increased by the presence of AgNPs to 49.3 ± 1.0 MPa. The decrease observed in the WVP values for the Cts/AgNPs system, in comparison to net Cts systems. The antibacterial activities of Cts/AgNPs BNCs at different ratio of AgNPs showed strong antibacterial activity against and Gram negative bacteria. These results indicate that Cts/AgNPs nanocomposites can be used in food packaging for certain bacterial inactivation and control.

References

1. M. Imran, S. El-Fahmy, A.-M. Revol-Junelles, S. Desobry, Cellulose derivative based active coatings: effects of nisin and plasticizer on physico-chemical and antimicrobial properties of hydroxypropyl methylcellulose films. Carbohydr. Polym. **81**, 219–225 (2010)
2. M. Darder, M. Colilla, E. Ruiz-Hitzky, Chitosan-clay nanocomposites: application as electrochemical sensors. Appl. Clay Sci. **28**, 199–208 (2005)
3. Q. Chaudhry, M. Scotte, J. Blackburn, B. Ross, A. Boxall, L. Castle, Applications and implications of nanotechnologies for the food sector. Food Addit. Contam. **25**(3), 241–258 (2008)
4. C.N. Lok, C.M. Ho, R. Chen, Q.Y. He, W.Y. Yu, H. Sun et al., Proteomic analysis of the mode of antibacterial action of silver nanoparticles. J. Proteome Res. **5**, 916–924 (2006)
5. J. Sawai, Quantitative evaluation of antibacterial activities of metallic oxide powders (ZnO, MgO and CaO) by conductimetric assay. J. Microbiol. Methods **54**, 177–182 (2003)
6. J. Sawai, S. Shoji, H. Igarashi, A. Hashimato, T. Kokugan, M. Shimizu, Hydrogen peroxide as an antibacterial factor in zinc oxide powder slurry. J. Ferment. Bioeng. **86**(5), 521–522 (1998)
7. J. Sawai, T. Yoshikawa, Quantitative evaluation of antifungal activity of metallic oxide powders (MgO, CaO and ZnO) by an indirect conductimetric assay. J. Appl. Microbiol. **96**, 803–809 (2004)
8. Q.L. Feng, J. Wu, G.Q. Chen, F.Z. Cui, T.N. Kim, J.O. Kim, A mechanistic study of the antibacterial effect of silver ions on Escherichia coli and Staphylococcus aureus. J. Biomed. Mater. Res. **52**(4), 662–668 (2000)
9. I. Sondi, B. Salopek-Sondi, Silver nanoparticles as antimicrobial agent: a case study on E. coli as a model for Gram-negative bacteria. J. Colloid Interface Sci. **275**(1), 177–182 (2004)
10. M.N. Angles, A. Dufresne, Plasticized starch/tunicin Whiskers nanocomposites mechanical behaviuor. Macromolecules **34**, 2921–2931 (2001)
11. J.R. Morones, J.L. Elechiguerra, A. Camacho et al., The bactericidal effect of silver nanoparticles. Nanotechnology **16**, 2346–2353 (2005)
12. D. Lee, R.E. Cohen, M.F. Rubner, Antibacterial properties of Ag nanoparticle loaded multilayers and formation of magnetically directed antibacterial microparticles. Langmuir **21**, 9651–9659 (2005)
13. C.N. Lok, C.M. Ho, R. Chen, Q.Y. He, W.Y. Yu, H. Sun et al., Proteomic analysis of the mode of antibacterial action of silver nanoparticles. J. Proteome Res. **5**, 916–924 (2006)

Chapter 37
Image Preprocessing Pipeline for Bright-Field Miniature Live Cell Microscopy Prototypes

S. Schöll, F. Mualla, B. Sommerfeldt, S. Steidl and A. Maier

Abstract One of the biggest technical challenges in live cell imaging is to keep the cells in a healthy state while imaging them. In fact, being able to observe living cells in their cultivation environment over time is a major step to understand and diagnose diseases. For this purpose, we are using a novel microscopic system composed of microscopy prototypes that can, in contrast to most available live cell microscopes, operate in an incubator. Each prototype observes one well of a 24 well plate over time. The in-incubator operability imposes manufacturing constraints on the size of our microscopic system and consequently degrades the delivered image quality. In order to get usable live cell images with the prototypes, a preprocessing pipeline was introduced. First, the exposure time is increased until the circular illumination is visible. Second, the illumination field is estimated using Gaussian smoothing and the center of the circular illumination is detected. Third, based on the illumination center, a 1 mm^2 region is cropped. Fourth, using the resulting image's standard deviation, a suitable exposure time is found in order to avoid under- or over-exposure. Finally, the illumination is corrected by subtracting the estimated illumination field and the image contrast is stretched. The prototypes and the pipeline are currently in use at the laboratory of our bioprocess engineering partners. The generated images and videos enable them to analyse the behavior of cells over time.

S. Schöll (✉) · F. Mualla · S. Steidl · A. Maier
Pattern Recognition Lab, Department of Computer Science, Friedrich-Alexander-Universität Erlangen-Nürnberg, Martensstr. 3, 91058 Erlangen, Germany
e-mail: simon.schoell@cs.fau.de

S. Schöll
ASTRUM IT GmbH, Am Wolfsmantel 2, 91058 Erlangen, Germany

S. Schöll · A. Maier
Erlangen Graduate School in Advanced Optical Technologies (SAOT), Friedrich-Alexander-Universität Erlangen-Nürnberg, Erlangen, Germany

B. Sommerfeldt
Institute of Bioprocess Engineering, Friedrich-Alexander-Universität Erlangen-Nürnberg, Erlangen, Germany

E. K. Polychroniadis et al. (eds.), *International Multidisciplinary Microscopy Congress*, Springer Proceedings in Physics 154, DOI: 10.1007/978-3-319-04639-6_37,

37.1 Introduction

To understand biological processes like diseases, it is essential to observe the interactions of living cells in space and time [1]. The basic requirement for reliable time-lapse imaging of cells is to keep the specimen in a healthy state while taking hundreds or even thousands of images over a long time period ranging from minutes to days [2].

A constant cultivation environment and a stable buffered medium are necessary for healthy staying cells [3]. Usually an incubator is used to ensure constant temperature, humidity, and carbon dioxide according to the cell needs. In contrast, for time-lapse imaging most standard microscopes are used outside the incubator and the cultivation conditions are satisfied by using incubation chambers. Our project partners are manufacturing novel miniature microscopy prototypes that are very small and able to operate inside the incubator. In this paper, an image preprocessing pipeline is presented that is able to deal with different prototypes and also with different stages of prototype development to generate useful images of cells out of the incubator.

In Sect. 37.2, the prototype systems are described. The preprocessing pipeline is presented in Sect. 37.3. Results and conclusions are presented in Sect. 37.4.

37.2 Miniature Bright-Field Microscope Prototypes

The illumination technique of the miniature prototype systems is bright-field. The basic principle of a bright-field microscope is shown in Fig. 37.1. Light is transmitted through the cells, collimated and focused by different lenses, and finally digitised by an image sensor (e.g. a charge-coupled device).

Manufacturing constraints for the miniature prototypes were necessary to operate inside the incubator. One feature is therefore that the prototypes are resistant to corrosion. Another feature is that they are small enough to observe single wells of a 24 well plate next to each other. Therefore the space for the hardware is limited to a square region of approximately $15 \times 15\,\text{mm}^2$.

The resulting image quality is decreased because of the aforementioned manufacturing constraints and also due to the fact that the hardware development is in progress. Different prototypes and different development stages of them necessitate flexible algorithms for image preprocessing. Those algorithms are shown in the next section.

37.3 Preprocessing Pipeline

One characteristic of the prototype systems is a distinct circular illumination in the acquired images. The position of the illumination center differs from prototype sys-

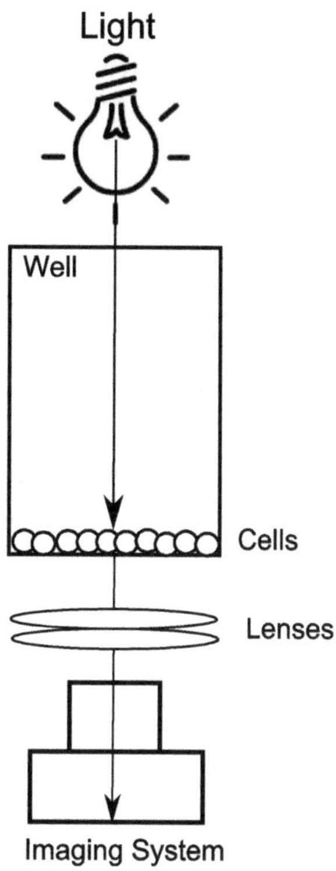

Fig. 37.1 Basic principle of a bright-field microscope. The novel miniature prototypes are using this principle to acquire images

tem to prototype system. After the first three pipeline steps (Sects. 37.3.1, 37.3.2 and 37.3.3) an image is reduced to a region around the center of the circular illumination.

Suitable exposure time and focus position vary severely for different prototypes, different cell cultures, and also for different well plates. In Sects. 37.3.4 and 37.3.5 the automatic adaption of exposure time and focus position is explained. The final pipeline step is the image enhancement that is composed of illumination correction and contrast enhancement.

37.3.1 Basic Illumination

The luminous exposure H is defined [4] as

$$H = E \cdot t, \tag{37.1}$$

where E is the illuminance in the image plane and t is the exposure time. In the miniature prototype systems the light source is a light-emitting diode (LED). A constant electrical power is assigned to the LED and therefore E is constant. The variable parameter for getting well illuminated images is the exposure time of the camera. The result of the first pipeline step is an image where the circular illumination is visible to enable the detection of the illumination center. The exposure time t_{basic} for this image is either manually selected or detected automatically. The automatic detection starts with an exposure time of 50 ms for taking an initial image. The exposure time is then increased in steps of 50 ms until the standard deviation of the image is greater than 15.

37.3.2 Center of Illumination

The first step to detect the center of the circular illumination is to estimate the illumination field. For this purpose, cells in the image are removed by a convolution with a Gaussian filter. The filter kernel is given by

$$G(x, y) = \frac{1}{\sqrt{2\pi\sigma^2}} e^{-\frac{x^2+y^2}{2\sigma^2}}, \tag{37.2}$$

where x and y are the horizontal and vertical distances to the origin of the kernel, and σ is the standard deviation. The variable parameter σ is experimentally set to 40 pixels and the kernel size to $6\sigma + 1 = 241$. In the estimated illumination field all pixels with maximum intensity are selected. The center of gravity of these pixels (x_{Center}, y_{Center}) is then calculated and considered as the illumination center.

37.3.3 Cropping

Images of the prototype systems have a resolution of 2592×1944 pixels. The hardware is optimised for a square image region of 1 mm^2 around the center of illumination. For the prototypes a region of 1 mm^2 corresponds to approximately 1400×1400 pixels. The start coordinates of the region of interest (ROI) are determined using the center of illumination (x_{Center}, y_{Center}):

$$\begin{aligned} x_{ROIstart} &= x_{Center} - 700 \\ y_{ROIstart} &= y_{Center} - 700 \end{aligned} \tag{37.3}$$

The determined ROI is shifted into the image if a part of it is outside. This adapted ROI is set at the camera and henceforward images are acquired with a reduced field of view of 1 mm^2.

37.3.4 Adjustment of Exposure Time

In this pipeline step, the basic exposure time t_{basic} is increased in steps of 50 ms until the standard deviation is maximised. Only the inner square of the image (700 × 700 pixels) is evaluated to consider the circular illumination and to avoid an overexposed image center. After this pipeline step, the camera takes images with a suitable exposure.

37.3.5 Adjustment of Focus

The maximum of the Tenenbaum Gradient (Tenengrad) [5, 6] focus measure is used to determine the focus position. The Tenengrad measure is given by the following equation:

$$F_{Tenengrad} = \sum_{x,y} ((I(x, y) * S_x)^2 + (I(x, y) * S_y)^2) \tag{37.4}$$

where I is the considered image, x and y are the spatial pixel coordinates, $*$ indicates the 2D convolution operator and S_x and S_y are the Sobel filter kernels:

$$S_x = \begin{pmatrix} 1 & 0 & -1 \\ 2 & 0 & -2 \\ 1 & 0 & -1 \end{pmatrix}; S_y = \begin{pmatrix} 1 & 2 & 1 \\ 0 & 0 & 0 \\ -1 & -2 & -1 \end{pmatrix} \tag{37.5}$$

After this pipeline step, the camera takes sharp images with a suitable focus position.

37.3.6 Image Enhancement

In the last preprocessing step, the image is enhanced. First, the illumination artifacts in the image are corrected by subtracting the illumination field from the image. The illumination field is calculated as described in Sect. 37.3.2.

Second, the contrast is stretched. The intensities of the image are therefore scaled to the whole range of gray values [0, 255]. To reduce the influence of outliers and noise, the lowest 1 % of intensity values are mapped to 0 and the highest 1 % are mapped to 255.

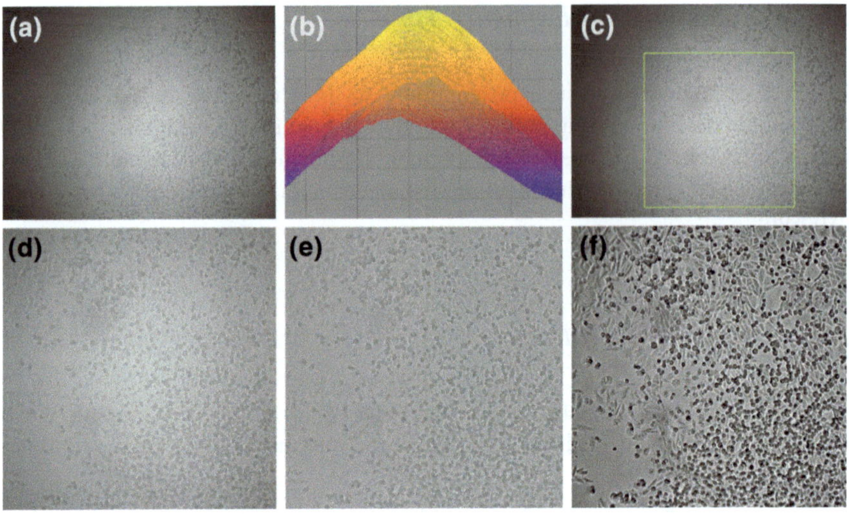

Fig. 37.2 An image from a prototype system (**a**) is used to estimate the illumination field (**b**). The center of the circular illumination is determined to automatically find a ROI around it (**c**). This ROI is cropped (**d**), the illumination artifacts are corrected (**e**) and the contrast is stretched for the final image (**f**) of the pipeline

37.4 Results and Conclusions

Example images of the pipeline steps are shown in Fig. 37.2. The final image (f) shows a considerable improvement in visual image quality compared to the input image (a). The presented preprocessing pipeline was used for three major development stages of the prototype systems and also for some minor stages. The preprocessed images allowed to use those systems in a similar way and to evaluate their results. In the meantime the pipeline was used for 51 systems. Currently 24 prototype systems are in use at a laboratory of our bioprocess engineering partners. The generated images and videos enable them to analyse the behavior of cells over time.

For the development of a novel hardware system a flexible and adapted preprocessing pipeline can be used to support the development process itself and is the basis for developing further image analyses tasks such as cell detection [7], pixel-wise cell/background classification [8] cell segmentation [9].

Acknowledgments The authors would like to thank the Bavarian Research Foundation BFS for funding the project COSIR under contract number AZ-917-10 and the industrial partners for the productive collaboration. Furthermore the authors gratefully acknowledge funding of the Erlangen Graduate School in Advanced Optical Technologies (SAOT) by the German Research Foundation (DFG) in the framework of the German excellence initiative.

References

1. R.Y. Tsien, Imagining imaging's future. Nat. Rev. Mol. Cell Biol. **4** (Suppl), SS16–21 (2003)
2. Q. Wu, F.A. Merchant, K.R. Castleman, *Microscope Image Processing* (Academic Press, Massachusetts, 2008)
3. D. Gerlich, J. Ellenberg, 4D imaging to assay complex dynamics in live specimens. Nat. Cell Biol. **5**(Suppl), 14–19 (2003)
4. Y. Ohno, Photometric calibrations. Natl. Inst. Stand. Technol. Spec. Publ. (1997). http://www.nist.gov/manuscript-publication-search.cfm?pub_id=104697
5. Y. Sun, S. Duthaler, B.J. Nelson, Autofocusing in computer microscopy: selecting the optimal focus algorithm. Microsc. Res. Tech. **65**, 139–149 (2004)
6. T.T.E. Yeo, S.H. Ong, R. Jayasooriah, Sinniah, Autofocusing for tissue microscopy. Image Vision Comput. **11**, 629–639 (1993)
7. F. Mualla, S. Schöll, B. Sommerfeldt, A. Maier, J. Hornegger, Automatic cell detection in bright-field microscope images using SIFT, random forests, and hierarchical clustering. IEEE Trans. Med. Imag. **32**, 2274–2286 (2013)
8. F. Mualla, S. Schöll, B. Sommerfeldt, J. Hornegger, Using the Monogenic Signal for Cell-Background Classification in Bright-Field Microscope Images, Proceedings des Workshops Bildverarbeitung für die Medizin 2013. (Springer, Heidelberg, 2013), pp. 170–174
9. R. Ali, M. Gooding, T. Szilágyi, B. Vojnovic, M. Christlieb, M. Brady, Automatic segmentation of adherent biological cell boundaries and nuclei from brightfield microscopy images. Mach. Vision Appl. **23**, 607–621 (2012)

Chapter 38
Measurement and Analysis of Three-Dimensional Shape Variations of Microalgae Based on a Digital Holographic Microscopy for Prediction of the Spread Point of Red Tide

Byung-Mok Kim, Kwang-Beom Seo, Jung-Sik Koo, Eun-Soo Kim, Eun-Seob Cho, Seok-Hyun Youn and Yung-Sang Seo

Abstract Thus far, a general approach to prevent the occurrence of the red tide phenomena and to minimize its resultant damage is to predict the spread point of the red tide. For that, conventionally a set of microalgae are periodically sampled from several shore areas and their numbers of microalgae per unit area are simply counted to predict the threshold level of the spread point of the red tide. However, this approach cannot foresee a decisive symptom for spreading of the red tide in advance. Accordingly, in this paper, we propose a novel approach for accurate prediction of the spread point of red tide by measuring and analyzing the three-dimensional (3-D) volume information of the microalgae with the digital holographic microscopy (DHM). This volume information of the microalgae can be converted into the thickness information through digital holographic reconstruction. With these thickness data, the volume variations of microalgae in time can be finally measured and analyzed. Experiments with several samples of microalgae may confirm the feasibility of the proposed method in the practical application.

38.1 Introduction

The red tide phenomena caused by the microalgae's hyperplasia has been widely occurred along the shores of many countries of the world. This occurrence of red tide may cause a critical damage to the cultured fishes and aquatic organisms. The

B.-M. Kim · K.-B. Seo · J.-S. Koo · E.-S. Kim (✉)
HoloDigilog Human Media Research Center (Holodigilog),
3D Display Research Center (3DRC), Department of Electronic Engineering,
Kwangwoon University, 20 Kwangwoon-ro, Nowon-gu, Seoul, Korea
e-mail: eskim@kw.ac.kr
URL: www.holodigilog.org, www.3drc.org

E.-S. Cho · S.-H. Youn · Y.-S. Seo
National Fisheries Research and Development Institute, 216 Gijanghaean-ro,
Gijang-eup, Gijang-gun, Busan, Korea

E. K. Polychroniadis et al. (eds.), *International Multidisciplinary Microscopy Congress*, Springer Proceedings in Physics 154, DOI: 10.1007/978-3-319-04639-6_38,

general reasons for the red tide phenomena seem to be the environmental factors such as sunlight, high water temperature, and external factor such as the nutrient salt flowed from the land and the geological factor. For that, in the conventional approach, a set of microalgae are periodically sampled from several shore areas and their numbers of microalgae per unit area are counted to predict the proliferation rate and the threshold level of the spread point of the red tide [1]. However, here in this method, only the population variations of the microalgae due to the change of several environmental factors [2], and two-dimensional variations of the microalgae's size have been studied for prediction of the spread point of proliferation of the red tide [3]. Therefore, this approach cannot technically provide us a crucial clue for predicting the right spreading time of the red tide in advance because the spreading phenomena of the microalgae can be seen only after the population of the microalgae has reached to a certain level. Accordingly, in this paper, we propose a novel approach to accurately predict the spread point of the red tide in advance by measuring and analyzing the biological three-dimensional (3-D) shape information of the microalgae by using the digital holographic microscopy (DHM) implemented on a shearing interferometer. That is, here in the proposed method, This volume information of the microalgae is converted into the thickness information by digital reconstruction of the hologram patterns. Then, with these data, the volume variations of microalgae in time can be analyzed. With experiments with several samples of microalgae, the feasibility of the proposed method in the practical application is to be discussed as well.

38.2 Theoretical Analysis of the Proposed System

(1) Optical setup of a shearing interferometer

The optical setup for the proposed system is described in Fig. 38.1. Here, a HeNe-laser ($\lambda = 632.8\,\text{nm}$) is employed for a light source and the laser beam emitting from it is headed for the sample and magnified by a microscope objective. Then, two divided beams are made by passing the laser beam through a glass plate with the thickness of 1 mm which is located at an angle of about 45° to the incident light beam. Afterwards, an interference phenomenon may occur between the two beams reflected from the front and the back surfaces of the glass plate. At this point, we can obtain the interference pattern which may contain the 3-D information of the object, which can be finally observed on the CCD camera.

(2) Numerical reconstruction

The numerical reconstruction procedure is based on the Fresnel diffraction theory [4]. However, the Fresnel diffraction theory may require that the distance between the object and the hologram must be sufficiently long compared to the size of the object or the hologram, which is called the Fresnel approximation condition. Moreover, with this Fresnel diffraction theory, object fields for very small and transparent objects cannot be correctly reconstructed. Therefore, for correct reconstruction of them, an

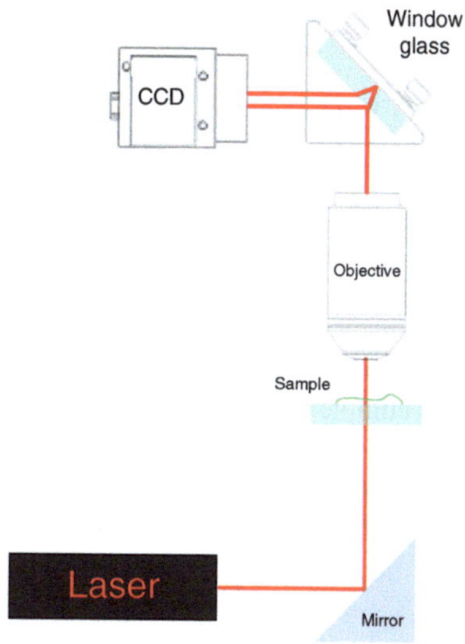

Fig. 38.1 Optical setup of a shearing interferometer

angular spectrum method seems to be useful because it can explain the wavefront propagation at any distance [5].

Here, the complex amplitude *U(x, y; 0)* at the hologram plane, *(x, y, 0)* is obtained by illuminating the hologram by the reference wave.

$$U_1(x, y; 0) = h(x, y)R(x, y) \tag{38.1}$$

where *h(x, y)* is the hologram transfer function and *R(x, y)* is the reference wave. Then, (38.1) can be represented as the angular spectrum with the Fourier transform [6].

$$\hat{U}_1(f_x, f_y, 0) = \int\int_{-\infty}^{\infty} U_1(x, y, 0) \exp[-i2\pi(f_x x + f_y y)]dxdy \tag{38.2}$$

where f_x and f_y are the spatial frequency in the *x* and *y* directions, respectively. By applying a process of spatial filtering to the spectrum, the reference beam and the conjugate image are removed and only the object spectrum can be selected by this filtering process as follows.

$$\bar{U}(x, y, 0) = \int\int_{-\infty}^{\infty} \textit{filtering}\ \hat{U}_1(f_x, f_y, 0) \exp[i2\pi(f_x x + f_y y)]df_x df_y \tag{38.3}$$

In (38.3), inverse-Fourier transformation of the filtered angular spectrum provides the complex amplitude at *(x, y, 0)* containing information about the object. Complex amplitude at the image plane which is located at a distance *d*, calculated from the filtered angular spectrum is given by

$$U(x, y, d) = \int\int_{-\infty}^{\infty} \bar{U}(f_x, f_y, 0) \exp[ik\sqrt{1 - \lambda^2 f_x^2 - \lambda^2 f_y^2 d}] \exp[i2\pi(f_x x + f_y y)] df_x df_y \quad (38.4)$$

The phases of the wave fronts with and without the object are calculated from the complex amplitudes of two reconstructed holograms $U_{with\,object}(x, y, d)$ and $U_{without\,object}(x, y, d)$, respectively. The phase difference between the holograms with the object and without the object can provide only the 3-D information of the object. Now, the phase of each hologram can be written as

$$\phi(x, y) = \arctan \frac{\mathrm{Im}[U(x, y, d)]}{\mathrm{Re}[U(x, y, d)]} \quad (38.5)$$

And the phase difference is written by subtraction

$$\Delta\phi(x, y) = \phi_{with\,object}(x, y) - \phi_{without\,object}(x, y) \quad (38.6)$$

This phase difference can be converted to the change of the optical path length of the object according to

$$\Delta\phi(x, y) = \frac{2\pi}{\lambda} \Delta n(x, y) \Delta L(x, y) \quad (38.7)$$

where $\Delta n(x, y)$ and $\Delta L(x, y)$ represent the changes of the refractive index and the thickness, respectively. These results mean that a change of the phase information can provide the corresponding change of the physical thickness of the object. Accordingly, in case the change of the refractive index and the phase information are known, the corresponding thickness information of the object can be easily calculated.

38.3 Experimental Results

Reconstructed 3-D images and the numerical thickness information of the microalgae can be obtained by using a shearing interferometer. Here, the isocrysis, which is transparent and circular-shaped, is used as the test sample of the microalgae for measuring its thickness information. Magnification of the objective lens has been scaled from 20 to 40 and 60. Figure 38.2 shows the interference pattern detected by using the CCD and the computationally reconstructed image. That is, Fig. 38.2a shows the interference pattern for a number of spread isocrysis. The reconstructed

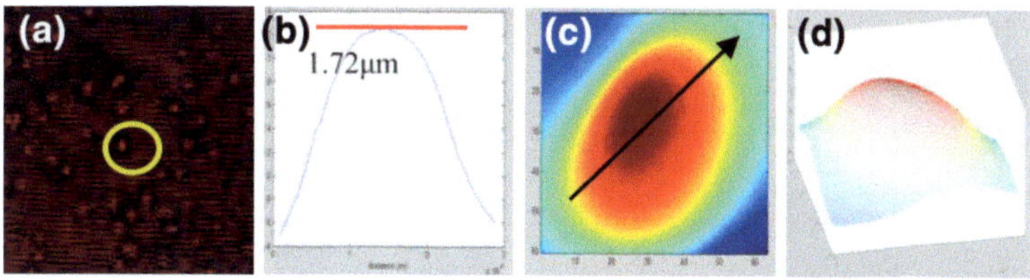

Fig. 38.2 Experimental results with a magnification objective scale of 40. **a** Interference pattern obtained with the CCD, **b** 1-D profile image of the isocrysis, **c** 2-D profile image of the isocrysis, **d** 3-D modeled image of the isocrysis

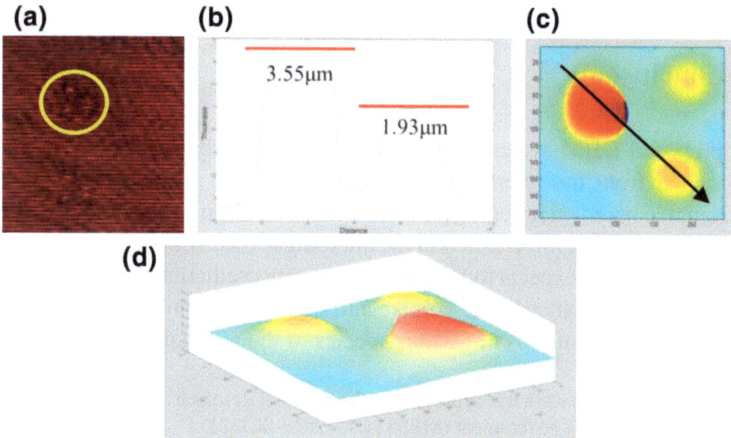

Fig. 38.3 Experimental results with magnification objective scale of 60. **a** Interference pattern obtained with the CCD, **b** 1-D profile image of the isocrysis, **c** 2-D profile image of the isocrysis, **d** 3-D modeled image of the Isocrysis

images for one microalgae cell are shown in Fig. 38.2b–d by using the numerical reconstruction method for conversion of the phase information into the thickness information. Here, the thickness of the microalgae is estimated to be 1. 72 μm.

Figure 38.3 also shows the interference pattern and its reconstructed images for three isocrysis cells which are closely located together. Here, Fig. 38.3b–d represent the numerically reconstructed images of three isocrysis cells, in which the thickness of each isocrysis cell are calculated to be 3.55, 1.93 and 1.86 μm respectively. In addition, several isocrysis cells located in the detection area are reconstructed and its quantitative thickness information is also obtained.

Figure 38.4 shows the reconstructed 3-D images of other microalgae with different shape compared to the isocrysis. From the results of Fig. 38.4, it is found that with thickness information of various types of microalgae, species of the microalgae can be quantitatively classified.

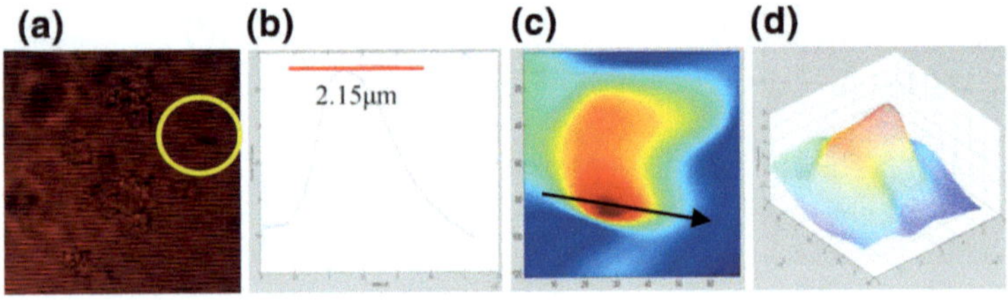

Fig. 38.4 Experimental results with the magnification objective scale of 20. **a** Interference pattern obtained with the CCD, **b** 1-D profile image of the microalgae, **c** 2-D profile image of the microalgae, **d** 3-D modeled image of the microalgae

38.4 Conclusions

In the proposed DHM system, 3-D images of the microalgae can be numerically reconstructed from the interference patterns picked up with the optical shearing interferometer-based on holographic microscopy, and from which quantitative thickness and volume shape information of the microalgae could be obtained. Accordingly, the proposed method expects to be used for accurate prediction of the spread point of the red tide just by analyzing the volume and shape variations of the microalgae and by recognizing 3-D patterns of various species of the microalgae in a near future.

Acknowledgments This work was supported by the National Research Foundation of Korea (NRF) grant funded by the Korea government (MEST) (No. 2013-067321).

References

1. E.S. Cho, S.H. Youn, Marine environmental and phytoplankton monitoring in Wando coastal water in august during the period of 2009–2010. J. Korean Soc. Mar. Environ. Saf. **18**(2), 95–100 (2012)
2. V. Sachithanandam, P.M. Mohan, R. Karthik, S. Sai Elangovan, G. Padmavathi, Climate changes influence the phytoplankton bloom (prymnesiophyceae: phaeocystis spp.) in North Andaman costal region. Indian J. Mar. Sci. **42**(1), 58–66 (2013)
3. E.S. Cho, Environmental and biological effects at Narodo, in the southern water of Korea, on bloom of Ichthyotoxic dinoflagellates cochlodinium polykrikoides. J. Life Sci. **15**(2), 225–230 (2005)
4. A. Dakoff, J. Gass, M.K. Kim, Microsocpic three-dimentional imaging by digital interference holography. J. Electron. Imaging **12**(4), 643–647 (2003)
5. L. Yu, M.K. Kim, Wavelength-scanning digital interference holography for tomographic three-dimensional imaging by use of the angular specturm method. Opt. Lett. **30**(16), 2092–2094 (2005)
6. J.W. Goodman, *Introduction to Fourier Optics* (Roberts and Company Publisher, Englewood, 2004)

Chapter 39
Fruit Trees Physiology and Breeding Programs Research Using Microscopic Technology

Kazem Arzani

Abstract The flower bud differentiation in bearing pistachio trees was examined using microtechnique and microscopic examination of samples for better pistachio orchard management. In addition, the pollen viability test was carried out using microscope technology in the controlled hybridization and breeding programs in sour and sweet cherry, almond, European and Asian pears, apricot, citrus and olives. This research was followed with the examination of pollen tube growth in the style using UV and florescent microscope in sweet cherry, European and Asian pears. Also, scanning electron microscopy (SEM) has provided a powerful technique in pomology research such as studying pollen morphology at TMU. SEM was used in order to investigate genotypic differences in pollen shape and exine structure in the species such as olive, apricots, Asian pear and pomegranate. In addition, the importance of stomatal density and its role in the pistachio trees gas exchange subjected to drought was studied and followed by evaluation of the existence of trichomes around the stomata in some pistachio rootstocks. Taken images were shown that stomata are often sunken and covered by trichomes to prevent water loss in the pistachio rootstocks subjected to drought. In the recent research the postharvest shelf life of peach fruit was examined using microscope technology in order to test the effect of some plant extracts in fruit quality attributes in the storage. Microscopic examination of the samples in wide range of physiological and breeding programs showed an essential part of research at TMU, lead to reliable and effective research results.

K. Arzani (✉)
Department of Horticultural Science, Tarbiat Modares University (TMU),
PO Box 14115-336 Tehran, Iran
e-mail: arzani_k@modares.ac.ir

E. K. Polychroniadis et al. (eds.), *International Multidisciplinary Microscopy Congress*, Springer Proceedings in Physics 154, DOI: 10.1007/978-3-319-04639-6_39,

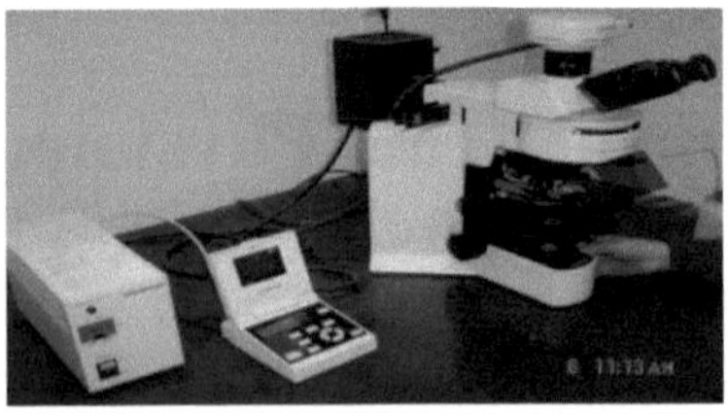

Fig. 39.1 Florescent microscope used for studying pollen tube growth in the style as well as pollen viability test assays [2, 5, 9, 10]

39.1 Background Information

There are a wide range of using microscopic technology in horticultural science, especially in fruit tree physiology and breeding research programs. This is essential in the basic as well as practical point of view to use microscopic technology in various stages of fruit tree growth and development. The objective is to review and discuss data on selected research results on various fruit tree species and future outlook of the research programs using microscopic technology at Tarbiat Modares University (TMU) and collaboration with the other universities and research institutes (Fig. 39.1).

39.2 Flower Bud Differentiation in Pistachio Trees and Pollen Viability Tests

The flower bud differentiation in bearing pistachio trees was examined using microtechnique and microscopic examination of samples for better pistachio orchard management [1]. In addition, the pollen viability test was carried out using microscope technology in the controlled hybridization and breeding programs in sour and sweet cherry [2], almond [3], European and Asian pears [4, 5], apricot [6], citrus [7] and olives [8]. This research was followed with the examination of pollen tube growth in the style using UV and florescent microscope in sweet cherry [2], European and Asian pears [5].

Results based on a microscopic examination of pollen tube growth showed that pollen grains were alive after pollination and that the pollen tube grew within the style after germination. Although pollen tube growth in the style was affected by genotypes, KS9 (♀) showed the lowest, whereas pollination of KS8 (♀) by KS11 (♂) showed the highest pollen tube growth (Fig. 39.2). In fruit trees the fruit set is affected by various factors, such as genetic and environmental conditions as well as the physiological status of the trees. In the present research the percentage fruit set in self pollinated cultivars was between 2.92 % in KS9 and 14.65 % in KS11, suggesting the existence of partial self incompatibility in KS9 and KS13. In these cultivars the percentage of fruit set was improved by cross pollination. Hence it is important to plant cross compatible cultivars for the establishment of new Asian pear orchards [9].

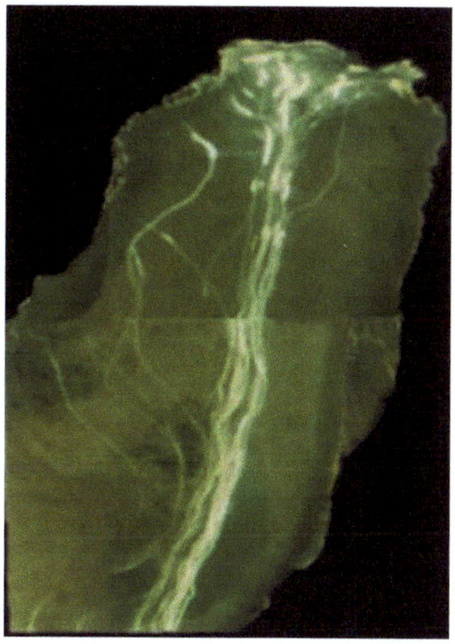

Fig. 39.2 Asian pear pollen tube growth in the KS8 style (♀) pollinated with KS13 (♂) (10 X) [9]

39.3 Pollen Morphology

39.3.1 Olives Pollen Morphology

Pollen grains of five olive cultivars *(Olea europaea* L.*)* were examined, using Scanning Electron Microscopy (SEM) for identification purposes. Four measurements, polar axis (P), equatorial diameter (E), P/E ratio and exine patterns were made. Forty pollen grains of each cultivar were viewed before finally selecting a representative pollen grain. Polar diameter ranged between 22.76 μm for the 'Mary' and 28.58 μm for the 'Shengeh' 'Zard' and 'Roghani' cultivars exhibited incomplete closure of the meshes. 'Zard' and 'Shengeh' cultivars had thick muri but 'Roghani', 'Mary' and 'Fishomi' possessed thinner muri. Scanning electron analysis of pollen grain revealed that specific differences including variation in size and form of the meshes are existing among cultivars. The differences in the exine pattern may help for the identification of some olive cultivars [8].

39.3.2 Apricots Pollen Morphology

Pollen grains of eleven Iranian apricot (*Prunus armeniaca* L.) cultivars were examined, using SEM. Pollens were collected in 2002-growing season from mature apricot trees grown in The Apricot Collection Orchard, Ismailabad Horticultural Research Station, Qazvin, Iran. For SEM evaluation, pollen grains were observed and photographed at 1000 X for whole grain and at 10,000 X for sectioning of exine pattern (surface topography). The length (L) and the width (W) of grain, L/W ratio, distance between two furrows and the width of ridges in elliptical pollen and the base, the altitude and the width of grain and ridges in triangular pollen were measured. Substantial variability in the size, shape and exine patterns among pollen grains was clearly distinguished. The apricot cultivars were easily separated into two groups according to their pollen grain shapes: (1) elliptical trizonocolpate (5 cultivars), and (2) obtuse-triangular (6 cultivars). The pollen grains of all studied apricot cultivars were medium in size: 51.32 × 25.51 μm (length × width) in elliptical shape, and triangular in shape: 39.03 × 31.22 μm (base × altitude). The resultant data of the 2-type pollens' shape were analyzed, using completely randomized design (CRD) with unequal replications. Moreover, multivariate statistical analysis was carried out to distinguish morphometric information, coming from measured parameters of the 2-shaped pollen grains. The projections of the loadings defined by the first two principal components allowed for the visualization of the position of apricot cultivars, demonstrating between-cultivar-pollen shape variations. Hence, the differences in both exine pattern and surface topography in particular ridges and pores characteristics can potentially be utilized as a diagnostic tool to aid in the identification of apricot cultivars (Fig. 39.3) [6].

39.3.3 Pomegranate Pollen Morphology

Pollen grains from 14 pomegranate (*Punica granatum*) cultivars were examined using scanning electron microscopy. Pollen grains were observed and photographed at 3,000× for whole grains and at 20,000× for exine patterns. The polar (P) and equatorial axis (E) of grains, P/E ratios, distances between the two furrows, and width of the pollen grain ridges were measured. The results indicated that the mean polar axis ranged between 22.63–25.86 μm, while the mean equatorial axis ranged between 12.69–14.47 μm.

Two groups were identified according to the polar axis and equatorial axis. Mean values of the P/E ratio varied from 1.75–1.94. In addition, pollen grains were placed in prolate (elliptical) shape group based on the P/E ratio. The studied cultivars were classified into four groups according to distance between furrow edges and into two groups according to width of ridges. All studied cultivars showed a similar foveolate ornamentation pattern. Correlation coefficients calculated between different traits of pomegranate pollen grains indicated that there were positive and significant correlations between polar axis and equatorial axis ($r = 0.675$), polar axis and distance between furrow edges ($r = 0.908$), P/E ratio and distance between furrow edges ($r = 0.671$). Hence, pollen morphology characteristics can be used for pomegranate cultivar identification (Fig. 39.4) [11].

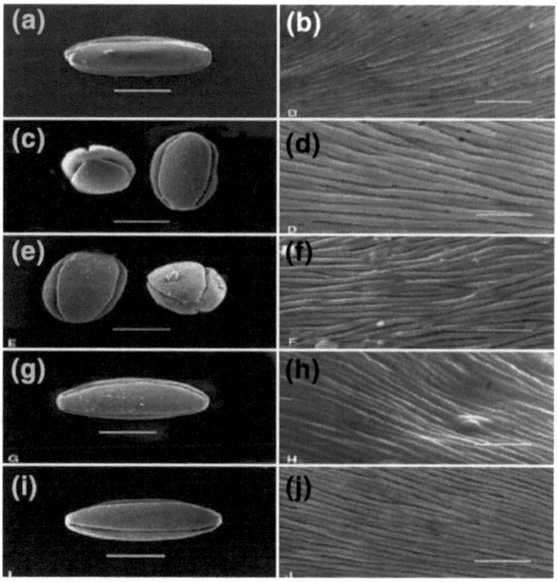

Fig. 39.3 Pollen shape and exine pattern in four elliptical trizonocolpate apricot cvs. 'Shasti' (**a**, **b**); 'Hosseini-Karimi'(**c**, **d**); 'Shamsi' (**e**, **f**); 'Nakhjavan' (**g**, **h**) and 'Noori' (**i**, **g**). Scale bars represent 20 μm for *left* photomicrographs (**a**, **c**, **e**, **g** and **i**), and 2 μm for *right* ones (**b**, **d**, **f**, **h** and **j**) [6]

39.4 Pistachio Rootstocks Stomatal Density

The importance of stomatal density and its role in the pistachio trees gas exchange subjected to drought was studied and followed by evaluation of the existence of trichomes around the stomata in some pistachio rootstocks. In this study epidermal cell in leaves of four Pistachio rootstocks (*Pistacia vera*, cv. 'Badami Zarand', 'Sarakhs', 'Ghazvini' and *Pistacia mutica*) grown under three irrigation levels (100, 65 and 30 % ET_c) were evaluated. Stomata density and stomata dimension (length and width) were determined by light microscope and scanning electron microscope (SEM), respectively. The results showed that irrigation, rootstock and their interaction had significant effects on stomata density and stomata length. Effect of irrigation on stomata width was significant while rootstock and interaction of irrigation and rootstock had no significant effects on stomata width. The highest stomata density and the lowest stomtal length and width were obtained in severe water stress. The highest and the lowest stomata density on the abaxial surface was belonged to *P. mutica* (530.30 no/mm^2) and *P. vera* 'Ghazvini (404.04 no/mm^2), respectively. The rootstocks *P. mutica* and *P. vera* 'Ghazvini' had the highest and *P. vera* 'Badami' and *P. vera* 'Sarakhs' the lowest stomata length of stomata. The rootstocks showed no significant difference in case of stomata width. The shape of epidermal cells also was different among studied rootstocks. All rootstocks had Anomocytic stomata. Among studied rootstocks only *P. mutica* had trichomes on abaxial surface. Presence of trichomes in this rootstock is a characteristic distinguishable. It was found a negative correlation (−0.432) between stomata density and stomata width ($P \leq 0.01$). There were not found correlation between stomata density and stomata

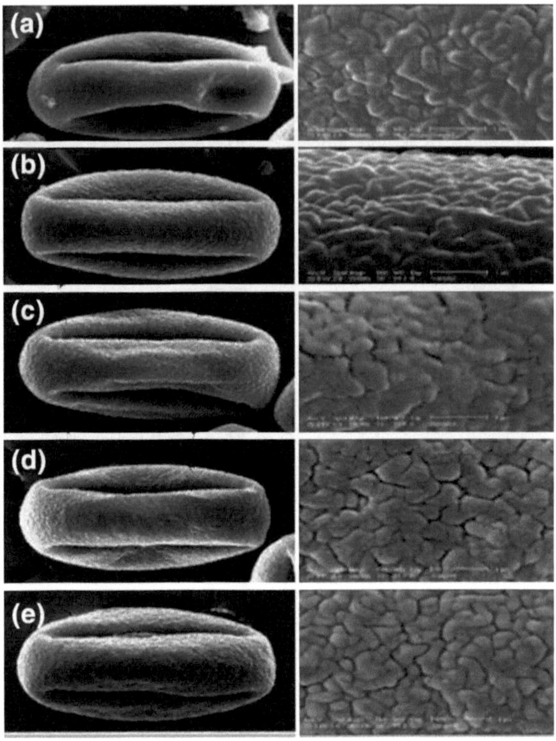

Fig. 39.4 Pollen shape (3,000×, *left panel*) and exine pattern (20,000×, *right panel*) of pomegranate cultivars. **a** 'Alake Malase Saveh'; **b** 'Bihaste Ardestan'; **c** 'Nabati'; **d** 'Agha Mohamad Ali' and **e** 'Naderie Natanz' [11]

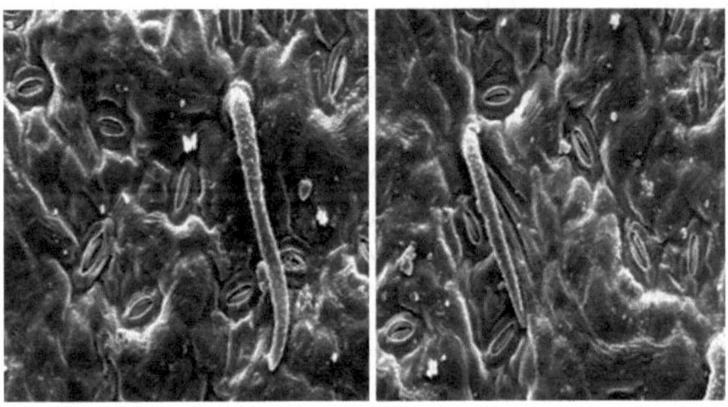

Fig. 39.5 Non-glandular trichomes on abaxial surface of P. mutica observed by SEM [12]

length, while in severe treatment of water stress was observed significant negative correlation between stomata density and stomata length (−0.675) at $P \leq 0.05$. *P. vera* 'Ghazvini' had the lowest number of stomata but wider stomata. Anatomical differerences may be useful as an initial screening method for classifying pistachio rootstocks of drought resistance (Fig. 39.5) [12].

In addition, in the recent research the postharvest shelf life of peach fruit was examined using microscope technology in order to test the effect of some plant extracts in fruit quality attributes in the storage [13]. In conclusion, microscopic examination of the samples in wide range of physiological and breeding programs showed an essential part of research at TMU, lead to reliable and effective research results.

Acknowledgments I would like to thank Tarbiat Modares University (TMU) for financial support and providing facilities.

References

1. A. Javanshah, K. Arzani, Y. Dehghani-Shouraki, P. Cappellini, Study of flowering in Pistaca vera L. cv. 'Ovhady'. Seed Plant J. **16**(4), 432–445 (2001) (In Persian)
2. M. Mahmoudi, K. Arzani, N. Bouzari, Pollination, pollen tube growth and determination of suitable pollinizer for sweet cherry (Prunus avium L.) cultivar 'Red Rezaeieh'. Seed Plant J. **23**(4), 571–585 (2008) (in Persian)
3. S. Alizadeh-Salteh, K. Arzani, A. Imani, Self-incompatibility studies of some Iranian late-blooming almonds and pollen source effect on some characteristics of Nuts. J. Nuts **3**(2), 37–44 (2012)
4. M. Ghazaeian, K. Arzani, A. Moeini, Study on pollen morphology of some Asian pear genotypes by scanning electron microscopy (SEM) and determination of suitable pollen germination medium. Iranian J. Hort. Sci. Technol. **8**(2), 113–124 (2007) (In Persian)
5. S. Mousavi, K. Arzani, H. Abdollahi, Evaluation of self-compatibility and self-incompatibility in some Asian pears (Pyrus serotina Rehd.) and European pears (Pyrus communis L.) and Native pears cultivars. Iranian J. Hort. Sci. **43**(3), 311–321 (2012) (In Persian)
6. K. Arzani, M.A. Nejatian, G. Karimzadeh, Apricot (Prunus armeniaca) pollen morphological characterization through scanning electron microscopy, using multivariate analysis. New Zealand. J. Crop Hort. Sci. **33**, 381–388 (2005)
7. N. Ahmadi, K. Arzani, A. Moeini, Study of pollen storage, germination and pollen tube growth in some citrus cultivars. Seed Plant J. **17**(4), 216–229 (2002) (In Persian).
8. T. Javady, K. Arzani, Pollen morphology of five Iranian olive (Olea europaea L.) cultivars. J. Agr. Sci. Technol. (JAST) **3**, 341–346 (2001)
9. K. Arzani, M. Koshesh-saba, F. Ghanati, Study on compatibility and pollen tube growth of some Asian pear (Pyrus serotina Rhed) cultivars. Acta Horticulturae **671**, 159–163 (2005)
10. M. Rasouli, K. Arzani, Effect of pollen source on pollen tube growth and qualitative and quantitative fruit characteristics of sweet cherry cv .'Zard Daneshkadea'. Iranian J. Hort. Sci. **41**(3), 309–318 (2010) (In Persian)
11. F. Varasteh, K. Arzani, Classification of some Iranian pomegranate (Punica granatum) cultivars by pollen morphology using scanning electron microscopy. Hort. Environ. Biotechnol. **50**(1), 24–30 (2009)
12. K. Arzani, M. Ghasemi, A. Yadollahi, H. Hokmabadi, Study of foliar epidermal anatomy of four pistachio rootstocks under water stress. IDESIA **31**(1), 101–107 (2013)
13. S. Alizadeh-Salteh, K. Arzani, R. Omidbeigi, N. Safaie essential oils inhibit mycelial growth of rhizopus stolonifer. Eur. J. Hort. Sci. **75**(6), 278–282 (2010)

Index

E. K. Polychroniadis et al. (eds.), *International Multidisciplinary Microscopy Congress*, Springer Proceedings in Physics 154, DOI: 10.1007/978-3-319-04639-6,

MIX
Papier aus verantwortungsvollen Quellen
Paper from responsible sources
FSC® C105338

If you have any concerns about our products,
you can contact us on
ProductSafety@springernature.com

In case Publisher is established outside the EU,
the EU authorized representative is:
Springer Nature Customer Service Center GmbH
Europaplatz 3, 69115 Heidelberg, Germany

Printed by Libri Plureos GmbH
in Hamburg, Germany